Thin Films

PVD for Microelectronics:
Sputter Deposition Applied
to Semiconductor Manufacturing

VOLUME 26

Thin Films

PVD for Microelectronics: Sputter Deposition Applied to Semiconductor Manufacturing

Ronald A. Powell
Director, Thin Film Technology
Novellus Systems, Inc.
Palo Alto, California

Stephen M. Rossnagel
T.J. Watson Research Center
IBM Corporation
Yorktown Heights, New York

VOLUME 26

ACADEMIC PRESS
San Diego London Boston New York Sydney Tokyo Toronto

This book is printed on acid-free paper. ∞

Academic Press
a division of Harcourt Brace & Company
525 B Street, Suite 1900, San Diego, California 92101-4495, USA
http://www.apnet.com

Academic Press
24–28 Oval Road, London NW1 7DX, UK
http://www.hbuk.co.uk/ap/

International Standard Book Number: 0-12-533026-X

Printed and bound in the United Kingdom
Transferred to Digital Printing, 2011

Contents

Preface

For more than a century, the physical vapor deposition (PVD) process known as sputtering has been applied to industrial thin film coating and, since the early 1970's, has been a key element of microelectronic fabrication. Currently, PVD is the established method of depositing metal contacts, barriers, and interconnects used in advanced silicon integrated circuits (ICs) such as microprocessor chips with clock speed greater than 500 MHz and DRAM memory chips storing nearly 1 Gigabit of information. These and other demanding applications of PVD technology have led to the development of sophisticated vacuum-integrated PVD production tools and a global market for PVD equipment in excess of $1B.

As the millenium approaches, the IC industry is faced with the economic and technical challenge of fabricating ultralarge scale integrated (ULSI) devices having minimum feature size 0.18 μm. Additional challenges are posed by the increase in Si wafer size from 200 mm to 300 mm, the replacement of Al alloy interconnects with Cu interconnects, and the related requirements for Cu diffusion barrier films and damascene processing. In order to meet the metallization challenge of ULSI devices, PVD will need to perform better than ever before.

In view of the large established base of PVD hardware and the many technical and cost advantages of sputter deposition, both users and suppliers have continued to push PVD technology to meet the challenge of coating and/or filling high aspect ratio features on ULSI devices. Activity to date has included modifications of conventional PVD hardware such as collimation (see chapter 3), extensions of PVD processing such as high temperature reflow and high pressure extrusion, as well as entirely new concepts such as directional deposition from ionized metal plasmas (see chapter 5) and the fusion of PVD and chemical vapor deposition (CVD) methods to take advantage of the best of both worlds (see chapter 9). All things considered, this an extremely exciting time to be involved in the development and application of PVD technology!

Given this resurgence of interest in PVD technology, the authors felt that an up-to-date monograph on the topic would be timely and well received. It is true that a number of excellent handbooks on general thin film technology have been published with specific chapter offerings on the underlying physics of sputtering, the design of magnetron plasma sources, and the application of PVD to a variety of industrial coating applications. However, virtually all of these treatments are now over 10 years out of date and do not include many of the exciting developments in PVD hardware and processing that have recently occurred, and that comprise much of the present book. Also, to our knowledge, no single volume on the application of PVD to microelectronics was available.

Another goal was to provide an historical and technical perspective on PVD that would be of value to persons who had either recently entered the field or who do not regularly attend scientific meetings where advanced PVD technology is discussed. For example, a receptive audience should be found among process engineers and technicians, product support personnel, and the sales and marketing staff of suppliers of PVD equipment and components such as vacuum pumps, pressure gauges, power supplies, robotic wafer handlers and gas delivery systems that are used in a modern, vacuum-integrated PVD cluster tool. We also hoped that the book would appeal to research scientists and R&D staff who were familiar with the technology of PVD but not with emerging trends in the field as applied to microelectronics.

To keep the book focused, we have only treated the application of PVD to silicon-based microelectronics. As a result, we do not discuss applications of PVD to digital electronics based on GaAs. Similarly, applications of PVD outside of conventional microelectronics, such as flat panel display technology or magnetic storage disk coating, are only mentioned briefly.

While the focus of *PVD for Microelectronics* is clearly on PVD, the book is *not* intended to be an in-depth, academic treatment of the subject (we have however given extensive references to more scholarly articles and monographs in each chapter). Instead, our aim was to present the reader with a modern overview of the field, covering a wide range of topics, and providing a blend of theoretical and practical knowledge. We have also paid attention throughout to the commercial implications of the technology, discussing such topics as cost-of-ownership and the historical growth in size of the IC and PVD equipment markets. Finally, we have tried to put a PVD spin on thin film issues to avoid duplicating information that is readily available in copious books and articles on the materials science of thin films used in microelectronics. The overall result is that this book is closer in spirit to a textbook than a treatise, and could serve as the basis for a shortcourse on PVD or as an adjunct to training on PVD equipment or processing.

A number of features differentiate this book from other technical treatments of PVD and increase its usefulness. For example, we have made extensive reference to articles appearing in trade magazines such as *Solid State Technology* and *Semiconductor International* to provide relevant material that is not always indexed and therefore may not be readily accessible. Comments that are either historically interesting (e.g. the origin of the word "sputtering") or deserving of special attention for pedagogic reasons (e.g. the quantitative difference between atomic and weight percent of an element in a sputter target) have been boxed and set off from the main text. Citations have been given in the most complete form possible and include the full title of the article cited, a listing of all authors, and inclusive page numbers as opposed to only the starting page number.

Material has been organized in a logical progression. Chapter 1 gives a broad-brush introduction to use of PVD in advanced microelectronic fabrication, with suggested sources of additional information on the field. Chapters 2, 3, and 4 then provide a conceptual understanding of the physics of sputtering, plasma discharges, and plasma sputter sources that set the stage for subsequent chapters on hardware (Chapter 5) and process (Chapters 6–8), including exciting hardware/process developments that have been developed to improve coating and/or filling of high-aspect-ratio features. Chapter 9 also concentrates on process, but is organized by material—beginning with mainstream applications such as Al alloys and Ti/TiN, and proceeding to advanced materials such as Cu and Ta/TaN barriers. Chapter 10 deals with theoretical modeling or simulations of PVD. Finally, Chapter 11 discusses sputter target technology. The material in Chapter 11 could have been included as part of an earlier chapter; however, we wanted to set this topic apart due to the critical role of the target in the overall PVD process.

PVD for Microelectronics is designed to be read as a coherent monograph and not as a collection of separate articles submitted by different authors or research groups. Nevertheless, to achieve a consistency of style and to take advantage of the authors' complementary knowledge of the field, each author was responsible for writing a given chapter as follows: the text and figures for chapters 1, 4, 5, 7, 9, and 11 were prepared by Ron Powell, while those for chapters 2, 3, 6, 8, and 10 were prepared by Stephen Rossnagel. It also should be noted that the views expressed in this book are those of the authors themselves and do not necessarily represent those of their employers past or present.

Finally, we would like to express our appreciation to the many individuals, suppliers, and institutions who contributed information or figures that appear in the book. Of special note, we wish to acknowledge

Chuck Wickersham (Tosoh, SMD) for providing useful background material for Chapter 11 and Daniel Lee (Novellus Systems) for the use of material contained in the PVD training manual he has developed.

Ronald A. Powell
Palo Alto, CA

Stephen Rossnagel
Yorktown Heights, NY

October 1998

Useful Conversion Factors and Constants

Length

1 Ångstrom (Å)= 10^{-8} cm
1 nanometer (nm) = 10 Å
1 micron (μm) = 10,000 Å = 10^{-4} cm
1 inch = 2.54 cm

Mass and Force

1 atomic mass unit (AMU) = 1.66×10^{-24} gm
1 pound (lb) = 454 gm
1 dyne = 1 gm/cm-sec^2
1 Newton (N) = 1 kgm/m-sec^2
1 Newton = 10^5 dynes

Energy

1 electron volt (eV) = 1.6×10^{-12} erg
1 eV/particle = 23.06 kcal/mole
1 joule (J) = 10^7 erg
1 watt (W) = 1 J/sec

Pressure and Vacuum-Related

1 standard atmosphere = 1.013×10^6 dynes/cm^2
1 standard atmosphere = 14.7 lb/in.^{2} = 14.7 psi
1 standard atmosphere = 760 Torr
1 Torr = 1.33×10^3 dynes/cm^2
1 Pascal (Pa) = 1 N/m^2= 7.5 mTorr
1 Torr = 133.3 Pa
1 micron = 10^{-3} Torr = 1 milliTorr (mTorr)
1 standard cubic centimeter (std. cc) = 0.76 Torr-liter
1 std. cc per minute (sccm) = 12.7 mTorr-l/sec
1 liter/sec = 2.12 ft^3/min
1 ft^3/min (CFM) = 0.47 liter/sec
1 Langmuir = 10^{-6} Torr-sec

Miscellaneous

1 Tesla = 10^4 gauss
mass of electron (m_e) = 9.11×10^{-28} gm
charge of electron (e) = 1.6×10^{-19} Coulomb
Avogadro's number (N_0) = 6.02×10^{23} molecules per mole
Boltzmann's constant (k) = 1.38×10^{-16} erg/°K
Stefan-Boltzmann constant (σ) = 5.67×10^{-8} W/m^2-K^4
permittivity of vacuum (ε_0) = 8.85×10^{-12} C^2/N-m^2
permeability of vacuum (μ_0) = $4\pi \times 10^{-7}$ N/A^2

Chapter 1 Introduction

> The phenomenon originally described as "cathodic disintegration" by Sir William Robert Grove in 1852 was renamed "spluttering" by Sir John Thompson in 1921. Spluttering refers to the rapid ejection of small particles, as in "frying bacon will splutter fat." In a scientific paper two years later, Thompson dropped the "l" from spluttering in favor of a less common variation, and it's been "sputtering" ever since.

1.1 The Role of PVD in Microelectronics

The physical process that we now call *sputtering* was first reported in 1852 by Sir William Robert Grove [1.1], who described the effect as "cathodic disintegration." Grove's apparatus (shown in Fig. 1.1) utilized a cathode made of silver-coated copper, but his manually pumped vacuum was sufficiently poor ($\approx$ 10 Torr) that the world's first sputter-deposited film was probably not silver but silver oxide. Moreover, it was possible to "disintegrate" the as-deposited film by reversing the electrical leads to the cathode and anode, in effect creating both the first sputter deposition system and the first sputter etching system. Subsequent scientific investigations by other workers in the late 19th and early 20th centuries led to an understanding of the basic physics of the sputtering process and resulted in a variety of industrial coating applications such as the deposition of metal films for mirrors (c. 1875) and the deposition of gold films on wax phonograph masters (c. 1930). By the time the first microelectronic device — the solid state transistor — was demonstrated publicly in 1948, sputter deposition was nearly 100 years old. Since that time thin film deposition by sputtering — i.e., by physical vapor deposition (PVD) — has become an established and essential part of integrated circuit (IC) fabrication technology and has given rise to a multibillion dollar, global PVD equipment market.

In the early years of semiconductor electronics, thin films of metals were typically deposited by electron-beam (e-beam) or hot filament evaporation. However, with the introduction of production-worthy DC magnetron sources in the 1970s, sputtering began to displace evaporation. DC magnetrons were capable of depositing high quality aluminum alloys of Al-Cu and Al-Cu-Si at deposition rates and cost per wafer comparable to evaporation. In addition, the improved step coverage and better control of alloy composition provided by PVD made it attractive for the production

FIG. 1.1 Sputtering was first observed in 1852 using this simple apparatus, making it the world's first PVD system [1.1]. The vacuum in the glass bell jar was manually produced using a hand-operated pump, and the working gas was introduced from a gas-filled bladder through a stopcock.

of advanced large scale integrated (LSI) devices such as the 16K DRAM. As a result, PVD quickly displaced e-beam evaporation for leading-edge applications.

There are a number of reasons why PVD has been so successful for microelectronic applications. First of all, sputtering can be used to deposit all of the conducting films currently used in interconnect metallization schemes, including low-melting-point metals such as Al ($T_{melt} \approx 660°C$) and refractory metals such as Ti ($T_{melt} \approx 1670°C$). Also, sputtering of important multicomponent alloys such as Al-Si-Cu and Ti-W can be deposited from a single alloy sputter target with the deposited film retaining the stoichiometry of the target. This was problematic with evaporation since the deposition rate of the respective alloy constituents depended on their individual vapor pressures. The PVD deposition rate is also well matched to the throughput needs of wafer fabrication ($\approx$ 40 wafers per hour), being about 1 μm/min for thick films (e.g., an 8000 Å Al alloy interconnect) and about 1000 Å/min for thin films (e.g., 500 Å Ti/TiN barrier/liner combination).

Critical film attributes — such as purity and microstructure, which affect electrical conductivity; surface roughness, which affects lithography; and film adhesion to oxides — have all proven acceptable for microelectronic applications. Because sputtering is done from an extended area target and not from a point source as in evaporation, shadowing is minimized and resulting step coverage is generally good ($> 50\%$) over features with relatively low aspect ratio (AR $< 0.5{:}1$).

Another reason for the success of PVD is that the global film uniformity from a properly designed magnetron source has kept pace with wafer size increases in the IC industry. Advanced PVD sources can deposit films with 3σ ("three-sigma") nonuniformity of 3–5% over 200-mm-diameter (8-inch) Si wafers in a production environment. Being primarily a physical deposition process whose underlying physics is well understood, sputtering lends itself to first-principles type of modeling or Monte Carlo simulation. Since PVD utilizes nontoxic targets and low pressures of inert gas ($\approx$ 1–10 mTorr of Ar), it is also in sync with increasing environmental concerns about the use and disposal of hazardous materials. PVD is also compatible with the established trend toward automated single-wafer, vacuum-integrated processing. Finally, PVD has demonstrated an acceptably low cost-of-ownership (CoO) consistent with the economic demands of production-line wafer fabrication.

Figure 1.2 shows the market for semiconductors and semiconductor process equipment since 1965 on a semilog scale. While equipment sales figures for a given year depend on the source (e.g., Dataquest reported capital equipment sales of $\approx$ \$30B in 1995 while VLSI Research reported $\approx$ \$20B), it is the trend that is most important. A significant PVD equipment market emerged in the late 1970s as sputtering began to replace e-beam deposition for mainstream IC metallization. During the period from 1980 through 1997, the PVD equipment market grew from \$100M to \$1.5B — a compound annual growth rate of $\approx$ 23%. This is reflected by the total sales of semiconductor processing equipment (lithography, doping, etching, deposition, annealing, etc.), which increased from about \$2B to \$30B over the same period, and by sales of semiconductors, which increased from about \$20B to \$150B; approximately 90% of today's market is for ICs as opposed to discrete devices. These semiconductors in turn are incorporated in personal computers and in other consumer electronic products valued at about \$800B. Finally, these electronic products sustain larger global industries such as automobiles and aviation. In fact, it has been estimated that all those industries ultimately dependent on semiconductors or electronics represent annual sales of approximately \$1,500B, or \$15T.

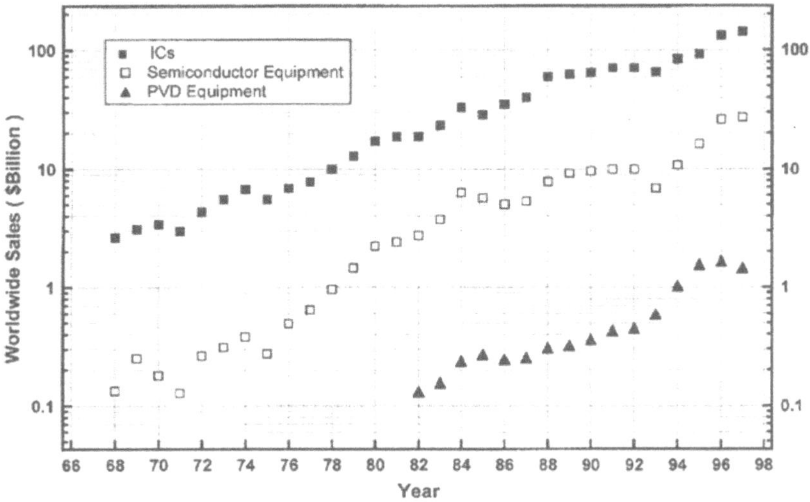

FIG. 1.2 Global sales of semiconductor process equipment and ICs that are fabricated using this equipment have continued to rise steadily, albeit with fluctuations, for more than 30 years.

The economics of the supplier–user interaction has made PVD technology part of a "food chain" within which levels are separated by about an order of magnitude. Using data for 1997, we see that the total semiconductor equipment market was about 15 times larger than that for PVD-specific equipment alone. This equipment in turn was used to produce ICs with sales 5 times larger, leading to electronic products valued about 5 times greater still. Ultimately it is the end user that drives all of these sales, so that when the market for such things as personal computers slumps, or is predicted to slump by economic forecasters, there is a ripple effect back down the chain.

> Strictly speaking, the term *physical vapor deposition* (PVD) can also be used to describe methods such as electron-beam evaporation, thermal filament evaporation, or molecular beam epitaxy (MBE) in which heated crucibles are used to produce vapors that condense at the wafer surface. In this volume, unless stated otherwise, PVD refers only to deposition by sputtering.

Since PVD emerged in the 1970s as a production-worthy technology for microelectronic fabrication, its major application has continued to be metallization and interconnection — i.e., the deposition of electrically connected, multiple levels of metal films. Although insulating films can be deposited by RF magnetron sputtering, methods such as chemical vapor deposition (CVD) and spin-on glass (SOG) technology dominate the deposition of the insulators that electrically isolate one level of metallization from the next and one metal line from an adjacent one on the same level.

Figure 1.3 presents a simplified cross section of an advanced IC and is intended to illustrate how PVD films are utilized for microelectronic fabrication. After the individual transistors are fabricated within the silicon surface, they are contacted and wired together locally to form specific functions (memory cells, logic gates, etc.) and then interconnected together globally to form a fully functioning integrated circuit on a chip. The number of front-end-of-line (FEOL) process steps needed to form active structures as compared to the number of back-end-of-line (BEOL) process steps needed to connect them has steadily decreased. In fact, it has been estimated that fully 60% of the process steps in making an advanced microprocessor are devoted to interconnection [1.2].

The reason for the dominance of BEOL can be understood from device scaling theory [1.3]. As the minimum feature size of a device is reduced to

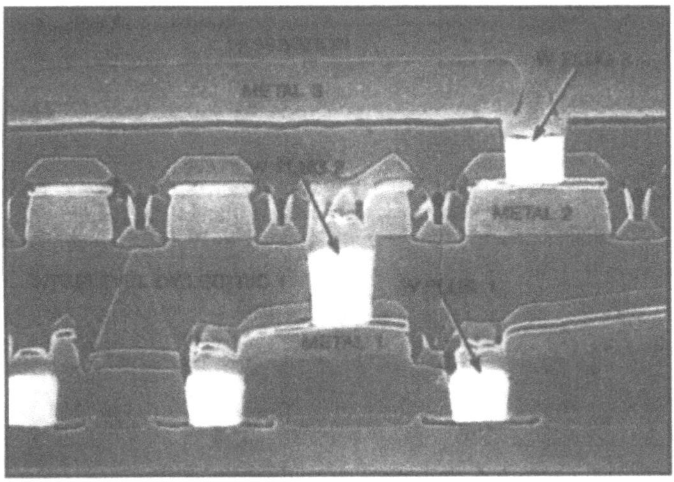

FIG. 1.3 Simplified cross section of an advanced IC (Intel Pentium chip) showing how PVD films are utilized. The IC shown uses CVD W plugs; however, device roadmaps show the eventual replacement of W by Al and/or Cu.

obtain increased device speed and device density, scaling theory shows that the cross-sectional area of the interconnect line needs to be decreased and its length increased. By the year 2001, the total length of interconnects on an advanced microprocessor may well exceed 2 km. A major design challenge has therefore become routing the lengthened interconnect lines so as to minimize RC time-constant signal propagation delays caused by their parasitic capacitance (C) and ohmic resistance (R). Interconnect delay is of increasing concern for advanced ultralarge scale integrated (ULSI) devices because even a small RC time delay associated with the dense wiring (e.g., pitch between adjacent lines $< 0.5\ \mu$m) can be a large fraction of the intrinsic clock cycle time (e.g., a 1 GHz frequency clock has a cycle time $\approx$ 5 nsec), which in turn limits the high-speed performance that was built into the chip. The device packing density of advanced ICs has become so large (e.g., $\approx 10^7$ transistors per cm^2 in a 64Mb DRAM) in fact that the area needed to sensibly route the interconnects now exceeds that of the Si chip itself.

To deal with this situation, a high-rise architecture is used in which multiple levels of metal are isolated by and interconnected through multiple levels of dielectric. Analogous to the framing of a house, this multilevel metallization (MLM) interconnect scheme results in a kind of "joist and stud" configuration with horizontal metal joists of rectangular cross section (the lines) connected by vertical metal studs with circular cross section (the contacts and vias). The chip area, A, required for multilevel wiring has been shown to depend on the number of levels, n, through the expression [1.4]

$$A^{1/2} = \frac{(PG^m)}{n} \tag{1.1}$$

where P is the pitch of the metal wires, G is the number of transistor gates to be connected, and m is analytically determined to be ≈ 0.2 for high-density wiring designs. Therefore, all things being equal, adding an extra level of metal to a three-level metal interconnect (i.e., increasing n from 3 to 4) is equivalent to increasing the chip area by a factor of $(4/3)^2 \approx 2$.

Regarding MLM nomenclature and terminology, the contact hole is the opening that connects the first level of metal (metal 1 or M1) to the Si device. Via holes, on the other hand, connect one layer of metal to the next through an interlevel or interlayer dielectric (ILD) — not to be confused with an intermetal dielectric (IMD), which is the insulator between adjacent metal lines in the same layer. ILDs are numbered as follows: The dielectric between the Si and M1 is referred to as ILD0, the dielectric between M1 and M2 is ILD1, etc. This nomenclature is followed up until the topmost layer of dielectric, which is referred to as the *passivating layer* and whose purpose is to provide physical and chemical protection of the underlying metal and device structures during final assembly of the chip and to prevent the diffusion of moisture and corrosive, mobile ions once the finished chip is operating.

Figure 1.3 presented a cross-sectional view of an MLM stack emphasizing the vertical layering of metal and dielectric, e.g., M1-ILD1-M2. Additional insight into PVD interconnect issues is provided by a plan view (Fig. 1.4) showing the metal wiring in a given layer with length L and cross-sectional, current-carrying area of $T \times W$. The RC time-constant delay introduced by this wiring scheme is then determined by (1) the resistance of the lines, R, (2) their pitch, P, which affects the lateral line-to-line capacitance, C_L, and (3) their vertical separation, which affects the layer-to-layer capacitance, C_V. Assuming the lines are very densely packed, their pitch P might be, say, twice the metal line width (i.e., $P = 2W$). Also, the vertical thickness of dielectric above and below a metal line

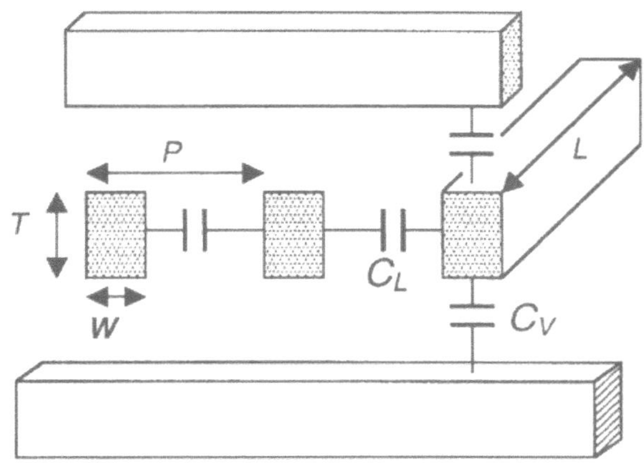

$$R = 2\rho L/PT$$

$$C = 2\,(C_L + C_V) = 2\,\varepsilon\varepsilon_0\,(2LT/P + LP/2T)$$

$$RC = 2\,(C_L + C_V) = 2\,\rho\varepsilon\varepsilon_0\,(4L^2/P^2 + L^2/T^2)$$

FIG. 1.4 Simplified plan view of wiring in an IC indicating how the line resistance and parasitic capacitance between metal lines contributes to the overall RC delay time.

will be close to the thickness of the line. With these assumptions, it is straightforward to estimate the RC delay [1.5] as follows:

$$R = \frac{2\rho L}{PT} \tag{1.2}$$

$$C = 2(C_L + C_V) = 2\epsilon\epsilon_0\left(\frac{2LT}{P} + \frac{LP}{2T}\right) \tag{1.3}$$

so that

$$\mathrm{RC} = 2\rho\epsilon\epsilon_0\left(\frac{4L^2}{P^2} + \frac{L^2}{T^2}\right) \tag{1.4}$$

where ϵ_0 is the permittivity of free space, ϵ (sometimes written as k) is the dielectric constant of the interlayer insulator, and ρ is the resistivity of the metal line.

It should be noted that while these equations can be used to gain insight into MLM issues, they are a highly simplified treatment of a mathematically complex problem. For example, the simple treatment used to obtain Eq. (1.3) assumes that the vertical capacitance C_V between two lines is proportional to their planar area $L \times W$ — similar to the elementary treatment of a semi-infinite, parallel plate capacitor with capacitor plate area $= LW$ and dielectric thickness T. This result would be valid when $W >> T$. However, when W is comparable or less than T, a rigorous treatment of the problem — which involves solving the second-order partial differential Laplace's equation — shows that C_V is proportional to log W and not to W. Hence, while narrowing the width of interconnect lines is expected to reduce interlayer capacitance, the gains will be much less than predicted by elementary theory once line width has been scaled down to the dielectric thickness, which today is $\approx 1\ \mu$m.

Equation (1.4) shows that interconnect delay is directly proportional to the product $\rho\epsilon$, which has driven the move toward higher-conductivity PVD metals (e.g., pure Cu with $\rho = 1.7\ \mu\Omega$-cm versus Al-0.5%Cu with $\rho = 3.0\ \mu\Omega$-cm) and/or lower dielectric constant insulators (e.g., fluorine-containing CVD oxide with $\epsilon \approx 3$ versus conventional CVD oxide with $\epsilon \approx 4$). Going to a lower dielectric constant insulator also reduces the AC power consumed by the chip since this power loss is directly proportional to ϵ. The L^2-dependence in Eq. (1.4) shows the value of using multiple levels of wiring that reduce the line length per layer, with a quadratic reduction in RC time constant associated with that layer of wiring. On the other hand, increased levels of metal wiring on a chip also require increased capital investment by the chip maker. Since the performance-driven trend in MLM materials away from Al/SiO_2 and toward Cu/low-k dielectrics will allow fewer levels of metal to be used for the same device generation, a considerable cost savings is expected as well. For example, it has been estimated by SEMATECH that \$1.3B in back-end-of-line capital equipment will be required to build a 10,000-wafer-per-week fab producing a high-end microprocessor (0.18 μm) having eight levels of metal and Al/SiO_2 wiring. However, by switching to Cu/low-k dielectric wiring, the same device performance could be achieved using only five levels of metal. The net result of this simplified device architecture is that the capital equipment investment can be reduced by \$500M.

Figure 1.3 indicates that a variety of films are used to engineer a contact or via plug having appropriate electrical and mechanical properties for the devices and circuits being fabricated. For example, a PVD "aluminum plug" might actually consist at the contact level of a thin bottom layer of Ti ($\approx$ 5 nm = 50 Å) to reduce contact resistance to the exposed Si (by chemically reducing native SiO_2 and also reacting to form a low contact resistance $TiSi_2$ film after annealing), a thin liner sleeve of TiN ($\approx$ 10 nm = 100 Å) to serve as a diffusion barrier between the Al and the Si, and the actual thick plug of Al-0.5%Cu ($\approx$ 1 μm). Similarly, a PVD "aluminum interconnect line" might actually consist of an engineered slab of several films (e.g., TiN/Al-0.5%Cu/Ti/TiN with the Ti/TiN on top), each chosen to enhance a desired property such as resistance to electromigration or to improve a subsequent process step such as a TiN antireflection coating (ARC), which is added to facilitate photoresist patterning with optical lithography.

As a practical matter, there are only two thickness ranges of interest for PVD films used in microelectronics. Roughly speaking, they are 50–500 Å for contacts, barriers, liners, and ARC layers, and 0.5–1 μm for contact plugs, via plugs, and interconnect lines. The ratio of height-to-width (i.e., the aspect ratio, AR) of features to be coated or filled with metal can range from zero for a planar surface to AR = 5:1 or even 10:1 for a contact hole in advanced devices. A common practice is to use closely packed lower levels of metal for local interconnections and thicker, higher conductivity, wider pitched metal patterns at upper levels for power supply buses and global interconnections. Given more or less the same interlayer dielectric thickness between levels, the via holes at upper levels then tend to be less steep than those at lower levels; this facilitates step coverage or filling with PVD metal. This is reflected by the fact that although fabrication of an advanced IC (e.g., a microprocessor with minimum feature size of 0.25 μm) requires on the order of 25 lithographic mask levels, only 5 of these masks will contain features with dimension ≤ 0.35 μm. On the other hand, as will be discussed in Chapter 5, the so-called thermal budget allowed for advanced chips leads to lower-temperature processing at the via levels than at the contact level, making it harder to achieve this architectural benefit.

Considering the fact that virtually any solid metallic or ceramic target can be sputtered, the number of materials sputter deposited in Si microelectronic fabrication is remarkably small — and many of these serve useful dual purposes (e.g., a TiN ARC layer also helps harden the Al interconnect slab against stress migration failure). Figure 1.5 shows that, on an industry-wide scale, the vast majority of PVD equipment in microelectronic production is used to deposit Al alloys (Al with ≈ 1% of Si and/or Cu), Ti, or TiN. This is followed by relatively low utilization of Ti-W alloys and refractory metal silicides such as WSi_x and $MoSi_x$. As MLM technology evolves, it is likely that Cu interconnects and Cu-compatible barriers such as Ta and TaN will also join the most-favored PVD materials list. However, the Cu interconnect may be deposited using PVD in combination with other deposition methods — e.g., a PVD Cu adhesion–seed layer followed by CVD Cu, or a PVD Cu "strike layer" followed by electroplated Cu (see Chapter 9).

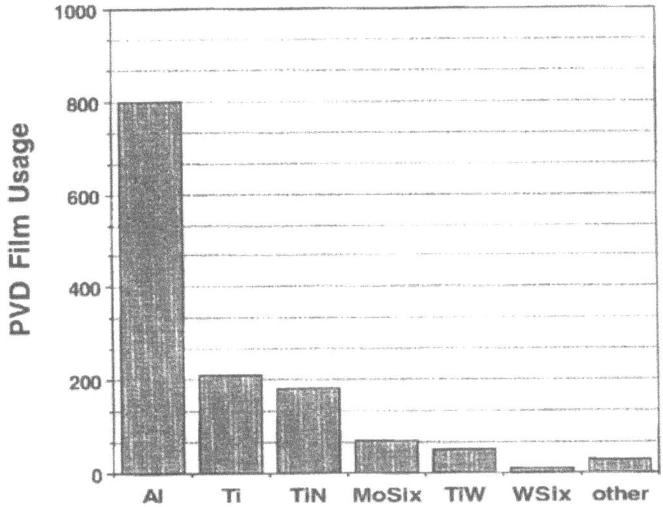

FIG. 1.5 Histogram showing the industry-wide use of PVD by film type. Whereas many different kinds of conducting films are deposited by PVD for advanced microelectronic applications, the dominant films are Al alloys, Ti, and TiN.

Thinking geometrically (see Fig. 1.6) one can regard PVD as having four major applications of interest in microelectronics: (1) a coating on the bottom of a contact hole, via hole, or trench; (2) a lining on the sides; (3) a plug that fills such a feature; and (4) a planar two-dimensional film. A blanket, conformal coating might simultaneously satisfy applications (1) and (2) and, if thick enough, (3) and (4) as well. The challenge for PVD is to cost-effectively deposit such films in features that have smaller dimensions and higher aspect ratios and to do this uniformly over larger-diameter wafers with film properties needed for advanced Si devices.

In this regard it is worth pointing out that while PVD has achieved a dominant position in microelectronic metallization, there are concerns that the step coverage of PVD films may not be good enough to conformally coat and/or fill high aspect ratio features (AR $\geq$ 4:1) in advanced ULSI devices (minimum feature size $\leq 0.25\ \mu$m). The fundamental reason for concern is that the material ejected from a sputter target has a broad angular distribution that results in a nonnegligible flux of low-angle material hitting the wafer surface. The deposited film associated with this flux can shadow the high-angle material that would otherwise coat the sidewalls and bottom of the feature.

The general situation is illustrated in Fig. 1.7, which shows how, for a sufficiently steep feature and/or thick deposition, PVD films can preferentially build up material at the top of the feature. Sometimes referred to as an *overburden* or *bread-loafing*, this material can reduce bottom coverage, thereby preventing uniform coating of the sidewalls. If the PVD film is thick enough, this overburden can bridge the diameter of the hole with metal and lead to an undesired keyhole-shaped void being produced

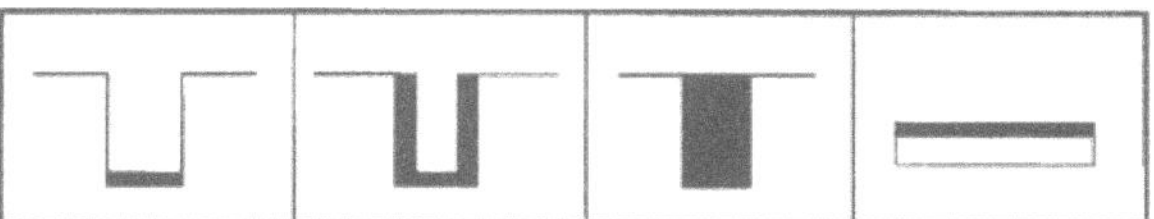

Geometry	3-Dimensional	3-Dimensional	3-Dimensional	2-Dimensional
Need	Bottom Coverage	Conformality	Filling	Planar Film
Application	Contact Clean	Barrier Liner or Adhesion Layer	Contact Plug or Via Plug	Interconnect

FIG. 1.6 PVD applications can be divided geometrically into those requiring a two-dimensional planar film and those requiring coating and/or filling of a three-dimensional feature.

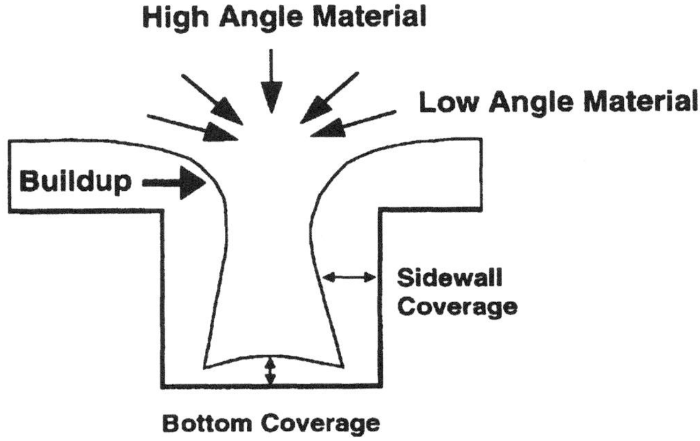

FIG. 1.7 A relatively large flux of low-angle deposition during conventional PVD can lead to preferential buildup of material at the upper edge of fine-geometry structures. This in turn limits bottom and sidewall coverage of thin films and may prevent void-free filling of the structure.

within the volume of an otherwise filled plug. The future role of PVD in microelectronics will be influenced strongly by our ability to control the directionality of the deposition and to improve the step coverage of the resulting films. Chapters 6–8 will treat this important topic in detail, discussing a number of recent developments in PVD source technology and processing.

1.2 PVD and the Interconnect Roadmap

It is almost obligatory in books about semiconductor technology to show the evolution of device complexity by plotting such things as minimum feature size, level of integration, die size, and clock speed versus year. One of the best-known and best-cited roadmaps is the *Semiconductor Industry Association's National Technology Roadmap*, referred to as the SIA Roadmap, which in 1994 extended historic device technology trends out to 2010 and identified device-driven and productivity-driven needs for key technologies such as interconnection that directly impact PVD. An updated version of the SIA Roadmap was released in 1997, which extended the industry's vision of the future through the year 2012 when, if the roadmap is followed, logic circuits will have minimum feature size of 0.035 μm, memory chips will hold 256 gigabits of memory, and silicon wafers will measure 450 mm in diameter. In addition to extending the 1994 roadmap by two years and one process generation, the 1997 SIA roadmap added a new "technology node" at 0.15 μm, between the 0.18 μm and 0.13 μm device generations on the earlier roadmap.

Because microelectronics is digital, the letter "K" used to indicate bit count per chip (e.g., a 256K DRAM = 256 kilobits) does not equal 1000 as in the metric system. Instead, the microelectronic K has a value of $1024 = 2^{10}$, which is very close to $1000 = 10^3$. Knowing that $2^{10} \approx 10^3$ and taking the logarithm of both sides is also an easy way to remember that $\log_{10}2 \approx 0.3$.

It is well beyond the scope of this book to present a comprehensive PVD roadmap (the reader is encouraged to review the actual SIA Roadmap or summarized versions [1.6]). Instead, in Figs. 1.8–1.10 we present greatly abbreviated roadmaps in graphical and tabular form based on both the SIA Roadmap and others (e.g., those generated by SEMATECH and the Semiconductor Research Corporation (SRC)) to set the stage for later chapters on PVD process and hardware development. Two caveats need to be made regarding technology roadmaps in general. The first is that a roadmap can extrapolate historic trends but cannot anticipate truly breakthrough technology. The second is that device details such as aspect ratio and number of metal levels are company-specific, so that a range of values

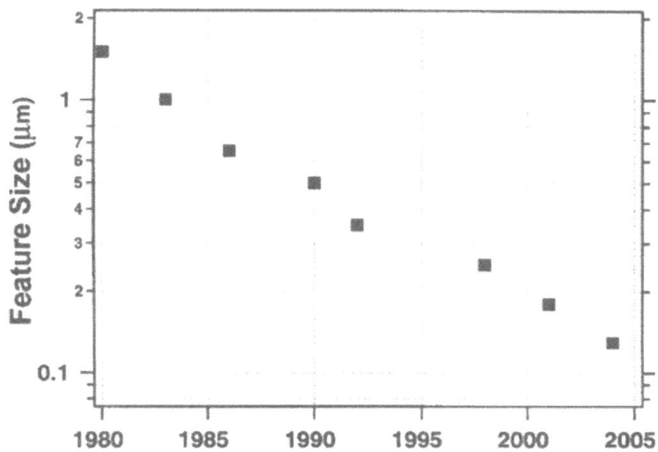

FIG. 1.8 **(a)**

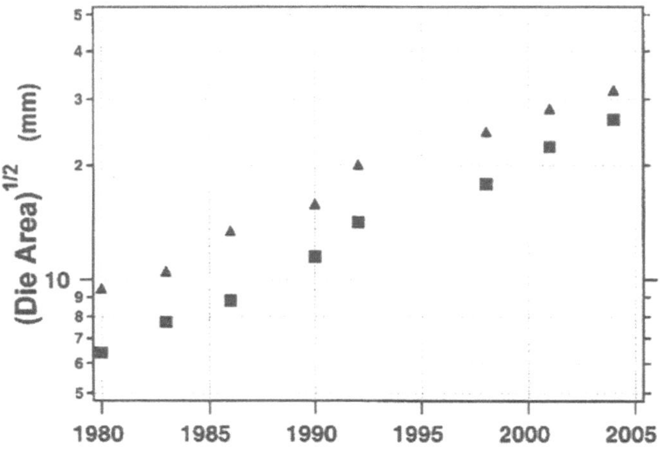

FIG. 1.8 (b)

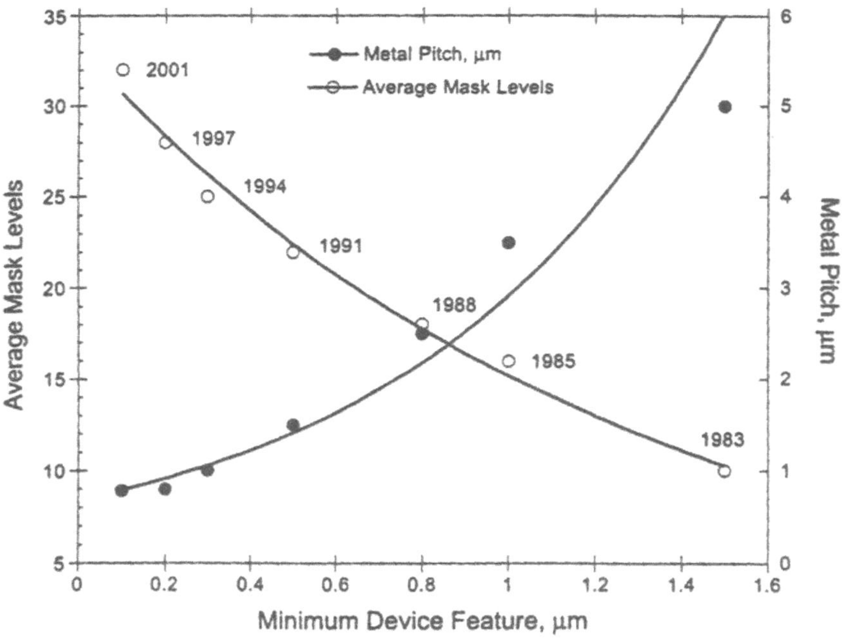

(c)

FIG. 1.8 Evolution of silicon device technology showing (a) decreasing feature size, (b) increasing die area, and (c) increasing complexity over time (Fig. 1.8c from ref. 1.7).

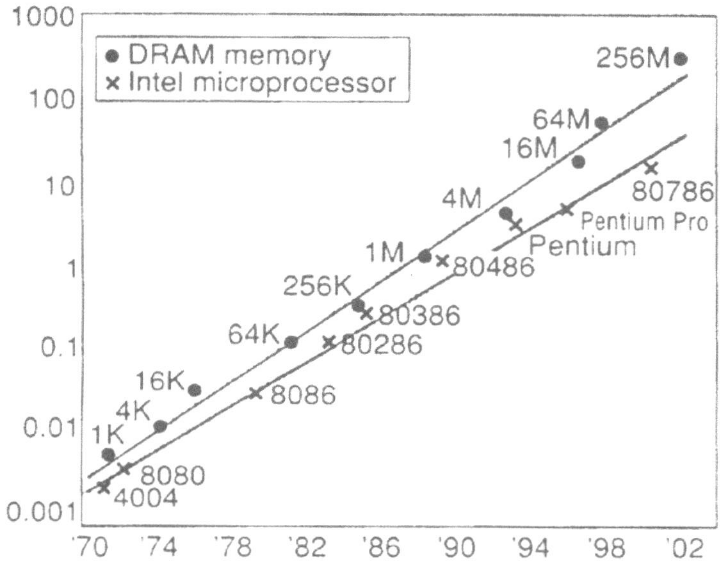

FIG. 1.9 Evolution of silicon IC sophistication is illustrated with product generations of Intel microprocessors (courtesy of VLSI Research Inc.).

is more appropriate for a given attribute at a given time. Nevertheless, several observations related to the application of PVD in microelectronics can be drawn from Figs. 1.8–1.10.

By 2001, when the geometry of devices in pilot production is 0.18 μm (equivalent to a 1-Gigabit = 1-Gb level of integration in a DRAM), leading-edge wafers will be 300 mm (12 inch) in diameter. Thus, not only will device geometry be smaller, but the area on which PVD films are to be deposited uniformly will be over 2 times greater than for a 200-mm wafer. Of special note is that the aspect ratio of features to be coated and/or filled with metal could become so high for both memory and logic that it will exceed the step coverage capability of PVD. The reason for the rise of aspect ratio is that vertical device scaling has not kept pace with lateral scaling. In particular, given the relatively high dielectric constant of the CVD oxides ($k \approx 4$) used for ILD, it is difficult to bring advanced metal interconnect levels closer than about 0.8 μm without increasing parasitic capacitance to the extent that signal propagation is seriously delayed. Therefore, when a device designer shrinks a via hole diameter, the aspect ratio is likely to go up since the vertical distance between metal-(n) and metal-(n + 1) cannot go down. The development of lower-k dielectric interlayers (e.g., k = 2)

Year of First Product Shipment →	1997	1999	2001	2003
Complexity (DRAM)	256 Mbit	1 Gbit	1 Gbit	4 Gbit
Transistor - minimum feature size - gate oxide equivalent thickness - junction depth at channel	 0.25 μm 4-5 nm 50-100 nm	 0.18 μm 3-4 nm 36-72 nm	 0.15 μm 2-3 nm 30-60 nm	 0.13 μm 2-3 nm 26-52 nm
Chip Size (average) - DRAM - microprocessor	 280 mm^2 300 mm^2	 400 mm^2 340 mm^2	 445 mm^2 385 mm^2	 560 mm^2 430 mm^2
Wafer Diameter (mm)	200 mm (8 inch)	200 & 300 mm (8 &12 inch)	300 mm (12 inch)	300 mm (12 inch)
Number of Metal Levels - memory (DRAM) - logic (microprocessor)	 2-3 6	 3 6-7	 3 7	 3 7
Contact/Via Aspect Ratio - memory - logic	 5.5:1 2.2:1	 6.3:1 2.2:1	 7.0:1 2.4:1	 7.5:1 2.5:1
Maximum Interconnect Length for Logic (meters/chip)	820	1,480	2,160	2,840
Particles - critical size - allowed particle density greater than size (per m^2)	 125 nm 125	 90 nm 125	 75 nm 125	 65 nm 125

FIG. 1.10 Technology Roadmap for Interconnect, based on the *SIA National Technology Roadmap* [1.6].

could help slow the growth of aspect ratio, but the approach is limited by the fact that the lowest dielectric constant "material" is free space with a dielectric constant of $k = 1$.

The trend for both memory (e.g., DRAMs) and logic devices (e.g., microprocessors) is clearly toward increased levels of MLM interconnect (six or more levels for logic), and since each level requires one or more engineered metal depositions, this trend also supports the continued growth of PVD. It should be noted that although a logic and a memory chip may be at the same device generation, the level of integration can be quite different. For example, a 256M DRAM chip and a microprocessor might both require the same 0.25-μm minimum feature size, but the microprocessor might have 10 times fewer transistors. The severe limits put on particles and other foreign matter by continued shrinking of minimum feature size and continued growth of die area will require PVD tools to provide the highest levels of vacuum, robotic, and process cleanliness. Finally, it should be underscored that the economic and support issues related to PVD equipment (cost-of-ownership, tool up-time, mean-time-to-repair the tool, etc.) will be every bit as important as the technical quality of the PVD process (blanket film properties, step coverage over high aspect ratio features, within-wafer uniformity, etc.).

1.3 Additional Sources of Information on PVD

Neither this book nor the references given at the end of each chapter are intended to be a comprehensive review of PVD for microelectronic applications. Fortunately, there are a number of ways of obtaining additional information about PVD process and hardware technology and of keeping abreast of developments in this fast-moving field.

Other than this volume, we are not aware of any monograph on the use of PVD in modern microelectronics. The book *Physical Vapor Deposition* [1.8] presents a good overview of the use of both e-beam evaporation and sputtering in industrial coating circa 1985, and useful review chapters on sputter deposition relevant to semiconductor fabrication can be found in handbooks such as *Thin Film Processes* [1.9, 1.10], *Thin Film Processes II* [1.11, 1.12], the *Handbook of Thin Film Process Technology* [1.13], and in the series on *VLSI Electronics: Microstructure Science* [1.14]. There are also a large number of books dealing with the general topic of thin films for semiconductor applications. Since the primary use of PVD in microelectronics is directed at interconnect and metallization technology, books on this topic often contain chapters on the materials science of relevant PVD thin films (e.g., Ti, $TiSi_2$, TiN, Al-Cu-Si alloys, etc.) and on their process integration. In addition to the handbooks mentioned earlier, the following books on thin film technology also include much valuable information: *Handbook of Multilevel Metallization for Integrated Circuits* [1.15], *Thin Film Deposition: Principles & Practice* [1.16], *Film Deposition by Plasma Techniques* [1.17], and *Principles of Plasma Discharges and Materials Processing* [1.18].

Manual or computer searches of scientific data bases under key words such as PVD, physical vapor deposition, sputtering, etc. will provide useful sources; a good starting point is the *Journal of Vacuum Science and Technology*, which has published many excellent papers and reviews on the physics and technology of sputtering. Several other scientific journals also have a strong history of reporting process, hardware, and device-related developments in PVD science and technology, including (in alphabetical order) *Applied Physics Letters, IEEE Transactions on Electron Devices, Japanese Journal of Applied Physics, Japanese Journal of Applied Physics Letters, Journal of Applied Physics, Journal of the*

Electrochemical Society, and *Thin Solid Films*. Given the trend toward electronic publishing, it should soon be possible to obtain journals such as these on CD-ROM to facilitate both searching and retrieving PVD-related information. In addition, it is now possible to use the Internet to get on-line information about microelectronic manufacturing in general and PVD in particular. While a number of search engines can be used to locate specific information, the web home pages of trade publications such as *Semiconductor International* (http://www.semiconductor-intl.com) and *Solid State Technology* (http://www.solid-state.com) are a convenient entry point for general information on microelectronic manufacturing. Also, suppliers of PVD hardware have worldwide web sites permitting 24 hour-a-day, real time access to company information, press releases, interactive catalogs, product brochures, data sheets and technical reports. Finally, large scientific societies such as the American Institute of Physics (http://www.aip.org/aiphome.html) have established home pages that allow visitors to browse through periodicals of member societies — such as the *Journal of Vacuum Science & Technology* published by the American Vacuum Society.

Many conferences regularly solicit papers on PVD for microelectronics and publish technical proceedings. For example, the yearly VLSI Multilevel Interconnection Conference (VMIC) has provided a forum for the discussion of advanced PVD processing technology since 1986; its published proceedings document many important developments in the field. Also of note are the yearly proceedings of the American Vacuum Society (AVS) National Symposium, the biyearly (spring and fall) Materials Research Society (MRS) Symposium, the biyearly (spring and fall) Electrochemical Society (ECS) Symposium, the International Electron Devices Meeting (IEDM), the International Conference on Metallurgical Coatings and Thin Films (ICMCTF), and the International Interconnect Technology Conference (IITC) which began in 1998.

Peer-reviewed papers in scientific journals and conferences are complemented by so-called trade publications such as *Semiconductor International* and *Solid State Technology*, both of which regularly include brief overview articles by their editorial staff and outside authors on process and hardware trends in sputter deposition for semiconductor, magnetic disk, and flat panel display fabrication. These articles are extremely useful and usually include topical comments from both PVD equipment suppliers and end users (see Fig. 1.11). Several trade publications also provide an annual buyers' guide, which includes a directory of vendors of products and services related to PVD — e.g., *Solid State Technology*'s "Resource Guide," *Semiconductor International*'s "Product Data Source,"

Title of Review Article	Citation
The Reasoning Behind "Cost of Ownership"	J. Secrest and P. Burggraaf, *Semiconductor Int'l*, pp. 56-60 (May 1993)
The Interconnect Challenge: Filling Small, High Aspect Ratio Contact Holes	P. Singer, *Semiconductor Int'l*, pp. 57-64 (Aug 1994)
Silicides for High Density Memory and Logic Circuits	J. Winnerl, Semiconductor Int'l, pp. 81-86 (Aug 1994)
New Interconnect Materials: Chasing the Promise of Faster Chips	P. Singer, *Semiconductor Int'l*, pp. 52-56 (Nov 1994)
Interconnect Metallization for Future Device Generations	B. Roberts, A. Harrus and R. L. Jackson, *Solid State Technology*, pp. 69-78 (Feb 1995)
Electrostatic Chucks in Wafer Processing	P. Singer, *Semiconductor Int'l*, pp. 57-64 (April 1995)
The Driving Forces in Cluster Tool Development	P. Singer, *Semiconductor Int'l*, pp. 113-118 (July 1995)
Straightening Out Sputter Deposition	P. Burggraaf, *Semiconductor Int'l*, pp. 69-74 (Aug 1996)
Filling Contacts and Vias: A Progress Report	P. Singer, *Semiconductor Int'l*, pp. 89-94 (Feb 1996)
Low k Dielectrics: The Search Continues	P. Singer, Semiconductor Int'l, pp. 88-96 (May 1996)
Design Challenges in Vacuum Robotics	E. Korczynski, *Solid State Technology*, pp. 62-70 (Oct 1996)
Depositing Diffusion Barriers	J. Baliga, *Semiconductor Int'l*, pp. 76-82 (Mar 1997)

FIG. 1.11 Articles from trade publications provide a capsule summary of PVD equipment and process development. Selected articles from *Semiconductor International* and *Solid State Technology* that address hardware and process issues of sputter deposition for microelectronics are used for illustration.

and *R&D Magazine*'s "Product Source." PVD system suppliers as well as suppliers of peripheral hardware and consumables (e.g., vacuum pumps, robotic arms, sputter targets) maintain technical reports and reprints of articles written by their staff. These reports can usually be obtained simply by contacting their customer support or marketing organizations.

Without question, the premier semiconductor fabrication equipment trade show is SEMICON-West, which is held in the San Francisco area and includes exhibits by all major suppliers of PVD hardware and often company-sponsored technical seminars highlighting a supplier's latest product or process introduction. The global nature of both IC device fabrication and the semiconductor process equipment market has led to similar SEMICON shows in many other parts of the world (e.g., SEMICON-Southwest, SEMICON-Japan, SEMICON-Korea, and SEMICON-Taiwan). All of these shows generally include technical sessions relevant to PVD with written proceedings; however, these proceedings are not always easy to obtain after the fact.

Short courses on PVD are regularly offered by a number of societies, such as the AVS and the MRS, and related process, hardware, and metrology workshops are sponsored by SEMATECH, the Semiconductor Research Corporation (SRC), and the Semiconductor Equipment and Materials Institute (SEMI).

Finally, a valuable source of information that is often overlooked is published patents. A patent begins with a brief overview of the "state of the art," which sets the stage for a discussion of what is new and different about the patented invention. Therefore, one can quickly get informed on a wide range of topics such as PVD cluster tool architecture, backside gas conduction cooling, magnetron sputter source design, electrostatic chucks for PVD, etc. by obtaining copies of key patents and reading the introductory sections that provide a brief, critical review of the technology.

References

1.1. W. R. Grove, "On the electro-chemical polarity of gases," *Phil. Trans. R. Soc.* **142**: 87–101 (1852).

1.2. E. Korczynski, "VMIC: Alternatives for interconnections," *Solid State Tech.*, 40 (Sept. 1995).

1.3. K. C. Saraswat and F. Mohammadi, "Effect of scaling on interconnections on the time delay of VLSI circuits," *IEEE Trans. on Electron Devices* **ED-29**(4): 645–650 (1982).

1.4. D. M. Brown, M. Ghezzo, and J. M. Pimbley, "Trends in advanced process technology — Submicrometer CMOS device design and process requirements," *Proc. IEEE* **74**(12): 1678–1702 (1986).

1.5. M. T. Bohr, "Interconnect scaling — The real limiter to high performance ULSI," *Solid State Tech.* 105–111 (Sept. 1996).

1.6. *The National Technology Roadmap for Semiconductors*, Semiconductor Industry Association, San Jose, CA, 1994 and 1997; see also summary of the 1994 Roadmap in *Solid State Tech.*, 42–53 (Feb. 1995).

1.7. T. Kodas and M. Hampden-Smith, *The Chemistry of Metal CVD*, chapter 1, VCH Publishers, New York, 1994.

1.8. *Physical Vapor Deposition*, 2nd ed., R. J. Hill, Ed., Temescal, a division of the BOC Group, Berkeley, CA, 1986.

1.9. David B. Fraser, "The Sputter and S-Gun Magnetrons," in *Thin Film Processes*, pp. 115–129, John L. Vossen and Werner Kern, Eds., Academic Press, New York, 1978.

1.10. Robert K. Waits, "Planar Magnetron Sputtering," in *Thin Film Processes*, pp. 131–173, John L. Vossen and Werner Kern, Eds., Academic Press, New York, 1978.

1.11. Stephen M. Rossnagel, "Glow Discharge Plasma and Sources for Etching and Deposition," in *Thin Film Processes II*, pp. 11–77, John L. Vossen and Werner Kern, Eds. Academic Press, New York, 1991.

1.12. Robert Parsons, "Sputter Deposition Processes," in *Thin Film Processes II*, pp. 177–208, John L. Vossen and Werner Kern, Eds., Academic Press, New York, 1991.

1.13. *Handbook of Thin Film Process Technology*, David A. Glocker and S. Ismat Shah, Eds., Institute of Physics (IOP) Publishing, Bristol, UK, 1995.

1.14. *VLSI Electronics: Microstructure Science*, Norman G. Einspruch, Series Ed. Academic Press, New York, 1990.

1.15. *Handbook of Multilevel Metallization for Integrated Circuits*, Syd R. Wilson, Clarence J. Tracy, and John L. Freeman Jr., Eds., Noyes Publications, Park Ridge, NJ, 1993.
1.16. Donald L. Smith, *Thin Film Deposition: Principles & Practice*, McGraw-Hill, New York, 1995.
1.17. Mitsuharu Konuma, *Film Deposition by Plasma Techniques*, vol. 10 in the Springer Series *Atoms + Plasmas*, G. Ecker, P. Lambropoulos, I. Sobelman, and H. Walther, Series Eds., Springer-Verlag, New York, 1992.
1.18. Michael A. Lieberman and Allan J. Lichtenberg, *Principles of Plasma Discharges and Materials Processing*, Wiley, New York, 1994.

Chapter 2 Physics of Sputtering

Thin film, vacuum-based deposition technologies fall into two basic catagories: physical vapor deposition (PVD) and chemical vapor deposition (CVD). PVD techniques include physical sputtering, which is the underlying topic of this volume, thermal evaporation [2.1, 2.1], and arc-based deposition [2.3, 2.4]. These techniques are generally atomic in nature, in that the films are deposited from single atoms or small clusters and any reactions that occur (such as oxidation or nitridization) occur at the film surface independently of the source process. This differs from CVD techniques, in which molecular species in the gas phase chemically react at a film surface, resulting in the formation of a condensed film as well as the emission of volatile by-products. CVD techniques will not be discussed in this volume.

Sputtering is a relatively simple process in which an energetic particle bombards a target surface with sufficient energy to result in the ejection of one or more atoms from the target. The sputter yield, Y, is just the ratio of the number of emitted particles to the number of bombarding ones:

$$Y = \frac{\text{(number of ejected particles)}}{\text{(number of incident particles)}} \tag{2.1}$$

Physical sputtering can result from bombardment with a variety of incident species. The most commonly used species is an inert gas ion (e.g., Ar^+, Kr^+), but sputtering can also result from the bombardment of other energetic ions, neutrals, electrons, and even photons. In general, the physical effects caused by bombardment with a neutral or an ion of the same species and energy will be identical. The ion is usually neutralized by pulling an electron from the near-surface region just prior to impact, and so it impacts the surface as a neutral. However, since the electrical current to the target due to ion bombardment is easily measured and it is quite easy to generate large fluxes of ions at controlled energies, virtually all applications of sputtering use ions as the bombarding particles. Because of the vast variety of possible effects that can occur, we will confine the discussion primarily to inert gas positive ion bombardment, with occasional divergences to neutrals or appropriate negative ions.

2.1 Sputtering

The sputtering process is one of relatively violent, kinetic collisions — first between the incident energetic particle and one or two substrate atoms, and then subsequent collisions between multiple atoms as the incident kinetic energy and momentum are distributed among many atoms (Fig. 2.1). Depending on the kinetic energy, E, of the incident ion, four different physical results can occur:

1. *Low Energy* ($0 < E <$ 20–50 eV). This regime is known classically (and somewhat inaccurately) as the *subthreshold region*. In this regime, it was thought that the incident ion had too little energy to dislodge and eject a target atom and that the resultant yield was zero. For many years, it was observed that sputtering seemed to have a threshold of about 40 eV for most materials, below which sputtering did not occur (Fig. 2.2). This was due to the dramatic fall-off in the yield as the ion energy decreased.

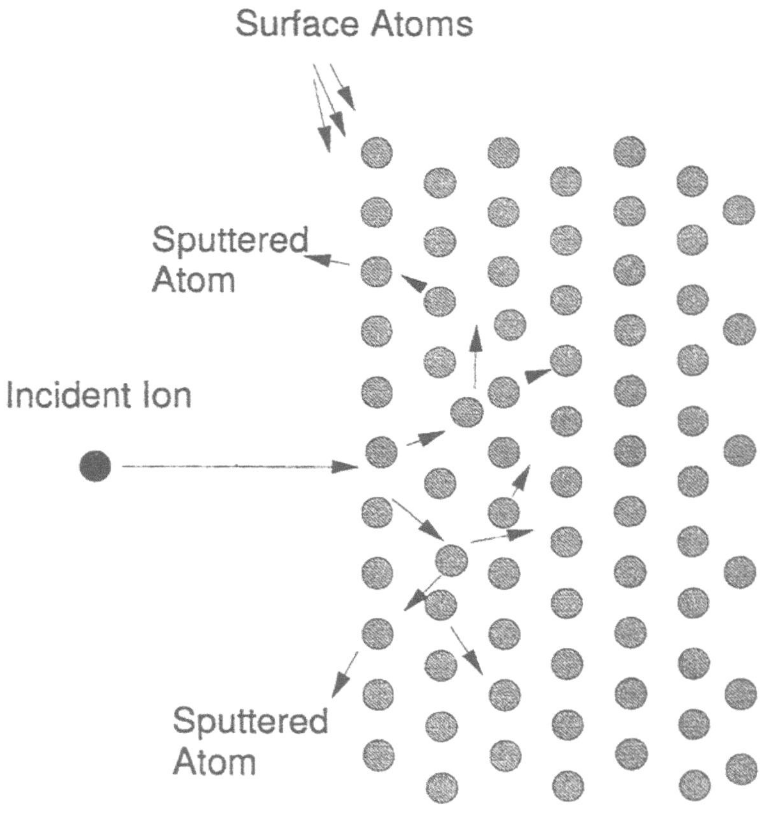

FIG. 2.1 Schematic of physical sputtering process.

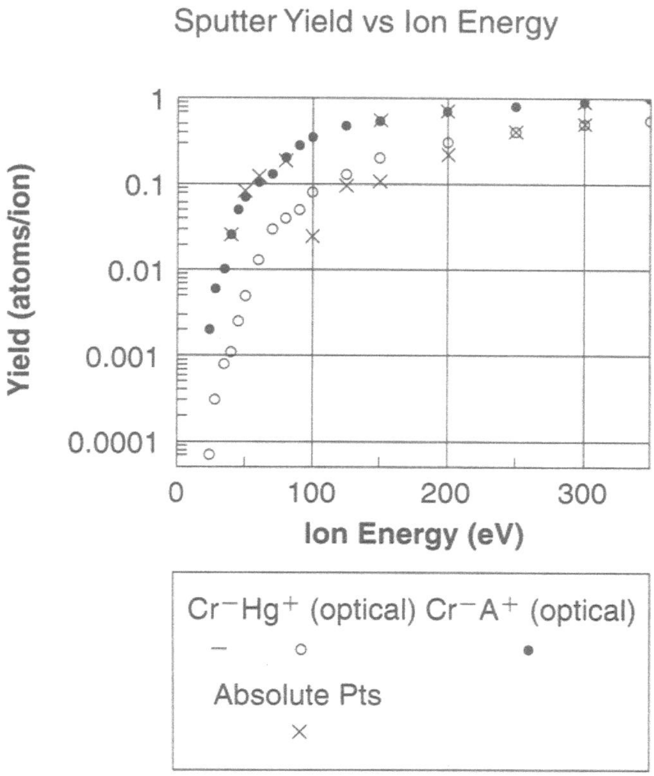

FIG. 2.2 Sputter yield for Cr sputtered with Ar and Hg as a function of ion energy at low energies [2.5].

Various models were developed that predicted thresholds of about 4 times the binding energy of the target material, which corresponded with energies in the 30 eV range.

Experimentally, though, more evidence has become available that suggests that sputtering can occur at energies below 4 times the binding energy. In high-density plasmas, such as those formed using ECR techniques, sputtering and film deposition at effective ion energies of below 15 eV are routinely observed. The required yields are in the 10^{-6} range, which is 2 to 3 orders of magnitude below the earlier measurements that suggested a threshold at higher energy. However, since the effective ion currents in an ECR tool may be many tens of amperes, even these very tiny yields can be quite significant.

EXAMPLE: With an ECR tool operated at 1 kW, the total ion flux within the source is on the order of 20 Amperes (at 50 eV/ion production rate). Most of this ion current lands on the chamber walls. With a sputter yield of 10^{-5} at perhaps 10–15 V of plasma potential, this leads to an erosion rate in a typical source (800–1000 cm^2) of about 0.005 atomic layers per second. While this seems small, since the material sputtered is randomly redeposited and can land on the dielectric window through which the microwave power enters the source, an electrically opaque film (approx. 10 nm) will be deposited in a little more than 1 hour of plasma run time. This film then reflects additional microwave power from entering the source. Obviously, even an extremely low level of sputtering can become crucial in these very high current tools.

There has been relatively little theoretical work on very low energy sputtering, perhaps because there are few applications. However, the previous concept of a true sputter threshold is really not that accurate. Under the right conditions, even an incident particle with very low energy (< 1 eV) might be able to dislodge an adsorbed surface atom.

2. *Moderate Energy* (50 eV $< E <$ 1000 eV). This range, sometimes known as the *knock-on sputtering regime,* covers most of the practical range of energies used for PVD technologies. In this range, the incident ion impacts a target atom, which recoils and strikes one or more atoms, which each then recoil, and the process continues much like in Fig. 2.1. However, this is a difficult process to predict and measure because it is keenly dependent on the exact collision point of the incident ion. The sequence of collisions will be completely different for each bombarding particle because each particle will hit in a different place with regard to the location of the surrounding atoms, and only a small fraction of the target atoms near the impact point will actually be dislodged as part of the collision chain.

This process must be evaluated practically by simply looking at the average of a large number of impacting particles. Various computer codes have been developed that follow the collision chains for many impacting ions. The most widely used program is called TRIM, and there are many variants [2.6–2.8] (see Section 10.1).

The sputter yield depends strongly on the incident particle's mass and kinetic energy as well as the substrate's mass and orientation. For many years it was thought that the substrate's temperature was important also. However, in the early 1980s a group in Julich, Germany, clearly showed that unless the temperature was very close to the melting point, it was not relevant to the sputtering process [2.9]. Conceptually, also, it would not make sense that energies on the thermal scale (0.1 eV) present in a warm substrate would have that much influence on sputtering events, which contain energies in the hundreds of eV range.

The yields for several materials of relevance to semiconductor applications are shown in Figs. 2.3 and 2.4. Sputter yields for many common materials used in semiconductor applications for several ion energies and inert gas species are given in Table 2.1 (adapted from [2.10]).

3. *High Energy* (1 keV $< E <$ 50 keV). This region, which is not relevant to semiconductor processing, is nevertheless a more well understood region. At these energies, the incident ion causes a dense cascade of secondary particles (target atoms) after the initial impact. Within this cascade

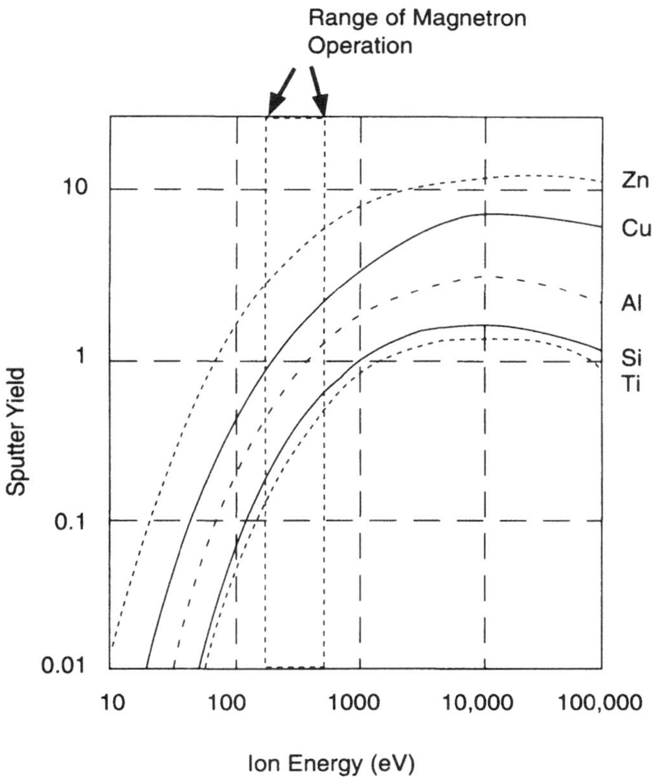

FIG. 2.3 Sputter yields as a function of ion energy for Ar^+ bombardment of common materials for ion energies up to 100 keV [2.10].

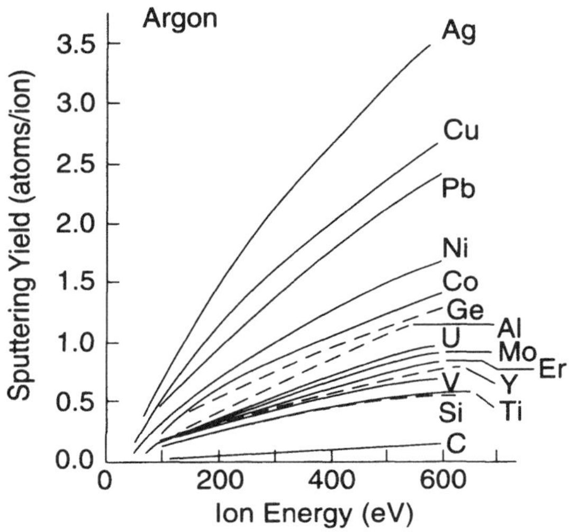

FIG. 2.4 Sputter yields as a function of ion energy for low energy: up to 600 eV (2.5).

volume, all of the bonds between atoms are broken and the region can be treated with a statistical mechanics-like approach. Since this energy region is more amenable to statistical calculation, the theory is well developed and accepted. Good reviews of this field have been published [2.11, 2.12], but the topic will not be covered in this volume.

4. *Very High Energy* ($E > 50$ keV). At these high energies, the incident ion can penetrate down into the target lattice many layers before causing a significant number of collisions. As a result, the affected volume is well below the surface and few if any atoms can be emitted. At these high energies, the incident ion is effectively implanted into the bulk of the target. This may be quite important to the electrical properties of the materials, particularly for semiconductors, but it is mostly irrelevant for physical sputtering.

Since sputtering is mostly a momentum and energy transfer process between the incident particle and the target atoms, the particular species used are very important. As shown in Fig. 2.4, the yields are different for different target materials using the same ion species and energy. There are two reasons for these differences. First, the binding energy will be different for each target material, and this is the barrier that a target atom must overcome to be emitted from the surface. There is a general trend toward

TABLE 2.1
SPUTTER YIELDS FOR SEMICONDUCTOR-RELATED MATERIALS FOR NE, AR, AND KR AT 200, 500, AND 1000 EV [2.10].

Ion	Ne	Ne	Ne	Ar	Ar	Ar	Kr	Kr	Kr
KE (eV)	200	500	1000	200	500	1000	200	500	1000
Be	0.14	0.54		0.17	0.57		0.14	0.51	
C				0.04	0.13				
Al	0.31	0.92		0.47	1.17		0.32	1.01	
Si	0.18	0.59		0.22	0.51		0.11	0.53	
Ti	0.26			0.25	0.57		0.16	0.51	
Ni	0.56	1.4	1.6	0.75	1.6	1.9	0.49	1.4	2.3
Cu	1.0	2.3	2.4	1.2	2.6	3.1	1.0	2.5	3.7
Zr	0.22	0.47		0.31	0.71		0.20	0.62	
Nb	0.18	0.43		0.26	0.65		0.19	0.59	
Mo	0.29	0.60	0.62	0.41	0.91	1.2	0.34	0.95	1.4
Ag	1.2	2.2	2.4	1.6	3.4	3.8	1.3	3.5	4.8
Ta	0.16	0.34		0.30	0.62		0.32	0.93	
W	0.15	0.35		0.30	0.64		0.36	1.0	
Pt	0.37	0.77		0.68	1.5		0.70	2.0	
Au	0.69	1.37		1.1	2.6	3.6	1.2	3.3	

higher sputter yields for materials with lower binding energies, and there is a general correlation between low melting point and low binding energy. This can be seen in Fig. 2.5, which plots the sputter yield for 1000 eV Ar^+ bombardment for a variety of materials as a function of the mass number of the target. However, sputtering is *not* a thermal process, so these correlations should not be taken too strongly.

A second reason for differences in yields is the efficiency of the momentum transfer process between the incident ion and the target atom. By conservation of energy and momentum, the energy transferred is related to the product of the masses of the two species divided by the square of the sum of the masses. This has a maximum value for two equal mass species, which implies that the highest sputter yields should be for cases of a target being bombarded by an ion of the same species. This situation is known as the *self-sputter yield.* It suggests, though, that the sputtering process will be rather inefficient and the yields relatively low for cases of a large mismatch between the incident and target masses.

The sputter yields for various inert gases on Si over a wide range of ion energies are shown in Fig. 2.6. In the high-energy regimes, there is a significant mass dependence to the yield. However, in the knock-on regime (< 1 keV), there is only a vague dependence of the yield on ion mass. This

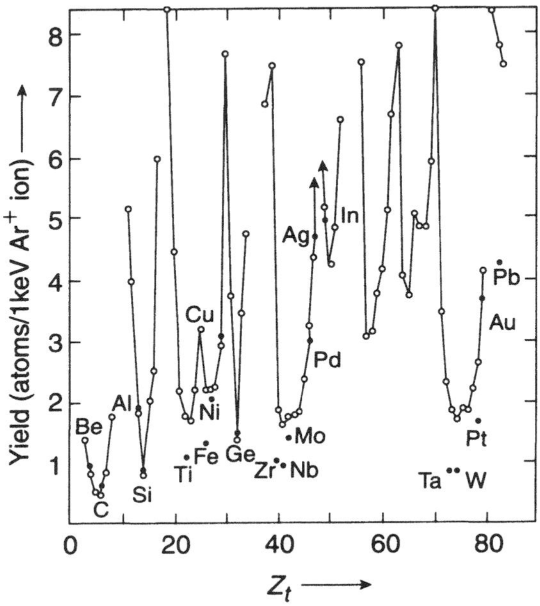

FIG. 2.5 Sputter yields for 1000 eV Ar^+ bombardment as a function of target mass number [2.12].

is particularly true for light-mass targets. It is routinely thought that going from Ar to perhaps Kr or Xe will result in a higher sputter rate and, for deposition applications, a higher deposition rate. From a yield point of view, this is only true for relatively high-mass target species with a mass much greater than 40 (Ar).

The angle of incidence of the bombarding ion can also have an effect on the sputtering process. This is shown schematically in Fig. 2.7. The incident ion at normal incidence affects the target in a regime roughly characterized by the spherical volume outlined as a dotted circle. A small fraction of this circle intercepts the surface, and this defines the area from which energetic, sputtered atoms might be emitted. As the incident angle goes to 45° or so, the volume affected by the impact is moved closer to the surface and, as a result, more atoms near the surface can be emitted by the collision process. The sputter yield in this case can easily exceed the case of 90°, normal incidence. However, as the incident angle becomes more grazing, eventually it is more likely that the incident ion will simply reflect off the sample surface, resulting in little energy deposition and very little sputtering. The angular dependence of the sputter yield, then, generally will be

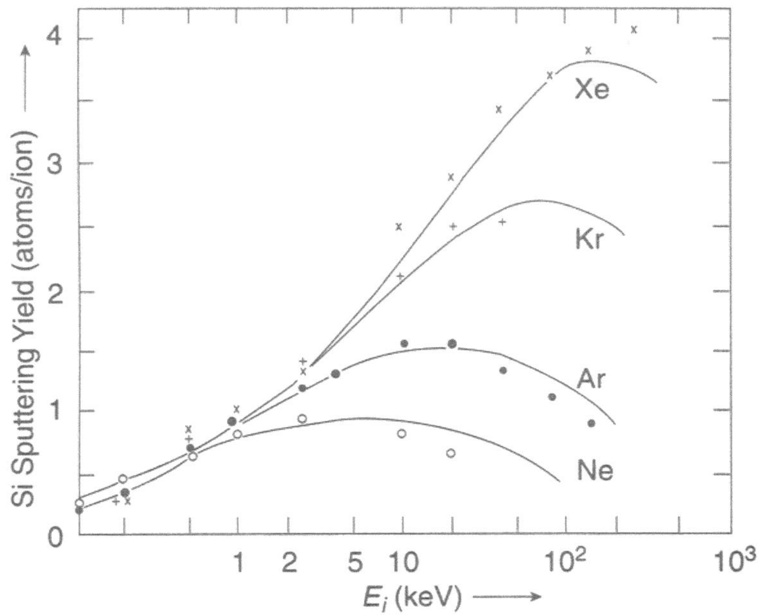

FIG. 2.6 Sputter yields for Si as a function of ion energy for several inert gas ions [2.12].

larger at angles near 45° than at 90°, and then will fall rapidly as 0° (grazing incidence) is approached (Fig. 2.8).

The dependence in Figure 2.8 is often described as a *cosine dependence.* This can be a little confusing depending on the frame of reference. If normal incidence (90° in the prior discussion) is converted to 0°, and near

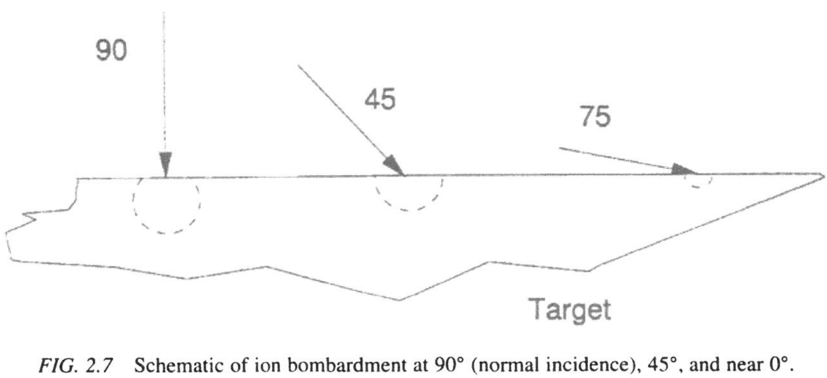

FIG. 2.7 Schematic of ion bombardment at 90° (normal incidence), 45°, and near 0°.

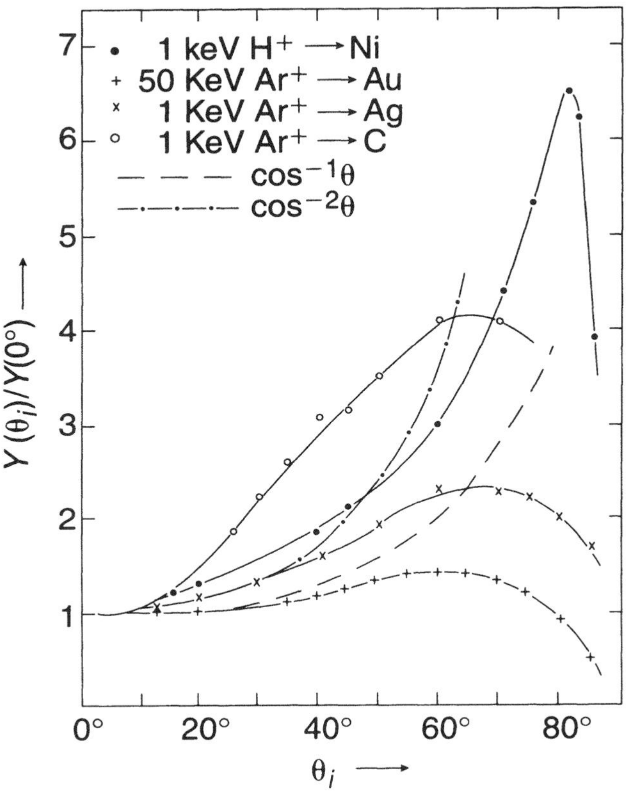

FIG. 2.8 The sputter yield as a function of the angle of incidence for the impacting ion [2.12].

grazing incidence (0°) is converted to 90°, then the yield scales roughly as 1/cosine of the angle from 0° up to about 50°. This is the origin of the cosine dependence of the sputter yield.

It is tempting at this point to infer that here is a way to increase the sputter emission rate from a target: If the surface were inclined at 45–50° from the ion direction, the yield would be increased nearly 2 times. However, there are two problems with this scenario. First, there is a differentiation between ions that come in the form of an ion beam and ions that come from a plasma. The ion-beam ions can be directed at will and the angle of incidence onto a surface is controllable simply by positioning the beam and sample. However, for plasma ion bombardment, which would be the case in an RF diode or a magnetron for example, the plasma sheath hugs the surface of the target and all ions are accelerated to 90° (normal incidence) to the surface regardless of the overall macroscopic shape of the target. It would be possible, though, to groove or texture the target surface in a plasma experiment such that the fine-scale surface is inclined at 45° to the surface normal. However, this requires that the grooves be much smaller than the sheath thickness.

Unfortunately, inclining the surface of the target to the incident ions by either means results in a reduction of the ion current density to the surface. Switching back to the reference frame where the sputter yield scales as 1/cos of the incident angle, the reduction in current density scales directly with the cosine of the angle. Therefore, these two terms cancel each other and generally lead to no enhancement.

2.2 Energy and Angular Distributions of Sputtered Atoms

Sputtering differs from evaporation in that the atoms are physically ejected from the target surface and as such can have significantly more kinetic energy. An example of this is shown in Fig. 2.9, which compares the velocity distribution of evaporated Cu atoms to sputtered atoms. Typically, the high-energy side of the sputtered-atom kinetic energy distribution follows a $1/E^2$-dependence. The peak in the kinetic energy distribution differs for

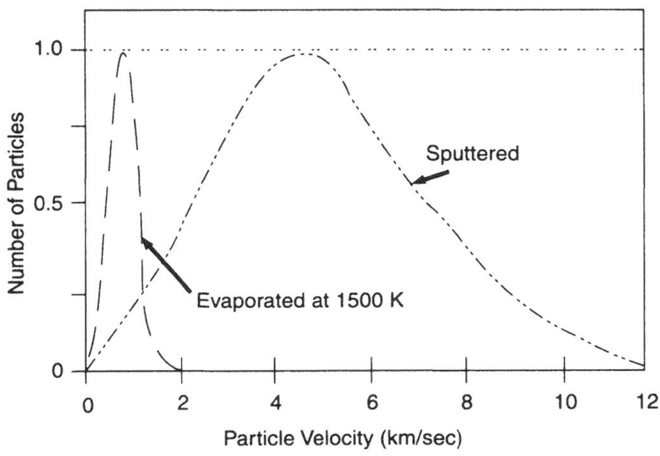

FIG. 2.9 The kinetic energy distribution for Cu atoms evaporated at 1600 K and sputtered with Ar^+ at 500 eV.

each ion-target system and is also slightly dependent on the ion's kinetic energy.

While the sputter emission of small clusters of atoms is relatively rare, such clusters should be expected to follow nominally similar emission characteristics in their energy spectrum, perhaps adjusted for the larger effective mass. Figure 2.10 shows work by H. Oechsner et al. measuring the energy spectrum of emitted Mo single atoms as well as atom pairs [2.13]. The kinetic energy of the Mo dimers is roughly one half that of the single atoms.

Perhaps more important than the exact distribution is the average kinetic energy of an emitted, sputtered atom. This will be a major component of the net energy arriving at the film surface during deposition. A chart of these average kinetic energies is shown in Table 2.2. Other significant components of energy that play a part in the deposition process come from the heat of sublimation — which is essentially the binding energy of the atom and is an intrinsic part of *any* PVD deposition process — and from other energetic processes related to the plasma. This can include photons from the plasma itself as two energetic neutral processes (which will be discussed below).

The angular distribution of sputtered atoms is generally described as a cosine distribution, which is accurate to first order for most situations. Traditionally, this is shown pictorially as in Fig. 2.11, which shows an

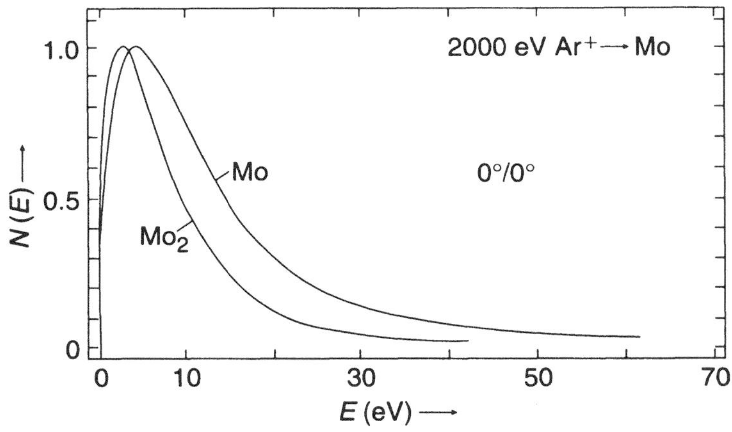

FIG. 2.10 Kinetic energy of Mo and Mo dimers for 2000-eV Ar^+ ion bombardment. The vertical scale has been normalized to show the comparative distributions. Typically, the emission of dimers is about 0.01 the magnitude of the level of the single atoms [2.13].

TABLE 2.2
THE AVERAGE KINETIC ENERGY FOR SPUTTERED-ATOM SPECIES COLLECTED FROM SEVERAL SOURCES.

Target	Ion/Energy	Ave KE (eV)	Peak eV	Reference
Be	Kr/1200	8	—	2.5
Al	Kr/1200	9	2	2.5
	Ar/900			2.31
Si	Kr/1200	10	—	2.5
Ti	Kr/1200	13		2.5
	Ar/900		3	2.31
V	Kr/1200	11	—	2.5
Cr	Kr/1200	13	—	2.5
Mn	Kr/1200	13	—	2.5
Fe	Kr/1200	14	—	2.5
Co	Kr/1200	12	—	2.5
Cu	Kr/1200	10		2.5
	Ar/500	10		2.25
	Ar/900		2	2.31
Ni	Kr/1200	17		2.5
	Ar/900		4	2.31
Ge	Kr/1200	13	—	2.5
Zr	Kr/1200	21	—	2.5
Mo	Kr/1200	21		2.5
	Ar/2000		7	2.13
Rh	Kr/1200	20	—	2.5
Pd	Kr/1200	16		2.5
Ag	Kr/1200	9	—	2.5
Ta	Kr/1200	33		2.5
	Ar/500	25		2.21
	Ar/2000		5	2.13
W	Kr/1200	34	—	2.5
Au	Kr/1200	21	—	2.5
Re	Kr/1200	39	—	2.5

impact point and an array of arrows at various angles. These arrows represent the relative fluxes in each direction and can be rotated about a vertical axis. The length of each arrow is related to the length (i.e., yield) at normal incidence times the cosine of the angle from 90°. Departures from cosine distributions occur as a function of incident ion energy. Generally, low energies change the distribution to a wider, less-normal-incidence distribution, described as *under-cosine* and higher energies have the opposite effect (*over-cosine*) (Fig. 2.12) [2.14]. These effects are fairly subtle, and the range of ion energies available in most practical plasma experiments (e.g., magnetrons) produce very little variation in the emission profile.

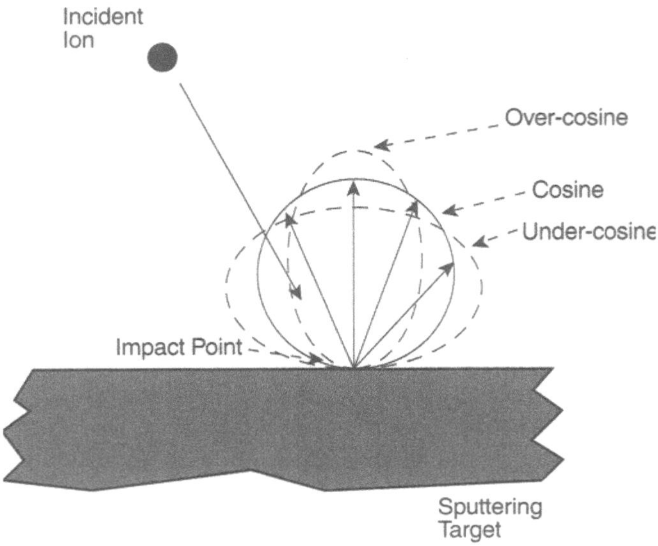

FIG. 2.11 Emission distributions for sputtered atoms.

The angle of incidence of the incident ion can have an effect on the emission dynamics. This was shown early on by Wehner and Rosenberg, who compared the emission distributions on Mo for smooth and rough surfaces (Fig. 2.13) [2.15, 2.16]. The rough surface showed no preferential direction, perhaps due to the intrinsic recapture of emitted atoms by the steep, rough surface. However, the smooth surface showed forward emission, consistent with a fairly shallow, low number of collisions, which retains some of the incident direction of the bombarding particle. Recent work by Doughty et al. has confirmed this work and extended it to Cu [2.17].

Forward sputtering is, of course, relevant to ion beam sputtering in which the incident ion's direction can be determined by design. However, in a plasma experiment, ions always impact the substrate surface at normal incidence, due to the planar electric field present over the sample surface. However, if the surface contains small features (perhaps on the micron scale), the incident ions (at 90°) may impact a slanting surface, resulting in the potential for forward sputter emission down into a feature or onto a nearby surface. This will become relevant in Chapter 8, which discusses ionized deposition.

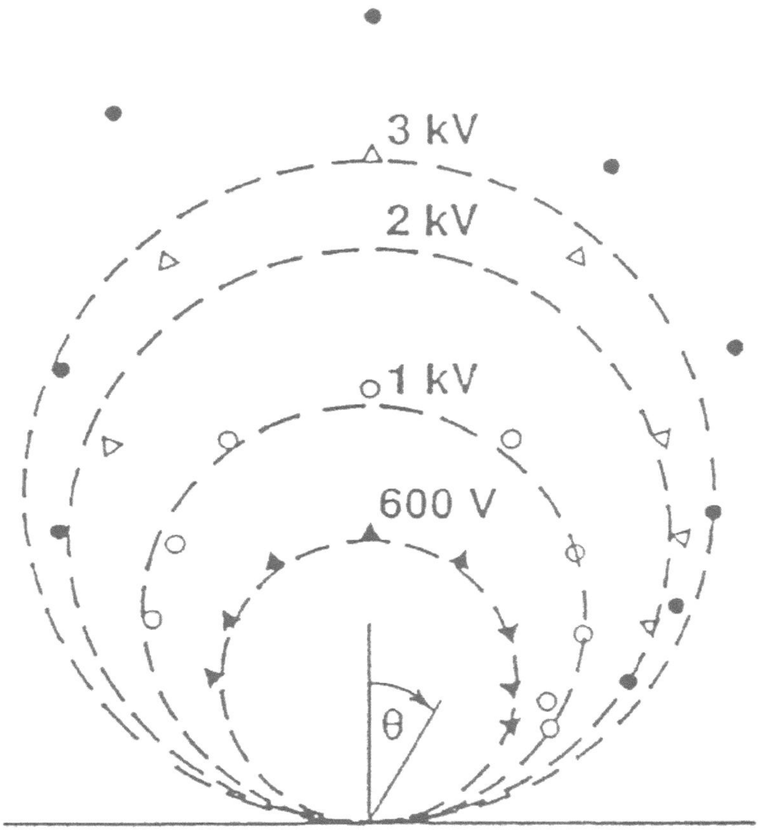

FIG. 2.12 Emission measurements as a function of ion energy [2.14].

Another general departure from a cosine emission distribution occurs for the case of single-crystal or oriented targets. First observed 40 years ago by Wehner, and described to this day as *Wehner spots*, the emission distributions have specific, preferred angles related to the underlying crystal structure [2.18]. This effect has been incorporated into target design in effect by at least one manufacturer as a means of developing a more-normal-incidence ejection profile [2.19]. While this last case may or may not be practical, the existence of preferred directions in the emission profile dependent on crystalline orientation indicates that at least some aspect of the

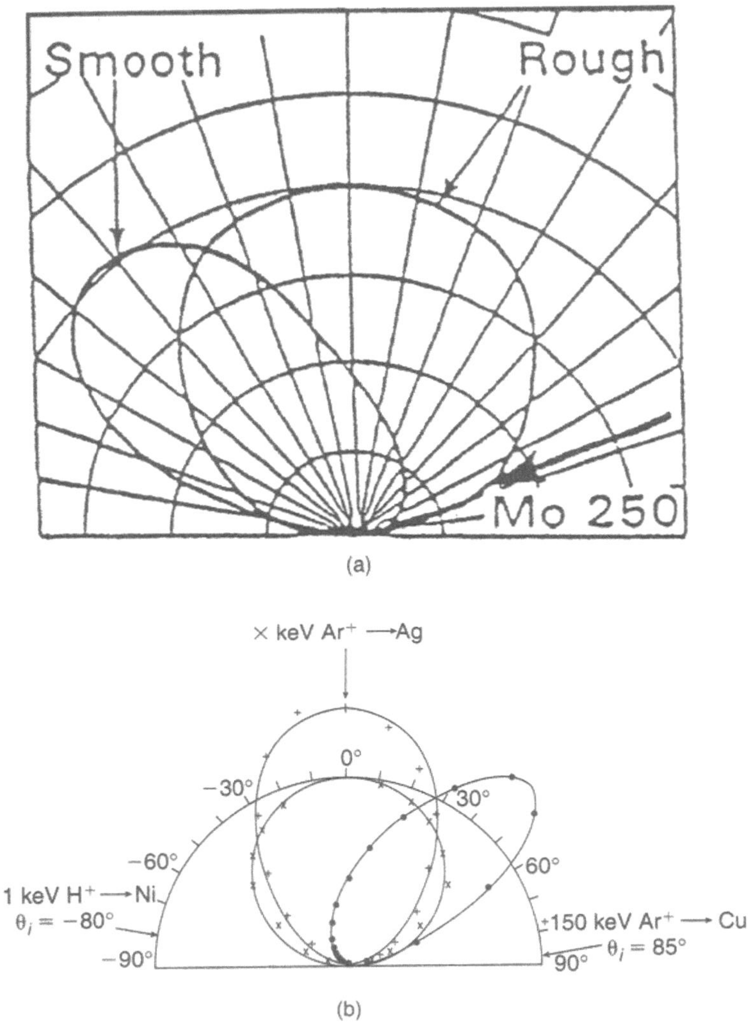

FIG. 2.13 (a) Emission angular distribution for 250-eV Ar^+ onto Mo at about 20° for smooth and rough surfaces [2.15], (b) emission distribution for various cases [2.12].

original lattice structure withstands the rather violent sputtering event on the target surface. This is further evidence of the lack of fully developed cascades in knock-on sputtering, which would lose any memory of their original structure or orientation.

2.3 Other Energetic Processes during Sputtering

There are two additional aspects of sputtering that may lead to significant effects on film deposition: reflected, energetic neutrals and negative ions. Both of these terms are slightly misleading but are in common usage.

Reflected, energetic neutrals are the result of energetic ion bombardment of the target. If the mass of the incident ion is equal to or less than the target atom mass, there is some probability of an elastic reflection of the ion from the surface. Since the ion is neutralized shortly before it impacts the surface, the reflected particle remains neutral and is unaffected by local potentials or sheaths. The reflected neutral can carry significant kinetic energy, often 20–40% of the incident ion energy. The angular distribution of these reflected atoms varies, but again to first order it might be considered roughly a cosine distribution.

The intrinsic problem with reflected neutrals is that they are very difficult to measure experimentally in the deposition system because they are uncharged. They can deposit considerable energy to the film surface and have long been thought to alter such physical properties as the film microstructure and stress. A long sequence of experiments by Dave Hoffman and John Thornton explored this situation and has been summarized by Hoffman [2.20].

The flux of energetic, reflected neutrals is strongly dependent on the ion-to-target mass ratio. If this number is very small, such as in the case of sputtering refractory materials like W or Ta, the reflected fluxes can approach the deposition rate, resulting in significant energy deposition along with the film atoms. For example, even though the kinetic energy of a sputtered Ta atom might be in the range of 25 eV [2.21], the average energy deposited during Ar^+ sputtering of Ta can approach 100 eV/Ta atom, resulting in significant sample heating and potential problems with stress and film microstructure.

This can also be inferred from a classic experiment by H. Winters [2.22]. In this experiment, a thin, carefully calibrated calorimeter was bombarded by a well-defined ion beam. The function of the calorimeter was simply to measure the temperature of the sample, from which the deposited energy could be calculated. Winters then compared the deposited kinetic energy as a fraction of the incident kinetic energy for ion energies of a few tens of eV up to 5 keV for various ion–sample combinations (Fig. 2.14). The data shows that for cases where the incident ion weighed much less than the target film, a sizable portion (20%) of the incident energy was not deposited but presumably was removed in the form of energetic, reflected neutrals. As the ion mass was increased to an amount to exceed the target mass, the

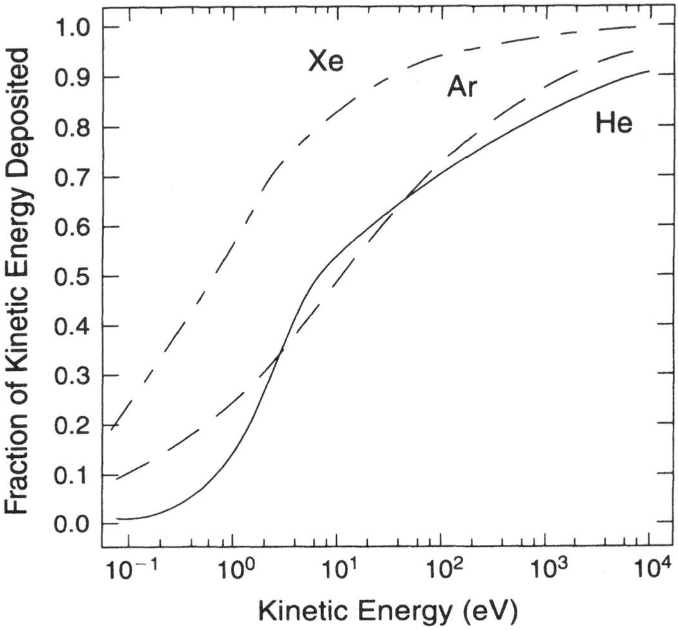

FIG. 2.14 Deposited kinetic energy fraction as a function of ion energy for He, Ar, and Xe onto Au as a function of ion energy up to 10 keV [2.22].

deposited energy moved closer to 100% of the incident energy. At this point, reflection was no longer present, although some energy was removed in the form of the kinetic energy of the sputtered atoms.

Negative ions can occur during the sputtering of materials that have components with high electronegitivity. A common example is oxygen. In many solid compounds containing oxygen, one component may be from the far left side of the periodic table, such as Ba, Y, Zr, Ti, and so on. These species readily give up an electron, which can then be attached to the oxygen atom, forming a negative ion. This negative O ion is then accelerated by the target sheath (to be discussed in Chapter 3) and enters the plasma at the target potential, which is typically many hundreds of volts. The negative O ion is then stripped of its electron in the plasma and continues on as a several-hundred-eV neutral [2.23].

Unfortunately, this neutral is moving directly toward the film location and can readily sputter the growing film. This resputtering effect may be minor, leading to small changes in film structure or composition. In cases of high levels of negative ion bombardment, the film structure or composition can be radically altered and the erosion rate can actually exceed the deposition rate, resulting in a etched substrate rather than a deposited film.

Negative ion effects are generally present when working with oxygen, although for cases such as Al, Ti, and Si, the effects are small. For cases such as ferroelectrics or pzieoelectrics (PZT, PLT, BST, etc.), the effect is quite strong, and it is extremely difficult to attain the correct film composition without a significant change (typically an increase in the level of the highest-sputter yield components) in the target composition.

2.4 Transport of Sputtered Atoms

Sputtered atoms must typically travel some distance (cms or more) before they can impact a sample surface to form a deposited film. The operating pressure for most sputtering applications ranges from 10^{-5} to 10^{-1} Torr, over which the mean free path for gas atoms varies from 500 cm down to 5 mm. This complicates the issue of atom transport. At low pressures, typically 1 mTorr or less, the sputtered atoms travel with few if any gas-phase collisions prior to deposition. This can be described as *ballistic* transport, or collision-free transport. At pressures in the tens of mTorr and above, the sputtered atoms are typically stopped by gas-phase collisions someplace between the target and the sample, and effectively become like any other gas atom and undergo *diffusive* transport. Another common term used to describe these slowed-down sputtered atoms is *thermalized*, implying that they have cooled down to the point of matching the gas temperature, which is typically a few hundred degrees C.

2.4.1 Ballistic Transport

In ballistic transport, the sputtered atoms have virtually no in-flight collisions and arrive at the film deposition surface with their original kinetic energy intact. This provides for a rather energetic deposition process, as the average kinetic energy can be 10 or more times the local thermal energy of the atoms at the film. The ballistically deposited atom can almost be thought of as implanting itself in the top layer or two of the film surface, and the deposition can be accompanied by the formation of defects and/or a sort of local annealing. Films deposited in the ballistic transport regime tend to be small-grained, dense, and often have relatively good adhesion. In addition, since the deposition is kinetic and not thermal, it is possible to manufacture unusual materials that might not be stable thermally simply by depositing the correct flux ratios. Ballistic depositions are generally characterized by small grain size and compressive stress.

Ballistic transport is also directional, at least within geometrical limits. Since there are no gas-phase collisions, the sputtered atoms travel in a straight line from the target to the sample (line-of-sight). This will allow for various means of directional filtering, such as collimation or long-throw sputtering, which will be discussed in Chapter 6. Ion beam sputtering, which will only be briefly discussed in Chapter 3, is also characteristic of ballistic sputter deposition because the operating pressures are well below 1 mTorr.

2.4.2 Diffusive Transport

As the pressure is increased, it becomes more likely that a sputtered atom may have a gas-phase collision with a background gas atom during transport. This starts to become significant at pressures above a few mTorr. In these collisions, significant kinetic energy (up to 50%) can be shared with the gas atom, resulting in both cooling of the sputtered atom and heating of the background gas. In addition, since the momentum- and energy-transfer cross sections are strongly energy dependent [2.24] (Fig. 2.15), as the sputtered atom slows down due to collisions, it becomes effectively larger. This effect is not physical (the atom does not grow); rather, it is related to the effective interaction time between the electron clouds of the two colliding atoms. As the atom slows down, there is more time for the electrostatic interaction and it is as though the particle increases in its effective size.

The end result is that as the sputtered atom slows down it becomes even more likely to have collisions, and it can quickly lose its initial kinetic energy in perhaps 5–10 collisions. This is known as *thermalization* and results in an effective temperature for the sputtered atom that is the same as the gas temperature, perhaps 1000°C or less. Measurements of average sputtered-atom temperature show a strong drop as the pressure is increased (Fig. 2.16) [2.25]. Thermalized deposition can be much different from ballistic deposition because the depositing atoms have virtually no kinetic energy [2.26]. In fact, thermalized depositions are much more similar to evaporated depositions, both in grain size and in stress. The grain size is typically larger and the stress more tensile. Since the deposition is more thermal, it may be less likely to deposit homogeneous alloys of unusual or immiscible materials.

Sputter deposition systems are usually intended for the rapid deposition of thin films, so it is relevant to see how efficient the transport process is.

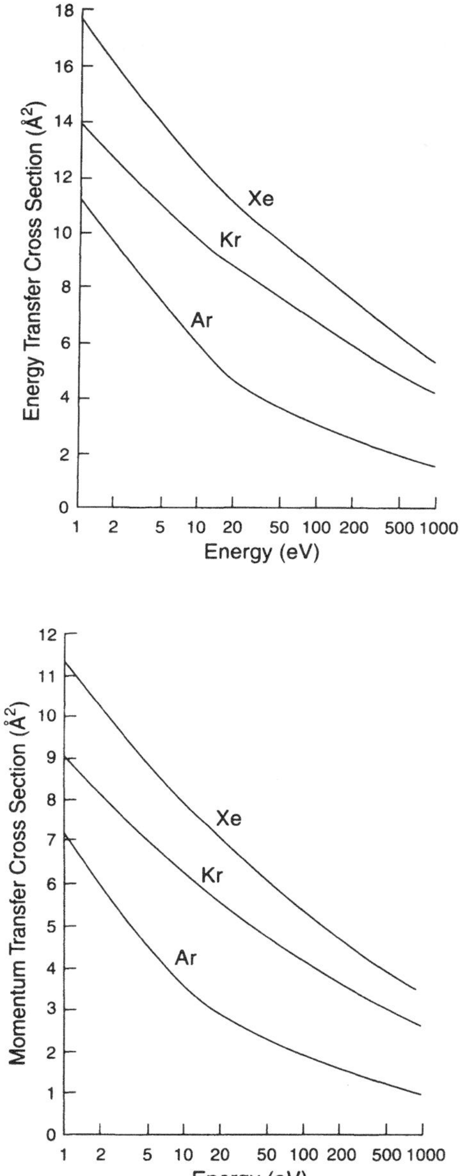

FIG. 2.15 Cross sections for energy (top) and momentum (bottom) transfer for various materials as a function of energy for Ar, Kr, and Xe [2.24].

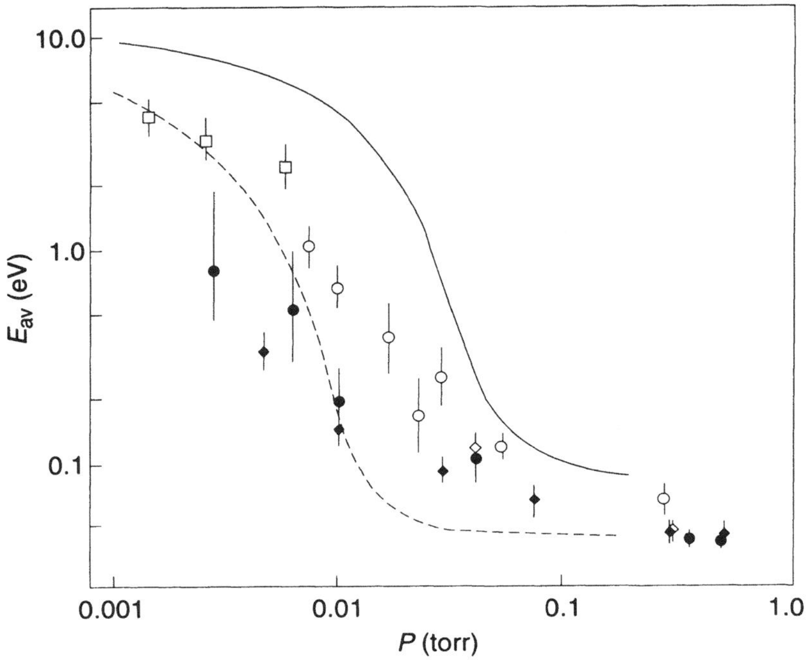

FIG. 2.16 The effective average kinetic energy of sputtered Cu as a function of system pressure. The different data points relate to changes in magnetic field and measurement position, and the solid lines are the result of modeling [2.25].

An atom sputtered from the cathode in a typical magnetron sputter tool (see Chapter 5) is likely to be deposited on one of three surfaces: the sample, the surrounding shields or fixtures, or potentially right back on the cathode itself. Obviously, the first case is most important. Deposition on the fixtures (shutters, shields, windows, etc.) results in lower net efficiency as well as cleaning concerns over time. Initially, deposition back to the cathode (redeposition) might not be considered that bad, in that the atoms are simply recycled by later sputtering. However, as will be seen in Chapters 4 and 5, most sputtering cathodes have "dead" areas, or areas where the erosion rate is fairly low. In these areas, there can be a net buildup of the sputtered material, which can result in either topographical problems on the source (bumps, nodules) or even peeling and flaking.

Sputtered-atom transport is rarely measured in direct terms. In many cases, systems are characterized by practical units, such as deposition rate per watt, which, if the exact sputter yield and cathode dimensions are known, may be extended back to more fundamental units. Generally this is not done, simply because users are interested in the actual performance of the system rather than the absolute units.

Transport can be defined as a probability of deposition, ranging from 0 to 1.0. A transport probability of 1.0 would mean that all of the atoms sputtered from some location arrived at the intended destination, and obviously a transport probability of 0.0 means that none of the sputtered atoms were deposited. Most practical systems will no doubt be someplace in between, simply because of the difficulty in managing either the emission or trajectory of the sputtered atoms.

Measurements have been published of sputtered-atom transport for a simple magnetron system in an open chamber [2.27]. The open chamber is necessary to remove the complication of the various shields, shutters, and fixtures that are usually present in most systems and act as collection points for material. The measurements used a magnetron sputtering source of diameter 20 cm mounted in a chamber of diameter 50 cm and length 30 cm. Samples were configured on the side areas (beside the cathode and on the walls) and also on a full-diameter sample plane that could be moved to different throw distances. It is necessary in this case to use a full-diameter sample plane to collect all the atoms that reach the sample location.

The results for several throw distances (5 to 14.5 cm), for pressures up to 30 mTorr, for Al and Cu, and for some different gases are shown in Table 2.3. The redeposition back to the cathode was inferred by locating samples on the various dead areas of the cathode and averaging the net deposition rate there over the entire cathode surface. This implies that the deposited atoms into the etch track (which cannot be measured) are recycled.

The results show several interesting points. As might be expected, the shortest throw distances result in the best transport, as do the lowest pressures. In general, though, the transport tends to be only moderately efficient: No more than 50% of the sputtered atoms typically make it to the sample plane. The best case is the sputtering of Al with Ne at low pressure and short throw distance. In this case, the mass of the gas is less than the mass of the sputtered atom, so gas scattering is reduced. It should be noted that the difference between Al transport in Ar and its transport in Ne at 5 mTorr even for the short throw distance of 5 cm (from 0.8 to 0.6) can *entirely* be associated with gas scattering. The results also show the significant impact of either increased sample throw distance or increased pressure. It is clearly most efficient to sputter at the lowest practical pressure and the shortest distance.

TABLE 2.3

TRANSPORT PROBABILITY FOR PLANAR MAGNETRON SPUTTER DEPOSITION ONTO THE SAMPLE PLANE, THE SIDE AREAS, AND BACK ONTO THE CATHODE. THE TOP GROUP IS FOR THE USE OF A CU CATHODE, THE LOWER GROUP IS FOR AL [2.27].

	P(Pa)	Sample plane	Magnetron plane	Side areas[a]
5-cm throw				
Kr	0.7	0.52	0.10	0.16
	2.6	0.45	0.18	0.17
	4	0.38	0.34	0.13
Ar	0.7	0.60	0.12	0.10
	2.6	0.46	0.26	0.12
	4	0.42	0.32	0.09
Ne	0.7	0.80	0.08	0.05
	2.6	0.56	0.16	0.10
	4	0.52	0.27	0.11
9.5-cm throw				
Kr	0.7	0.35	0.18	0.20
	2.6	0.27	0.35	0.24
	4	0.22	0.39	0.20
Ar	0.7	0.44	0.13	0.10
	2.6	0.45	0.35	0.15
	4	0.36	0.40	0.17
Ne	0.7	0.40	0.10	0.20
	2.6	0.42	0.36	0.18
	4	0.40	0.34	0.09

Throw 1000 W	*P*(Pa)	Sample plane	Magnetron plane	Side areas[a]
5 cm	0.7	0.63	0.031	0.16
	2.6	0.49	0.11	0.20
	4	0.53	0.14	0.22
9.5 cm	0.7	0.48	0.031	0.24
	2.6	0.47	0.13	0.24
	4	0.45	0.18	0.18
14.5 cm	0.7	0.39	0.045	0.25
	2.6	0.35	0.16	0.30
	4	0.31	0.18	0.35
200 W				
5 cm	4	0.53	0.23	0.13
3000 W				
5 cm	4	0.48	0.09	0.24

[a] The side areas include only those areas adjacent to the magnetron cathode, parallel to the cathode surface. It does not include all wall areas where deposition was too small to be measured.

2.4.3 Gas Rarefaction

In parallel to the thermalization process of cooling the sputtered atoms, the gas temperature can increase significantly. Since sputtering chambers are fairly open and have only modest gas flows, significant gas heating results in a local rarefaction of the gas density, as hot gas atoms leave the near-target region faster than cooler gas atoms arrive from the perimeter. Gas rarefaction effects were first observed in a dynamic mode known as the *sputtering wind*, in which convection-like flows were observed in the background gas within the chamber [2.28]. Later work showed a significant rarefaction of the average gas density — down to as low as 15% of the original density — due to the heating effect of the sputtered atoms (Fig. 2.17) [2.29]. Rarefaction may be important in scaling issues, in that high-rate sputtering (and as a result, more rarefaction) may have similar

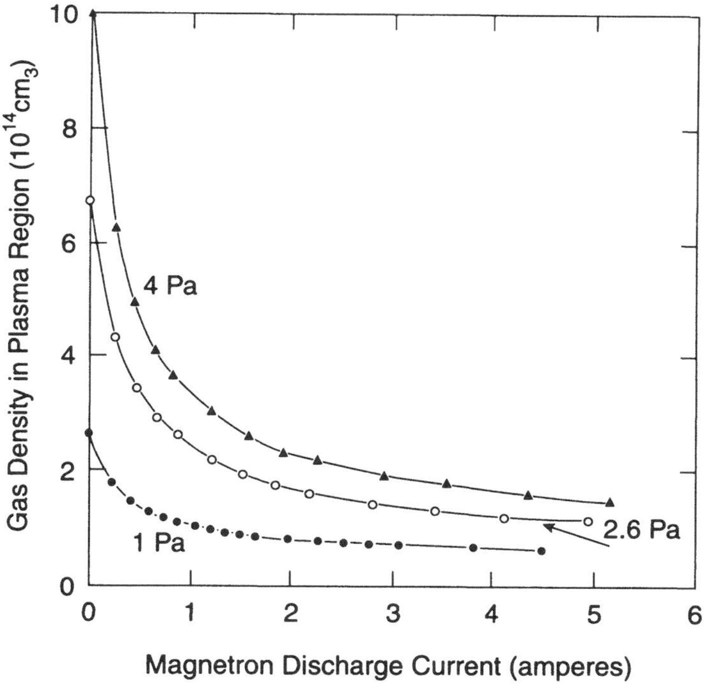

FIG. 2.17 Gas density in the region in front of the sputtering target as a function of ion (discharge) current to the target for 4 Pa (30 mTorr), 2.6 Pa (20 mTorr), and 1 Pa (7.5 mTorr). The cathode diameter was 150 mm, and the measurement position was 5.3 cm from the cathode face, on axis [2.29].

characteristics to low-pressure sputtering. Thus, a process developed at a low sputtering and deposition rate at a moderate pressure may be completely different as the deposition rate is scaled up and the effective gas density is reduced. This effect will also affect chemical effects, as seen in reactive sputtering [2.30].

References

2.1. C. Deshpandey and R. Bunshah, "Evaporation Processes," in *Thin Film Processes II*, J. L. Vossen and W. Kern, Eds., Academic Press, New York, 1991.

2.2. R. Glang, "Vacuum Evaporation," in *Handbook of Thin Film Technology*, L. I. Maissel and R. Glang, Eds., McGraw-Hill, New York, 1970.

2.3. D. M. Sanders, "Vacuum Arc-Based Processes," in *Handbook of Plasma Processing Technology*, S. M. Rossnagel, J. J. Cuomo, and W. Westwood, Eds., Noyes Publications, Park Ridge, NJ, 1989.

2.4. *Handbook of Vacuum Arc Science and Technology*, R. L. Boxman, P. J. Martin, and D. M. Sanders, Eds., Noyes Publications, Park Ridge NJ, 1995.

2.5. G. K. Wehner and G. S. Anderson, "The Nature of Physical Sputtering," in *Handbook of Thin Film Technology*, L. I. Maissel and R. Glang, Eds., McGraw-Hill, New York, 1970.

2.6. W. Eckstein, "Energy distributions of sputtered particles," *Nuclear Instru. & Meth. in Phys. Res.* **B18**: 344 (1987).

2.7. J. P. Biersack and W. Eckstein, "Sputtering studies with the Monte Carlo program TRIM.SP, *Appl. Phys.* **34**: 73–94 (1984).

2.8. D. N. Ruzic, "Fundamentals of sputtering and reflection" in *Handbook of Plasma Processing Technology*, S. M. Rossnagel, J. J. Cuomo, and W. Westwood, Eds., Noyes Publications, Park Ridge NJ, 1989, page 70.

2.9. K. Besocke, S. Berger, W. O. Hofer, and U. Littmark, "A search for a thermal spike effect in sputtering" *Radiation Effects*, **66**: 35 (1982).

2.10. H. R. Kaufman and R. S. Robinson, *Operation of Broad Beam Ion Sources*, Commonwealth Scientific, Alexandria, VA, 1987.

2.11. P. Sigmund, p. 9 in *Sputtering by Particle Bombardment I*, R. Behrisch, Ed., *Topics in Applied Physics* **47**, Springer-Verlag, Berlin, 1981.

2.12. P. Zalm, "Quantitative Sputtering," in *Handbook of Ion Beam Processing Technology*, J. J. Cuomo, S. M. Rossnagel, and H. R. Kaufman, Eds., Noyes Publications, Park Ridge, NJ, 1989.

2.13. H. Oechsner, "The Application of Postionization for Sputtering Studies and Surface or Thin Film Analysis," in *Handbook of Ion Beam Processing Technology*, J. J. Cuomo, S. M. Rossnagel, and H. R. Kaufman, Eds., Noyes Publications, Park Ridge, NJ, 1989.

2.14. Y. Matsuda, Y. Yamamura, Y. Ueda, K. Uchino, K. Muraoka, M. Maeda, and M. Akazaki, "Energy dependence of angular distributions of sputtered particles by ion beam bombardment at normal incidence," *Jpn J. Appl. Phys.* **25**: 8–11 (1986).

2.15. G. K. Wehner and D. Rosenberg, "Angular distribution of sputtered material," *J. Appl. Phys.* **31**: 177–179 (1960).

2.16. G. K. Wehner, "Momentum transfer in sputtering by ion bombardment," *J. Appl. Phys.* **25**: 270–271 (1954).

2.17. C. Doughty, S. Gorbatkin and L. A. Berry, "Spatial distribution of Cu sputter ejected by very low-energy ion bombardment", *J. Appl. Phys.*, vol 82 (1997) pp 1868–1875.

2.18. G. K. Wehner, "Sputtering of metal single crystals by ion bombardment," *J. Appl. Phys.* **26**: 1056–1057 (1955).
2.19. Tosoh Inc., Grove City, OH.
2.20. D. W. Hoffman, "Perspective on stresses in magnetron-sputtered thin films," *J. Vac. Sci. & Tech.*, **12A**: 953–961 (1984).
2.21. S. M. Rossnagel, C. Nichols, S. Hamaguchi, D. Ruzic, and R. Turkot, "Thin, high atomic weight refractory film deposition for diffusion barrier, adhesion layer and seed layer applications," *J. Vac. Sci. & Tech.* **B14**: 1819–1827 (1996).
2.22. H. F. Winters, H. Coufal, C. T. Rettner, and D. S. Bethune, "Energy transfer from rare gases to surfaces: Collisions with gold and platinum in the range 1–4000 eV," *Phys. Rev. B* **41**: 6240 (1990).
2.23. J. J. Cuomo, R. J. Gambino, J. M. E. Harper, J. D. Kuptsis, and J. C. Webber, "Significance of negative ion formation in sputtering and SIMS analysis," *J. Vac. Sci. & Tech.* **15**: 281–287 (1978).
2.24. R. S. Robinson, "R energetic binary collisions in rare gas plasmas," *J. Vac. Sci. & Tech.* **16**: 185–188 (1979).
2.25. L. T. Ball, I. S. Falconer, D. R. McKenzie, and J. M. Smelt, "An interferometric investigation of the thermalization of copper atoms in a magnetron sputtering discharge," *J. Appl. Phys.* **59**: 720–724 (1986).
2.26. G. M. Turner, I. S. Falconer, B. W. James, and D. R. McKenzie, "Monte Carlo calculation of the thermalization of atoms sputtered from the cathode of a sputtering discharge," *J. Appl. Phys.* **65**: 3671–3679 (1989).
2.27. S. M. Rossnagel, "Deposition and redeposition in magnetrons," *J. Vac. Sci. & Tech.*, **A6**: 3049–3054 (1988).
2.28. D. W. Hoffman, "A sputtering wind," *J. Vac. Sci. & Tech.* **A3**: 561–566 (1985).
2.29. S. M. Rossnagel, "Gas density reductions in magnetrons," *J. Vac. Sci. & Tech.* **A6**: 19–24 (1988).
2.30. W. D. Sproul, P. J. Rudnick, C. A. Gogol, and R. A. Mueller, "Advances in partial-pressure control applied to reactive sputtering," *Surface and Coatings Tech.* **39/40**: 499–506 (1989).
2.31. Wolfgang Hofer, "Angular, Energy and Mass Distribution of Sputtered Particles," in *Sputtering by Particle Bombardment III*, pp. 15–90 R. Behrisch and K. Wittmaack, Eds., Springer-Verlag, Berlin, 1991.

Chapter 3 Plasma Systems

The applications of sputtering and sputter deposition used in semiconductor processing all utilize plasmas as the source of the energetic, bombarding ions. A typical plasma is a partially ionized gas of Ar, Ne, Kr, or Xe, or a mixture of an inert gas and a chemically reactive gas such as oxygen or nitrogen. The ionization levels of these plasmas are low, and typically only one gas atom in 100–10,000 is ionized and the rest are neutral.

Plasma technology is characterized by the use of multiple sets of units as well as misnamed usage of units. Depending on the author and the contemporary trend at the time of publication, plasma-related publications may use SI units (kilograms, joules), cgs units (centimeters, grams, etc.), or other hybrid terms appropriate at the time. For the purpose of this book, we will use the terms most commonly used in the late 1990s, which unfortunately are a hybrid of various sets. For energy terms, we will use eV (electron volts), in which $1\ \text{eV} = 1.6 \times 10^{-19}$ joules. For densities, we will use particles/cm^3. A gas density, for example, of $1 \times 10^{15}/\text{cm}^3$ is equivalent to a chamber pressure of 30 mTorr or 4 pascals. We will use gas pressure units of mTorr, which are 10^{-3} Torr, where 760 Torr equals one atmosphere. (The SI unit of pascal never really caught on in the late 1970s and early 1980s. The pascal pressure unit is equal to 7.5 mTorr. The unit mTorr has also been known in past as a micron.) For dimensions, we will use centimeters, and for the various masses, we will use the atomic weights in grams. For calculations involving the masses of atoms or ions, we will try to convert most equations into mass units of AMU (Ar = 40 AMU, Xe = 131 AMU), which are easily found in the periodic table. Finally, for temperatures, we will work with degrees K, although electron temperatures (described below) use an energy unit, eV.

Even though it is somewhat tedious, we will try to restate the exact units as we go along to help those who might be searching for specific terms rather than reading the complete text. This is, perhaps, not the correct scientific approach to the purity of various systems of units, but it is consistent with the rather chaotic (and sloppy) usage within the various groups working with, publishing, and selling plasma technology.

The most common method for producing a plasma for sputtering applications is to place a moderate voltage (hundreds to a few thousand volts DC or RF) between two metal electrodes in a vacuum system (Fig. 3.1). Under the appropriate conditions of applied voltage and gas density, electrons may gain enough kinetic energy to ionize background gas atoms, and the gas may break down and a plasma can be formed. The condition for breakdown between two electrodes is a function of gas density; too low a

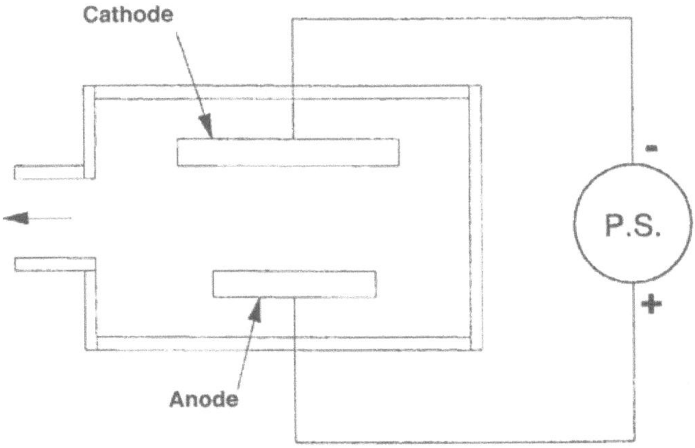

FIG. 3.1 Typical diode plasma system.

density prohibits a cascade-like breakdown, and too high a pressure has too many gas atoms, which can damp down the plasma formation due to too many collisions. This can be shown pictorially on a Paschen Curve (Fig. 3.2).

For most sputtering applications, the cathode will serve as the source of the sputtered atoms: Ions from the plasma will bombard the cathode, also known as the target, with sufficient energy to cause sputter emission of cathode atoms. The typical parameters of a system like that in Fig. 3.1 are a gas pressure in the 0.1 to 50 mTorr range, electrodes with dimensions of a few to several tens of centimeters, and a separation of a few centimeters to perhaps 20 cm between the cathode and the sample.

The vacuum systems used for sputtering usually have ultimate, or "base" pressures of below 10^{-5} Torr (0.01 mTorr) and often as low as the 10^{-9} Torr range. The chambers are constructed of steel or aluminum, and the vacuum seals are either copper gaskets or viton o-rings (or a similar material). Originally, most sputtering was done in diffusion-pumped chambers with a partially closed baffle between the diffusion pump and the chamber since the diffusion pump would not operate well at several mTorr. Most systems in current usage are pumped with either cryopumps or turbopumps, and many do not have pressure baffles between the sputtering chamber and the pump. In this case, the plasma is formed and the cathode is sputtered at the true chamber base pressure. If a baffle is used, it can often degrade the base pressure 1 to 2 orders of magnitude. This last feature is often not documented well in many articles and trade publications.

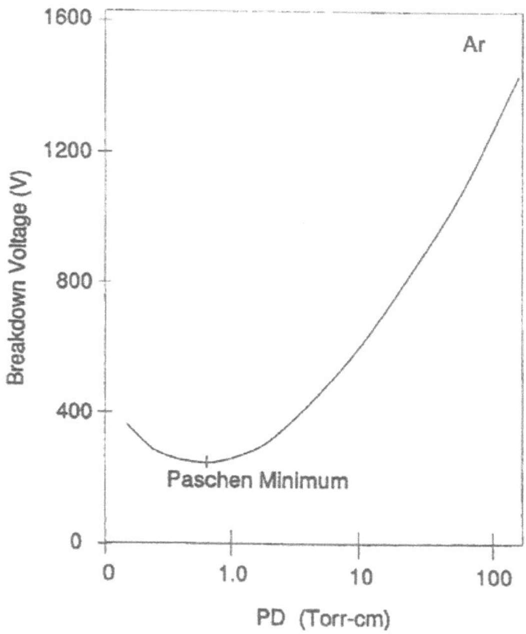

FIG. 3.2 The breakdown voltage for plasma formation between two electrodes as a function of the product of the pressure and the electrode spacing. This curve is known generically as a Paschen Curve.

3.1 Diode Plasmas

The two-electrode plasma system is known as a diode and is characterized by a flux of ions that impact the cathode (the negative electrode) along with a flux of electrons that move toward the anode. The specific direction of the current flow leads to the term *diode*. The bombardment of the cathode with ions causes the emission of secondary electrons from the cathode, which are then accelerated toward the anode and can gain enough energy to ionize more gas atoms. The flux of secondary electrons is the primary means of energy input into the plasma. From charge balance, it is necessary that each secondary electron create at least 1/gamma electron-ion pairs in the plasma, where gamma is the secondary electron yield.

The secondary electron yield depends on both the bombarding ion species and the substrate. The values can be quite low, and a sampling of this data is shown in Table 3.1 [3.1]. For a given material, the secondary

TABLE 3.1
SECONDARY ELECTRON YIELDS FOR VARIOUS SEMICONDUCTOR-RELATED MATERIALS [3.1].

Metal	Ar^+ (low E)	Ar^+ (100 eV)	Ar^+ (1 keV)	Reference
Al			0.10	3.1
Ag		0.05	0.07	3.1
Cu		0.05	0.3	3.1
Mo		0.07	0.1	3.2
Mo	0.12	0.12	0.12	3.1
Ta	0.01	0.015		3.1
Pt	0.03	0.03		3.1
W	0.1	0.01	0.10	3.1
Si (100)	0.02	0.03	0.04	3.1
Ni (111)	0.3	0.04	0.7	3.1
Ge (111)	0.3	0.04	0.5	3.1

electron yield is rather insensitive to the incoming ion's energy over a broad range from near 0 eV up to 1 keV or so (Fig. 3.3). The secondary electron emission process at these energies is an Auger-like process that is independent of the ion's kinetic energy. At higher energies (above 1 keV or so), the secondary electron yield increases linearly with ion energy. In this

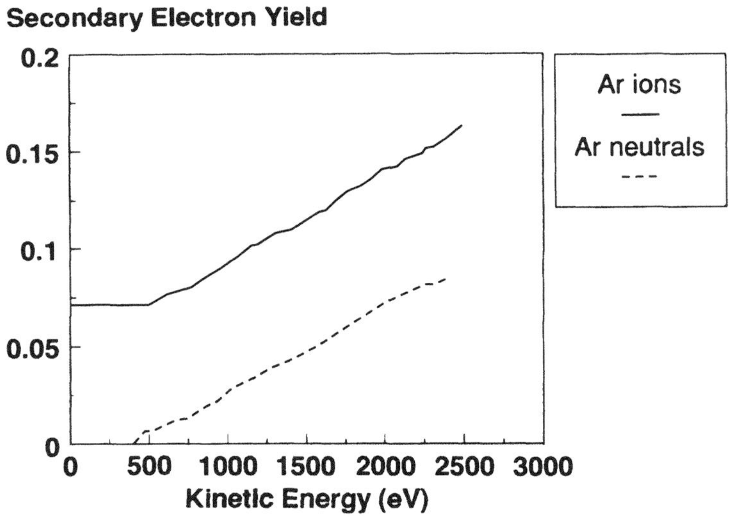

FIG. 3.3 General energy dependence of the secondary electron yield for energies up to a few keV (adapted from [3.2]).

energy regime, the yield is composed of the Auger component as well as a kinetic component. A classic test of this is the comparison of the secondary yields for ions and neutrals of the same species. This test, first done by Medved *et al.* in 1963, shows the additive effect of these two components for ions and the lack of the Auger process for neutrals [3.2].

A quick, crude calculation from the secondary electron yield puts an estimate on the voltages needed to operate a discharge. If the secondary electron yield for Ar^+ on some cathode is 0.05, then each secondary electron is responsible (in some form) for the generation of 20 Ar ions. Since the ionization potential for Ar is around 16 eV, the minimum energy necessary for the secondary electron to make 20 ions is 320 eV, requiring that the minimum discharge voltage be around 320 V. The real situation in the plasma is immensely more complicated, and this simple estimate should not be considered too binding. However, it does point out the inverse relationship between secondary electron yield and discharge voltage: If the yield of secondaries is high, the required discharge voltage is likely to be smaller than a similar case with a low secondary yield.

Surface contamination can change the secondary electron yield. The presence of oxygen on the surface often increases the secondary electron yield by 20 to 100%. Therefore, an oxidized cathode might have a lower operating voltage than a clean cathode. This implies that when starting with a dirty cathode, the operating or discharge voltage of a plasma system may be observed to increase with time until the oxided layer is sputtered off the cathode. Conversely, if oxygen is added to a sputtering system (see Section 3.8), the result may be to decrease the discharge voltage as the cathode becomes lightly oxidized.

Many common diode plasma systems actually have only one obvious electrode, which is the cathode, and the chamber walls function as the anode. Since the chamber walls are grounded for safety reasons, the cathode potential is then many hundreds of volts negative.

Plasmas have specific regions of interest for sputtering applications. The two most relevant regions are the bulk of the plasma and the sheath, or dark space, located between the cathode and the bulk of the plasma. The plasma itself is a conductor, and as such its potential is fairly constant across its width. Virtually all of the voltage drop between the cathode and the anode occurs in the thin dark space, or sheath, at the cathode.

Current flow across the cathode sheath is limited by space charge effects, which relate to a self-shielding effect of a stream of charged particles. The current density across the sheath, j, the voltage across the sheath, V, and the sheath thickness, d, are related by the Child-Langmuir Law (3.3):

$$j = \frac{4}{9}\left(\frac{2e}{M}\right)^{1/2} \frac{V^{1.5}}{4\pi d^2} \tag{3.1}$$

where M is the ion mass, V is the voltage, and d is the acceleration distance. The current density in most plasma systems is in the range of milliAmps per square centimeter (mA/cm^2). This equation in a useful form becomes

$$j\left(\frac{\text{mA}}{\text{cm}^2}\right) = 5.5 \times 10^{-5} \cdot V^{1.5} \cdot d^{-2} \cdot W^{-0.5} \tag{3.2}$$

where V is the voltage (in volts), d is the sheath width in cm, and W is the atomic weight of the ion in AMU.

> Example: The maximum, space-charge-limited current between two grids spaced 0.2 cm apart with a voltage across them of 1000 V for Argon (40 AMU) is (d = 0.2, V = 1000, W = 40) j = 2.41 mA/cm^2. If Xe was used in place of Ar (W = 131), the maximum current density drops to (d = 0.2, V = 1000, W = 131) j = 1.33 mA/cm^2.

For electrons, the conversion leads to

$$j\left(\frac{\text{mA}}{\text{cm}^2}\right) = 2.3 \times 10^{-3} \cdot V^{1.5} \cdot d^{-2} \tag{3.3}$$

and the units are, again, volts and cm for V and d, respectively.

By comparison, with the same voltage difference and d-spacing, the maximum electron current density (i.e., the space charge value) is orders of magnitude larger than the ion current density value. This is evidence of the much greater mobility of electrons in a plasma compared to the ions. The electron space charge current is effectively ignored in most cases and the fundamental limit is the ion current. Also, this means that in a sheath at the edge of a plasma — where the ions are passing from the plasma to the cathode and the electrons are moving from the cathode to the plasma — there are virtually no electrons in the sheath. Their flux is only a few percent of the ion flux anyway, and they move very rapidly through the sheath. One result of this is that the sheath is dark because there are few electrons available to excite the atoms and ions present. Strictly speaking, the sheath is not completely dark, but the emitted light levels are significantly below that of the bulk plasma.

The Child-Langmuir Law is relevant for virtually all cases of current flow either across a sheath or between two grids or apertures and can be easily derived from Poisson's equation [3.4].

Plasmas also have a characteristic length, known as a Debye length, which is given by

$$L = \left(\frac{kT_e}{4\pi n_e e^2} \right)^{1/2} \tag{3.4}$$

where n_e is the electron density, e is the electron charge, and kT_e is the electron temperature (described below). The Debye length can be thought of as a self-shielding distance for an electrical disturbance of a plasma. If a small potential is applied to some surface in a plasma, the movement of charge will be such that most of the potential is shielded within one Debye length. In a practical example, if a grid or an array of holes was present on a metal surface in a plasma, electrically the grid would appear as a solid surface if the Debye length was larger than the hole radius. If the Debye length was significantly smaller than the holes, then the plasma would penetrate the holes. Converting the Debye length into useful units, the above equation can be rewritten as

$$L(\text{cm}) = 743 \left(\frac{T_e}{n_e} \right)^{1/2} \tag{3.5}$$

where T_e is the electron temperature in eV and n_e is the electron density in electrons/cm^3.

> Example: A typical high-density plasma has a density of 10^{12}/cm^3. If the electron temperature is 2 eV, the Debye length is 1×10^{-3} cm, or 10 microns. A lower-density plasma of 10^9/cm^3 with an electron temperature of 5 eV would have a Debye (self-shielding) length of 0.05 cm.

There are three primary species of particles in a plasma: ions, electrons, and gas atoms. In the bulk of the plasma, the electron and ion densities, n_e and n_i, are identical. Experimentally, techniques may measure either the electron density or the ion density, although measurements of the electron density are more common and more accurate. For this reason, usually the electron density is given as the plasma density even though it means exactly the same as the ion density. The electrons, which are much more mobile than the ions due to their low mass, become thermally equilibrated and develop a "temperature" or energy dependence that can be characterized by a Maxwell–Boltzmann distribution (Fig 3.4). This is given by the relation [3.4]

$$f(v) = A \exp\left(\frac{-1/2(mv^2)}{kT}\right) \tag{3.6}$$

with

$$A = n\left(\frac{m}{2\pi kT}\right)^{1/2}$$

where m is the electron mass and v is the electron velocity. The peak in this distribution is generally called the electron temperature, although the average energy is 3/2 of the peak in the distribution. Even though this is actually an energy, the "k" in kT is assumed, and the electron temperature is given in electron volts (eV). A typical value for a processing plasma might be 2 eV, and rarely would the electron temperature exceed 10 eV.

The use of terms and units in plasma technology can be somewhat confusing and ambiguous. The electron temperature is almost always described in energy units (eV, where 1 eV = 1.6×10^{-19} joule) and virtually never in the form of degrees K. A simple rule of thumb is that a 1-eV

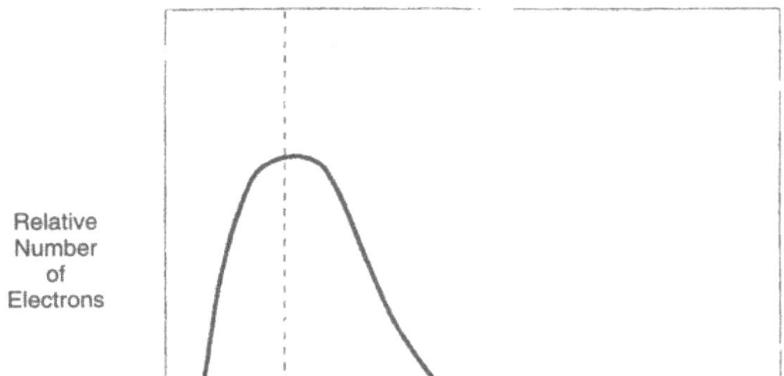

FIG. 3.4 Maxwell–Boltzmann electron energy distribution.

plasma temperature is about equal to 11,600 K. Another ambiguous term with plasmas is the description of ion or electron energies in terms of volts, rather than in energy units such as eV or joules. Since all electrons have a single charge and virtually all of the ions also have a single charge, when these particles are accelerated over some voltage, they attain the same number of eV in kinetic energy as the acceleration voltage. So, their kinetic energy in eV is equal to the acceleration voltage in volts, and the general tradition is to simply describe the energy in terms of volts.

The ions in a plasma are significantly colder than the electrons. This is due to the large numbers of collisions that each ion has with the background gas atoms. The cross sections for these elastic collisions as well as collisions where an electron can be transferred (known as charge exchange collisions) are large. Since there are 100 to 10,000 neutral gas atoms for each ion, the gas temperature dominates the ion temperature, and this stays at a few hundred degrees C, which is equivalent to 1/10 of an eV or less.

3.2 Plasma Potential

Since the electrons are so light and so energetic compared to the ions, many aspects of plasma calculations or models assume that the ions are virtually immobile. In the bulk of the plasma, since the electrons are so mobile, the loss rate for electrons from the plasma edge is greater than the loss rate for ions. This results in a slight positive charging of the plasma, setting up a small positive potential that retards the rate of electron loss to be the same as the ion loss rate. The *plasma potential* is virtually always positive in this situation and is typically positive on the order of the electron temperature in voltage. Virtually all processing plasmas have a fairly uniform plasma potential, and it is almost always a few volts positive of the most positive potential exposed to the plasma, which is either the anode or the grounded chamber walls that function as the anode.

3.3 Floating Potential

An electrically isolated object immersed in a plasma receives a flux of both electrons and ions. Since current flow away is inhibited, the net current flow to the object must be zero. However, since electrons move so much more rapidly than ions, the object may develop a net negative charge and potential, which limits the flow of electrons to be equal to the ion flow.

This negative potential is known as the *floating potential* and is typically 2 to 3 times the electron temperature (in voltage).

> Note: While the plasma potential and the floating potential are related to the electron temperature, there are other factors that preclude a direct calculation of the electron temperature from measurements of these potentials. In general, low electron temperatures correlate with small plasma and floating potentials. The topic of direct plasma probe measurements is well treated in [3.1, 3.4, 3.6].

3.4 Flux to the Sheath

The bulk of the plasma is at a uniform potential, and only small, local variations in potential or density can occur. In many ways, the plasma functions as a gas of charged particles that are randomized by collisions. Near the edge of the plasma, though, small potentials may extend into the plasma. This potential is on the order of one-half of an electron temperature (in volts, now) and typically perturbs the edge regions of the plasma (on the order of a Debye length or two) in the region between the dark space of the sheath and the bulk of the plasma. In this region, ions can be weakly attracted by this potential difference and drawn to the edge of the sheath. The ions attain what is known as the ion acoustic velocity, which is

$$V_{\text{(ion acoustic)}} = \left(\frac{kT_e}{M}\right)^{1/2} \tag{3.7}$$

where M is the ion mass. This can be converted into useful units as

$$V_{\text{(ion acoustic)}} = 9.8 \times 10^5 \left(\frac{T_e}{W}\right)^{1/2} \tag{3.8}$$

where T_e is the electron temperature in eV, W is the mass of the ion in AMU, and the velocity is in cm/sec.

> Example: For a 2-eV plasma of argon (40 AMU), this leads to $V = 2.2 \times 10^5$ cm/sec.

This is called *Bohm presheath diffusion* and was first understood by Bohm in the late 1940s. The flux to the sheath edge is then just the ion density (which is also the electron density) times the ion acoustic velocity, typically with a factor of 0.6 [3.5]:

$$j = 0.6 n_e V_{\text{(ion acoustic)}} \tag{3.9}$$

which, converted into mA/cm^2, becomes

$$j\left(\frac{\text{mA}}{\text{cm}^2}\right) = 8.9 \times 10^{-11}\, n_e \left(\frac{T_e}{W}\right)^{1/2} \tag{3.10}$$

where n_e is the density in electrons/cm^3, T_e is the electron temperature in eV, and W is the atomic weight of the ion in AMU.

Example: For a Ar ($W = 40$ AMU) plasma density of 10^{11}/cm^3 and an electron temperature of 2 eV, the current density to the sheath edge is $j = 2$ mA/cm^2.

This ion flux is then accelerated by the sheath potential and bombards the cathode/target surface or the anode surface. The net ion energy for the cathode/target is the difference between the plasma potential (typically a few volts positive) and the cathode potential, which is generally several hundred volts negative (Fig. 3.5). In practice, the plasma-potential contribution is usually ignored for the cathode. For the anode (or the chamber walls if they are the anode), the ion's energy is just due to the plasma potential. This bombardment is generally ignored in practical systems, although it can be important in high-density plasma etch tools.

3.5 DC and RF Plasmas

Historically, DC diode plasmas were first used for sputtering and sputter deposition applications. DC diodes are limited in practical terms either by low ion currents and high voltages or by operating pressures that are too high to allow significant deposition rates. A key problem is that the cross section for ionization of the background gas by the secondary electrons emitted from the cathode is small and effectively decreases as the electron energy increases (Fig. 3.6), [3.1]. Since the cross section is small, many secondaries pass through the plasma region and impact the anode/wall

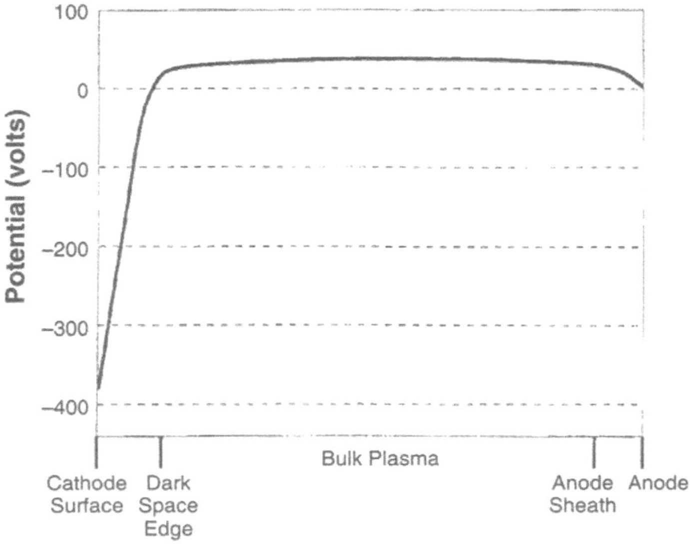

FIG. 3.5 The potential as a function of distance between the cathode and the anode. It should be noted that the position of 0 volts (ground) is not specified and can be set wherever desired. In most sputtering tools, the anode is grounded.

without making a sizable number of ion-electron pairs. From the earlier discussion, the loss of some fraction of the secondary electrons (50%, for example) due to the low collision probability results in the requirement that the energy (or voltage) given to the remaining secondaries be increased by two to keep the same net ion-electron pair production rate. However, as the voltage (energy) is increased, the cross section for ionization of the background gas becomes even smaller, so the efficiency of using the secondaries can drop even farther. The net result is a low plasma density and hence a small discharge current to the cathode at a rather high voltage. This results in a low sputtering rate and a poor deposition rate of sputtered atoms.

One solution to this problem has been to increase the gas density to increase the number of gas-phase, ionizing collisions. This works to a point, but the gas pressures required are often quite high — many hundreds of mTorr. At these pressures, the mean free path of the sputtered atoms is on the order of millimeters or less, and all of the sputtered atoms

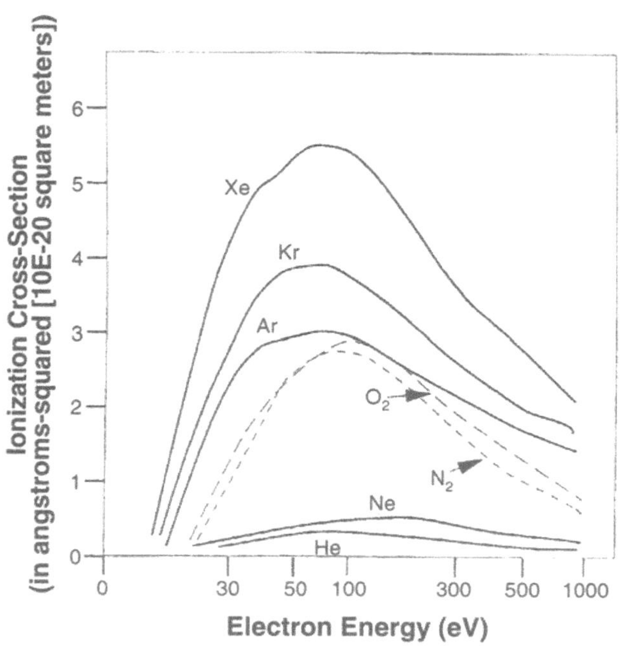

FIG. 3.6 Electron-impact ionization cross sections as a function of electron energy [3.1].

are effectively stopped very close to the cathode. They can then move about by diffusion, but the cathode face provides a very large sink for the diffusing particles, and the net number deposited onto a nearby surface can be very low.

Another problem that occurs with DC diode plasmas occurs when a reactive gas, such as oxygen, is added to the gas mixture during sputtering. This might be done, for example, to deposit films of aluminum oxide when sputtering a cathode made of Al. Unfortunately, the cathode can become rapidly oxidized in this case, and as a result the net DC current through the insulating, oxidized layer can be quite small. This means that the deposition rate will be too small to be useful.

Therefore, although DC diode plasmas can be easily made and operated, from a practical point of view they are rarely (if ever) used for the deposition of films. In terms of semiconductor fabrication, DC diode plasmas are not used in any production cases.

3.6 RF Plasmas

Using an RF potential (typically 13.5 MHz) in place of DC on a cathode completely solves the insulating-oxide problem and helps to solve the low deposition-rate problems discussed above. An RF diode system, shown in Fig. 3.7, is quite similar to the DC diode with the exception of the addition of impedance-matching networks between the power supply and the electrodes. In an RF diode, it is more common that the second electrode (the anode in the DC case) is actually powered by the power supply and not just grounded.

The potential on the RF-powered electrode oscillates positive and negative. Because of the much higher mobility of the electrons, the RF-powered electrode can pick up a much greater electron current than ion current for one cycle, and this begins to charge the electrode negative. The capacitor in the impedance network (to be discussed below) blocks this DC potential from the power supply. On each succeeding RF cycle, the electrode charges more negatively, which causes the effective ion collection time during each cycle to increase and reduces the electron collection time. The net result after several cycles is a negative DC potential on the electrode at a value that can approach one-half the applied RF peak-to-peak voltage. For very short fractions of each RF cycle, the

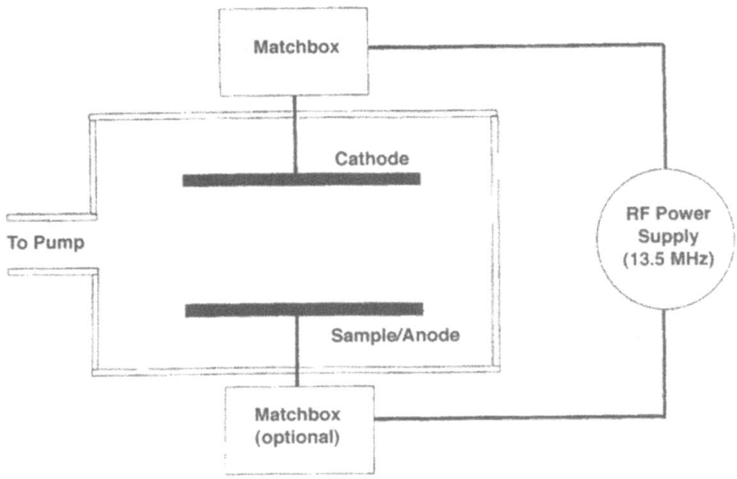

FIG. 3.7 RF diode system with nongrounded sample electrode.

electrode goes positive and collects electrons. For the remainder of the cycle, the electrode is negative and repels the electrons, collecting ions instead at an average potential roughly equal to the net DC negative bias (Fig. 3.8). The average DC bias is usually equal to one-half of the applied RF peak-to-peak voltage.

Since there is no net current flow (the ion current and the electron currents cancel out), there is no charging of an insulating surface; thus oxidized or nonconducting surfaces can be sputtered at modest rates. In addition, because of the rapidly changing potentials in the plasma due to the RF-powered electrode, secondary electrons as well as other plasma electrons are mildly trapped within the plasma and can cause additional ionization, compared to the DC diode case [3.6]. This results in a higher plasma density at the same pressure and a larger ion current to the cathode. It also results in a lower applied voltage (compared to a DC diode at the same applied power), which means that the secondary electrons will have a higher probability of being used efficiently for ionization.

RF diodes are routinely used for sputtering and sputter deposition of insulating materials — such as silicon dioxide, aluminum oxide, and other oxides — as well as magnetic materials such as Ni and Fe. The etching and deposition rates are modest (tens to a few hundred angstroms/minute). RF diode tools can be scaled up to rather large electrode size, and tools with diameters of 1 meter are routinely used for batch processing. Unfortunately, much like the DC diode case described above, increasing the RF

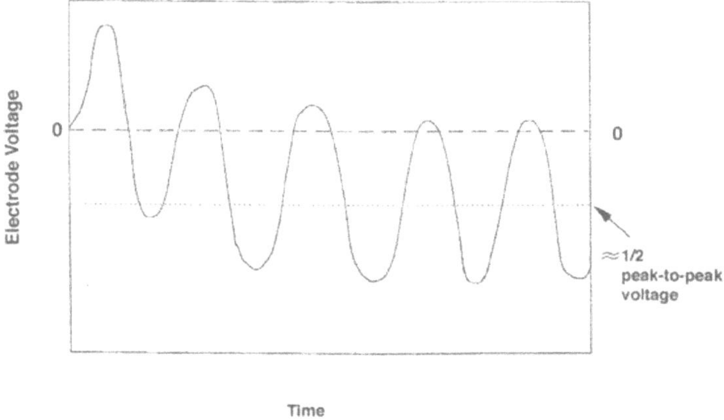

FIG. 3.8 Applied RF voltage as a function of time to the powered electrode.

power to the cathode does not linearly increase the sputtering and deposition rate. This is, again, due to the decrease in the ionization cross section as a function of increases in electron energy. Therefore, as the power is increased, more and more secondary electrons pass through the plasma without collisions and can strike and heat the anode/sample. The practical limit for RF diode systems appears to be a few kilowatts, and as such, RF power supplies larger than 3 kW are uncommon.

3.7 RF Matchboxes

RF plasmas have a complex impedance that has capacitive, inductive, and resistive components. Most power supplies, though, are designed to be most efficient powering a simple 50-ohm load. The role of the matching network — often called the *matchbox* — is to optimize the impedance of the plasma-and-matchbox system to maximize the amount of power that can be delivered to the plasma. The most common network is the "L" network, shown in Fig. 3.9 [3.7]. In this matchbox, the incoming power is split upon entering the matchbox. One side goes through a variable capacitor to ground, known as the load or shunt capacitor. The second side goes through a multiple-turn inductor, a second capacitor (the series or tuning capacitor), and then to the plasma electrode. The circuit is then completed through either the chamber walls (ground) or sometimes through the second electrode, which may be impedance-matched to ground.

Operationally, the role of the matchbox is to match the net impedance of the plasma side of the circuit to the shunt capacitor within the matchbox. Under these conditions, half of the applied power goes into the plasma and the remainder goes through the shunt capacitor. The power reflected back to the power supply will be a minimum at this point, and most matchboxes contain feedback control circuits that systematically adjust the two capacitors to minimize the reflected power.

Matchboxes are engineered for the appropriate impedance and power levels for each given system. At powers of a few hundred watts, it is possible to use variable capacitors constructed of interspersed plates in air. As the applied RF power approaches 1 kW, it becomes necessary to use vacuum-gap capacitors, which typically are rated to 5 to 6 kV and have capacitances of 10 to 2000 pF. The inductor in the matchbox is often a silver-plated Cu coil, which might be 3 to 7 turns. For powers greater than 1 kW, this coil is configured for water cooling. The capacitors are generally motor-driven, and a feedback circuit is organized to minimize the reflected power to the power supply. The inductor is generally fixed and permanently mounted.

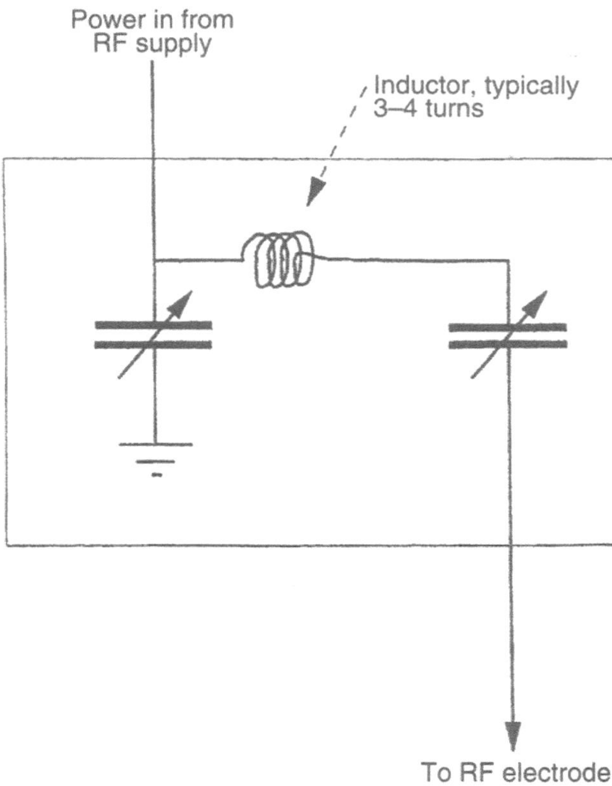

FIG. 3.9 Common L-type matching network.

3.8 Magnetic Fields

The other significant way to alter the plasma is to impose a magnetic field into the plasma region. The force on a charged particle moving in a magnetic field is just

$$F = q\mathbf{v} \times \mathbf{B} \tag{3.11}$$

where q is the particle charge, $\mathbf{v}$ is the velocity, $\mathbf{B}$ is the magnetic field, and the term $\mathbf{v} \times \mathbf{B}$ is the vector cross product of the velocity and the magnetic field. If the charged particle is moving in the same direction as the magnetic field, this cross product is zero and there is no magnetic force. If the particle is moving at right angles to the magnetic field, the particle is constrained to move in a circular path, as though it were circling a magnetic field line (Fig. 3.10). The net result for most particles is movement in a circular or helical path in which the helix axis is the same as the magnetic field direction. This can greatly increase their net path length within the plasma chamber. The radius of curvature for this motion is mass dependent and is given as

$$r = \frac{mv}{eB} \tag{3.12}$$

where e is the electron charge, B is the magnetic field, m is the particle (electron or ion) mass, and v is its velocity ($v = (2KE/m)^{1/2}$, where KE is the kinetic energy). Since the ion mass is many orders of magnitude larger than the electron mass, the effect of the magnetic field on ions is minimal and only the electrons respond in any practical way.

In practical units for electrons, r (cm) $= 2.4(KE)^{1/2}/B$, where KE is the kinetic energy of the electron in eV and B is the magnetic field in gauss. In practical terms for ions, r (cm) $= 104\ (W)^{1/2}\ (KE)^{1/2}/B$, where W is the weight of the ion in AMU (Ar = 40), KE is the kinetic energy of the ion in eV, and B is the magnetic field in gauss.

> Example: The radius of curvature for a 100-eV electron in a magnetic field of 100 gauss (0.01 T) is 0.3 cm. For an Ar (40 AMU) 100-eV ion in the same 100-gauss field, $r = 66$ cm. At energies of 1000 eV, the electron radius is 0.7 cm and the Ar^+ radius is 203 cm.

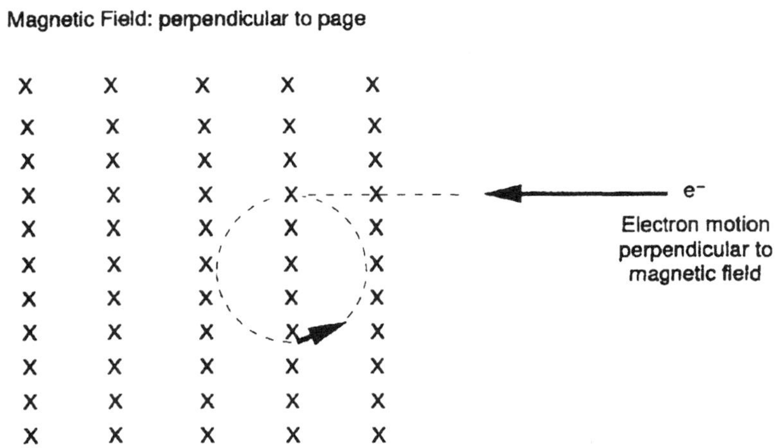

FIG. 3.10 Motion of an electron in a magnetic field.

Forcing the electrons in a plasma to move in helical, rather than straight, paths results in a great increase in the probability that an electron will have a collision with a gas atom, leading to either exciting or ionizing the atom. This effect can be used to great advantage to form a very dense, low-impedance plasma.

It should be noted here that the frequency of rotation of the charged particle, which is called the *cyclotron frequency*, is proportional to B/m, where m is the mass of the electron or ion. Since the ion's radius is too large to be relevant in sputtering tools, the electron's cyclotron (angular) frequency can be easily found to be

$$\omega = \frac{eB}{m} \tag{3.13}$$

In Hertz units (1/sec), the oscillation frequency is

$$f = 2.8 \times 10^6 B \tag{3.14}$$

where B is the magnetic field in gauss.

Example: The cyclotron frequency at a magnetic field of 5 Gauss is 13.5 MHz; at 875 gauss, f = 2.45 gHz.

The cyclotron frequency does not depend on the energy of the oscillating electron, only the magnetic field. However, the radius scales with the square of the kinetic energy.

There are three general configurations for imposing a magnetic field on a plasma. The first is simply to impose a magnetic field that is perpendicular to the cathode surface (and the sample surface). This forces the secondary electrons emitted from the plasma to move in helical paths and increases the plasma density. This approach is rarely used because (a) it is hard to get a uniform magnetic field (it is usually peaked on the system axis), and (b) there are better alternatives.

The second two magnetic-enhancement techniques require the use of a magnetic field that is parallel to the cathode surface. The secondary electrons in this case undergo a more complex motion and end up being trapped in the plane of the magnetic field, fairly close to the cathode. The trapping process uses an effect known as an **E** × **B** drift, which is another cross product, and moves the electrons on a trajectory perpendicular to both the electric field (vertical) and the imposed magnetic field (horizontal). This drift is sketched out in Fig. 3.11.

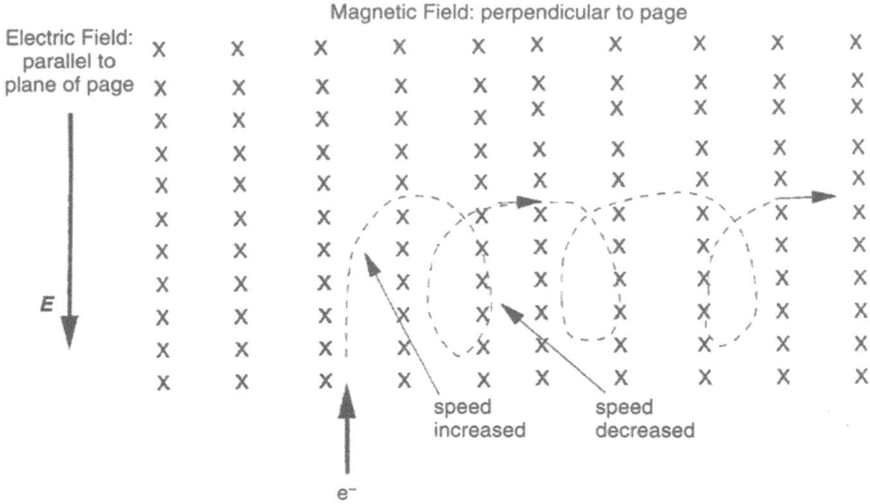

FIG. 3.11 Schematic of an **E** × **B** drift effect.

The **E** × **B** drift effect results in a pile-up of electrons at one side of the electrode system: sort of a plasma-phase Hall Effect. While this would not be desirable, the second type of magnetic-enhancement tool uses this effect combined with a moving magnetic field. The field is set up by oscillating electromagnets (Fig. 3.12), which cause the net magnetic field to rotate around the vertical axis of the tool. The moving magnetic field spreads the **E** × **B**-trapped electrons around the cathode surface and can lead to good uniformity. This effect is the basis of the AME5000 tool [3.8].

The third magnetic-enhancement system also uses the **E** × **B**-trapping effect, but instead of moving the magnetic field, it uses a magnetic field that is not uniform around the system. As shown in Fig. 3.13, the magnetic field in this case is configured radially and parallel to a round, planar cathode surface. The result is a closed-loop **E** × **B** drift path for the secondary electrons. The secondary electrons are trapped in this ring close to the cathode and can lead to very high levels of ionization of the background gas. This geometric design is known as a *magnetron.*

The formation of the closed-loop path for the **E** × **B**-drifting secondary electrons is what defines a magnetron. The simplest geometric design is that shown in Fig. 3.12, which is called generically a circular planar magnetron. As a measure of the trapping efficiency of this class of cathode,

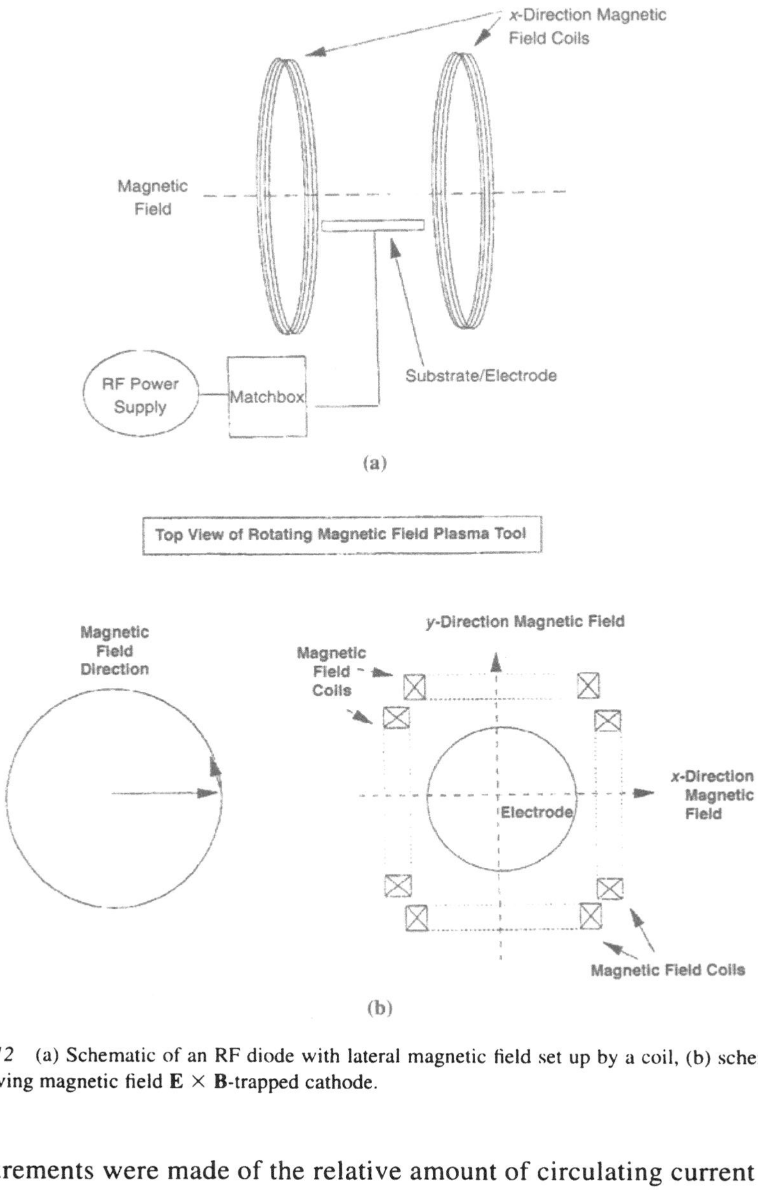

FIG. 3.12 (a) Schematic of an RF diode with lateral magnetic field set up by a coil, (b) schematic of a moving magnetic field **E** × **B**-trapped cathode.

measurements were made of the relative amount of circulating current and were compared to the net discharge current [3.9]. An example of that work for a small, 15-cm-diameter cathode is shown in Fig. 3.14, where the circulating current was inferred by measuring the magnetic field it induced.

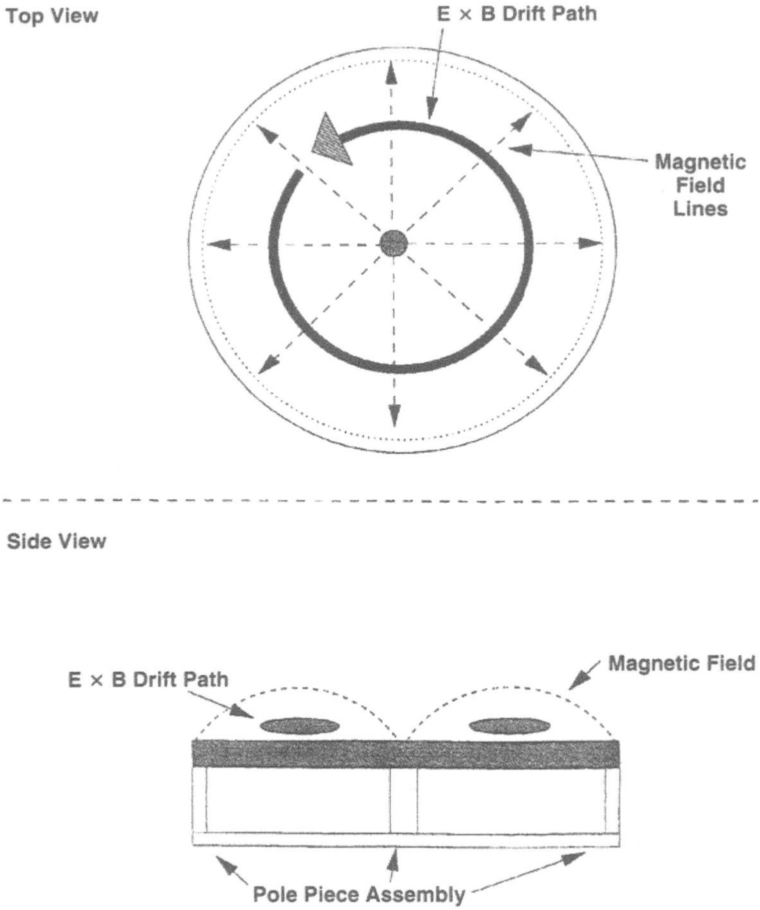

FIG. 3.13 Top and side view schematic of a circular planar magnetron cathode.

The slope of that graph indicates that the electrons can be considered to go around the **E** × **B** loop path from 3 to 8 times before they are lost to the walls of the system. This amount of trapping is indicative of an electron mobility consistent with Bohm diffusion, rather than classical electron diffusion in a magnetic field [3.9].

The circular planar magnetron is the most widely used example of the magnetron cathode. However, since the overall requirement is simply that

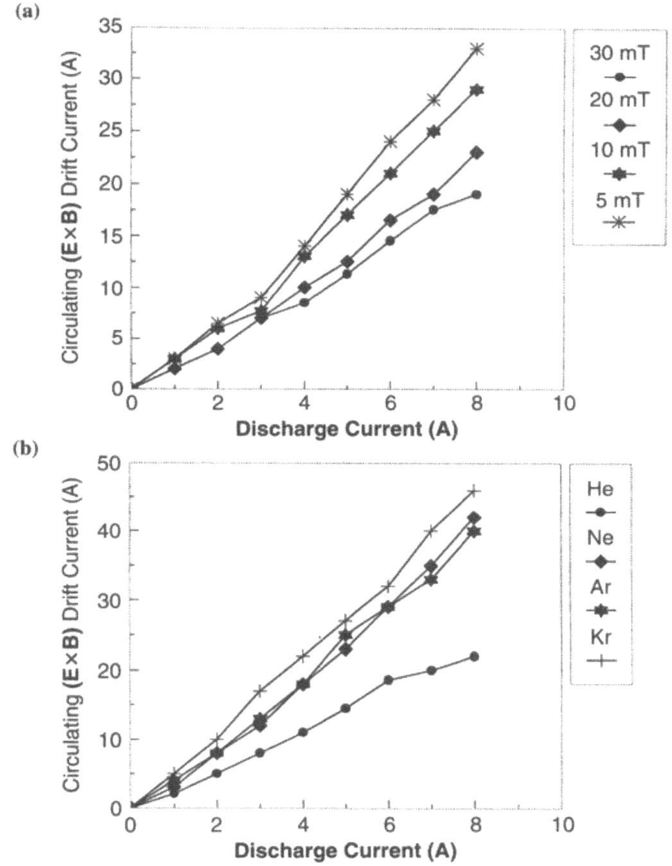

FIG. 3.14 Circulating **E** × **B** drift current in a 150-mm-diameter circular planar magnetron device as a function of discharge current: (a) as a function of chamber pressure, (b) as a function of working gas [3.9].

the **E** × **B** drift path form a closed path, there are many other geometrical permutations. A common example is the rectangular planar magnetron, shown in Fig. 3.15. This geometry is similar to the circular planar one, but is simply stretched in one direction, forming an **E** × **B** drift path somewhat like a racetrack oval. The magnetic fields in a rectangular magnetron are usually 10–20% stronger at the ends to compensate for curvature-driven loss processes for the electrons. Dimensionally, there is no real limit to the

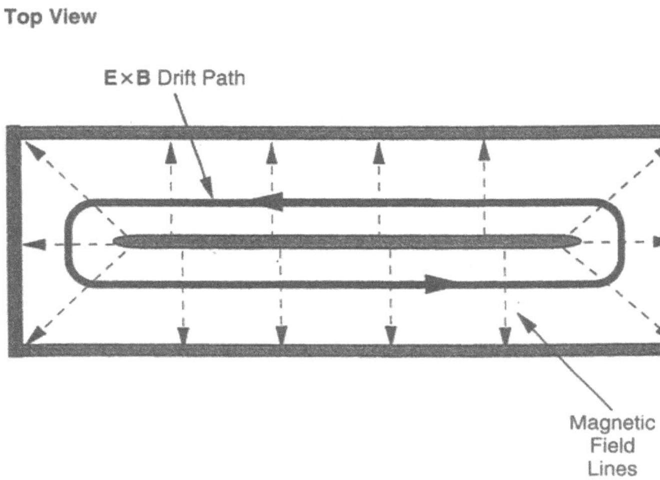

FIG. 3.15 Rectangular planar magnetron.

length of the rectangle, and sources have been constructed several meters in length. Typically this class of magnetron is used in a linear system, where samples pass by the long sides of the cathode. Since this type of source design scales up so easily, this is usually the kind of magnetron used to coat very large surfaces, such as architectural glass or large rolls of plastic sheeting.

A three-dimensional variant on the circular planar magnetron uses a beveled, cone-shaped cathode surface, as shown in Fig. 3.16. This class of magnetrons, known as the S-Gun class, was first developed by Peter Clarke [3.10]. The magnetic field is oriented such that it is parallel to the beveled cathode surfaces, and the **E** × **B** drift path is adjacent to the cathode surface. Larger versions of this class of sources use two nested loops, independently powered, for better deposition uniformity control.

The final geometrical variant of the magnetron uses a cylindrical geometry in which the magnetic field is uniform and the cathode surface is arranged in the form of a cylinder with the magnetic field aligned with the axis. This comes in two variations, one with the plasma on the external face of the cathode cylinder and the second with the plasma on the internal face (Fig. 3.17). The external-plasma version is known as the cylindri-

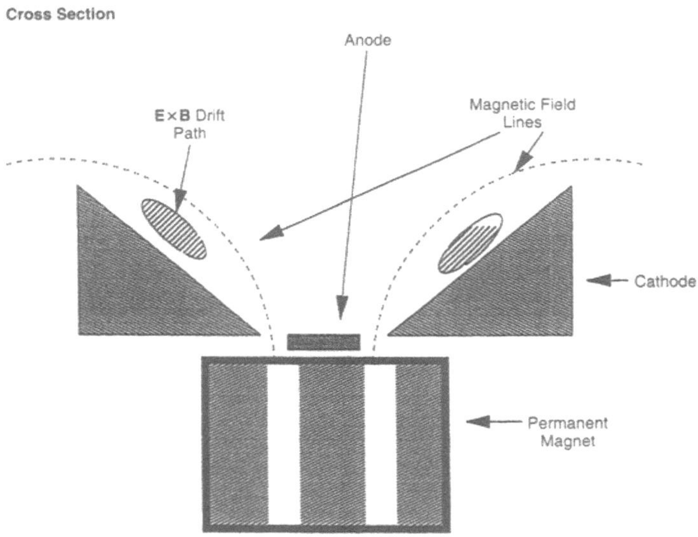

FIG. 3.16 S-Gun class of magnetron geometries.

cal post magnetron, and the internal version is known as the hollow cathode magnetron. The essential part of these designs is, again, a closed-loop path for the secondary electrons, which is now a cylindrical band in each geometry. The external-plasma version of the cylindrical magnetron is often used to coat large areas, whereas the internal version is usually used to deposit metal on fibers passing along the axis of the cathode.

There are additional variants of all of these design classes of magnetrons. For example, wedge-shaped cathodes have been designed that have a serpentine **E** × **B** drift path around the surface of the wedge to provide uniform erosion of the cathode. Another geometrical variation is the rotating cylindrical magnetron, which uses an **E** × **B** drift path that looks somewhat like a Hawaiian lei wrapped around a cylinder, which is then motor-driven to provide uniformity [3.11]. A widely used variation of the circular planar magnetron has a small heart-shaped **E** × **B** drift path, which is motor-driven behind the cathode to circle the cathode face and provide uniform erosion. This design is the basis for most current-day semiconductor wafer deposition systems and will be discussed at greater depth in Chapter 4.

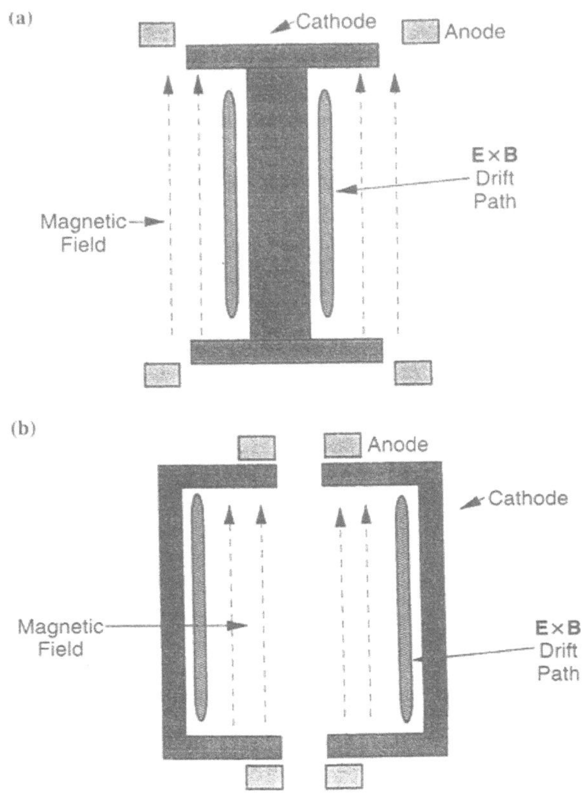

FIG. 3.17 (a) Cylindrical post magnetron, (b) hollow cathode magnetron.

3.9 Reactive Sputter Deposition

Reactive gas species, such as oxygen or nitrogen, are routinely added to the inert gases used during sputtering for the purpose of depositing oxide or nitride films. In the case of the DC diode, this was found (as discussed above) to strongly alter the cathode surface, resulting (particularly for the case of oxygen) in a much-reduced discharge current. In the case of magnetron sputtering, due to the very high levels of metal sputtered from the cathode, an interesting behavior can be observed.

A typical reactive deposition system is shown in Fig. 3.18. The system contains a sputtering source (usually a magnetron), gas inlets for inert and

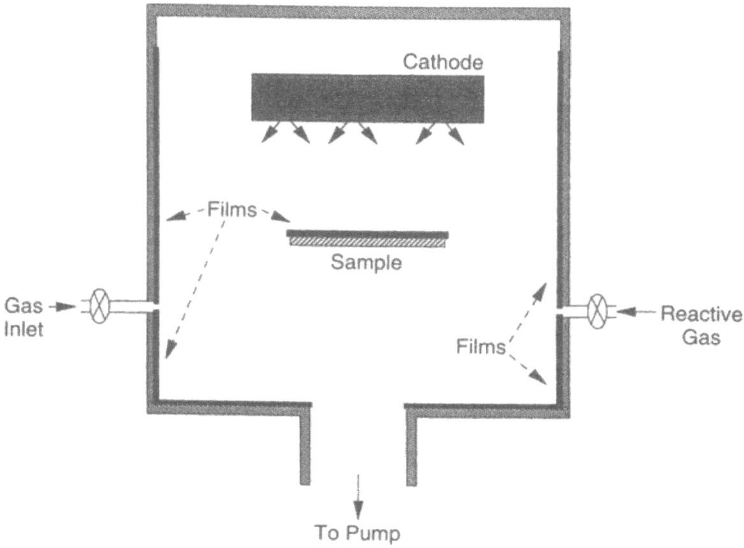

FIG. 3.18 Reactive sputter deposition system.

reactive species, a pumping system, and various surfaces such as the walls or the sample on which films are deposited. As a starting place for this explanation, consider the operation of the magnetron at a fixed DC power with an Al target in a background of inert gas without any reactive species present. At this point, the films deposited around the chamber are metallic and the deposition rate is relatively high.

If a small amount of a reactive gas species is added — oxygen, for example — very little change is observed. This is because the oxygen is rapidly adsorbed by the freshly sputtered Al films on the walls and the sample. Chemical analysis of these films would show a mostly metallic film with a low level of oxygen contamination. If the flow of the oxygen is increased (Fig. 3.19), very little change is observed in the system: The deposition rate is similar, the discharge voltage on the cathode is unchanged, and there is no change in the chamber pressure. The only real change is that the films are becoming increasingly more oxygen-rich.

If the oxygen flow is increased farther, eventually the deposited films will reach a terminal oxidation level and will not be able to absorb any additional gas. For Al, this occurs at the formation of Al_2O_3. Any additional flow of oxygen cannot be picked up by the already-saturated films, and

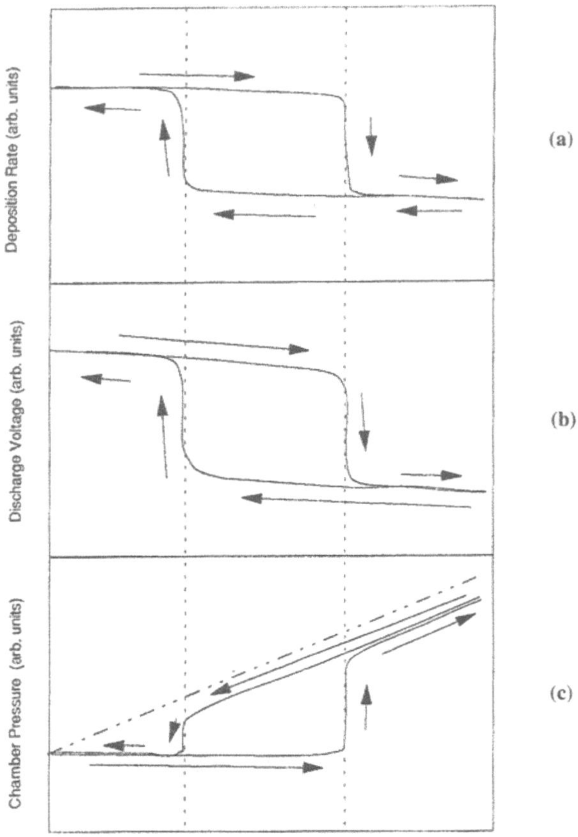

FIG. 3.19 Reactive sputtering hysteresis curves. (a) Deposition rate, (b) discharge voltage, and (c) chamber pressure, all as a function of reactive gas flow during the sputtering of a metal target.

oxygen now becomes part of the working background gas. This leads to oxidation of the Al cathode, which has significant and unfortunate consequences. The most significant effect is that the sputter yield of the oxidized Al target is much, much lower (up to 25 times) than that of clean Al. This means that the amount of Al atoms emitted from the cathode drops radically, and as a result, the capability for this sputtered Al film to absorb oxygen reduces markedly. This is a runaway type of effect. As the cathode starts to become oxidized, the Al emission rate drops, which leads to an increasing amount of free oxygen in the working gas, which causes additional oxidation of the cathode.

In short, at this critical flow, the system undergoes an irreversible transition from a metallic cathode operating at high rates to an oxidized cathode operating at very low rates. The deposition rate (Fig. 3.19a) drops rapidly and the chamber pressure (Fig. 3.19c) increases rapidly. It also turns out that for many gas-cathode systems, the discharge voltage changes significantly also, due to changes in the secondary electron yield going from a clean to a compound surface (Fig. 3.19b).

Another way to characterize the hysteresis problem is to examine the dependence on the overall system pumping speed. In a classic experiment, it was shown that by increasing the pumping speed, and at the same time increasing the reactive gas flow to attain the same operating pressure, the hysteresis effect became less pronounced and was virtually eliminated at the highest flows (Fig. 3.20) [3.26]. This shows that if the pumping effect of the film can be overshadowed by the pumping speed of the system pump, then changes in the wall pumping speed (i.e., oxidation of the wall) will have little net effect.

At flows above the critical flow, there are few changes in the plasma. The chamber pressure increases linearly, but the deposition rate and voltage do not change significantly. If the flow is reduced back to the critical flow point, the system does not revert back to the clean metallic state. This is due to the presence of the compound layer on the cathode, which reduces

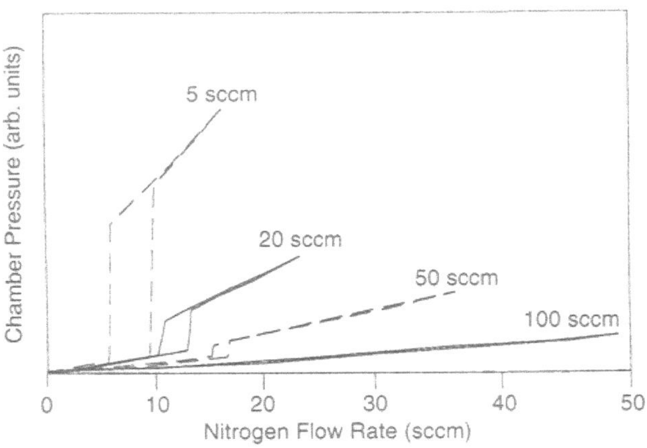

FIG. 3.20 The chamber pressure hysteresis plot for reactive sputtering as a function of chamber flow rate [3.26].

the amount of metal that can be emitted. It is necessary to go to a significantly lower reactive gas flow to get to the point where the films on the walls and the sample can take over as the primary sink for the reactive gas species. The overall flow dynamics of this system show a response much like the hysteresis response of magnetic materials, and these reactive gas flow plots are called hysteresis curves.

The fundamental problem with reactive sputter deposition is that the best films (i.e., the most completely oxidized films) are deposited at a flow just below the critical flow point. In addition, since the cathode is not yet oxidized at this point, the deposition rate is high. The basic problem is that as the critical flow is reached, the system changes rapidly into the oxidized-cathode mode, and once this transition has begun, it is very difficult to control. It should be noted, though, that in most cases it is easier to control the nitride transition than the oxide transition. There are a number of subtle reasons for this: The sputter yield of the cathode does not change as much when the cathode is nitrided (for many materials) compared to when it is oxidized; the secondary electron yield of the cathode may not go up (which reduced the voltage) and may actually go down, which results in higher cathode voltage and potentially increased sputter yields; and finally many of the nitride reactions of interest (for example, TiN and AlN) are less spontaneous than their equivalent oxygen reactions. In fact, many of the nitride reactions require energy to form, which is markedly different from the case of oxidation. This allows better control of the process using simply the power supplies.

Until recently, reactive sputter deposition of oxides was fairly difficult on an industrial scale due to these intrinsic control issues [3.27]. Although it was always possible to operate in the oxidized-cathode mode, the rates were low enough to be impractical. In addition, since a nonconducting oxide was being deposited around the chamber, there were problems with both arcing on the cathode edge and coating the anode with oxide, resulting in a "hidden anode" that inhibited current flow and led to more arcing.

The solutions to controlling reactive deposition lie in two areas: the control of the reactive gas flow and the time-dependent control of the cathode potential. The gas flow problem, which is the desire to operate just below the critical flow, can be solved by using a fast-feedback control system for the gas supply [3.12]. The voltage to the cathode can be pulsed in the 20–200 kHz range such that for a relatively small fraction of the cycle, the cathode is operated at a positive potential. Much like the case of the RF diode, this reduces charging and arcing effects, and allows better control of the deposition process. This has allowed, for example, the deposition of high-quality aluminum oxide at a rate of 78% of the pure metal deposition rate [3.13]. This is about 25 times faster than when using conventional RF power in an oxide mode. As a parallel to this pulsing approach, it is also possible to use two adjacent cathodes, one of which functions as the anode for the other. These cathodes are then switched in polarity in the 100 kHz range to reduce arcing and charging [3.14, 3.15].

3.10 Practical Plasma Issues in PVD Tools

The mechanical design of magnetron cathodes will be covered in Chapter 4, and some aspects of the overall tool design will be discussed in Chapter 5. There are some plasma issues that are relevant to introduce here, such as shielding and tooling, plasma breakdowns, gas density issues, etc.

The plasma in a magnetron system, as well as in most other types of plasma tools, expands to fill the volume of the tool: There is no real confinement or containment mechanism. This can be misleading with a magnetron system, where it appears that the plasma is all tightly connected to the cathode surface and the rest of the tool is mostly dark.

In a magnetron tool, the plasma density 'in' the etch track ($\mathbf{E} \times \mathbf{B}$ drift path) can be 100–1000 times denser than the surrounding regions. A measure of this can be to apply a small negative bias to the sample holder and see how much current it draws. When sputtering with a few kW of discharge power on the cathode (perhaps 5 A at 500 V), substrate currents of only tens or a hundred milliamps are measured. This means that a strong substrate bias power is going to be difficult to obtain; there simply are not enough ions at the sample to make much of a difference. There is a class of magnetrons, called unbalanced magnetrons, which are designed to have a less well confined plasma near the cathode and, as a result, much higher substrate bias capability [3.16]. However, unbalanced magnetrons have not been applied to semiconductor manufacturing technology and will not be discussed here.

The weak chamber plasma does result, though, in concerns about cathode currents and arcing. Most magnetron cathodes are configured with a ground shield, or the same function may be designed into the mount for the cathode. A ground shield is a grounded piece of metal located near the edge of a magnetron cathode and serves to keep the chamber plasma away from the sides of the cathode. This eliminates sputtering of the edge of the cathode and perhaps the backing plate, as well as shorting or arcing across the insulator used to hold the cathode in the chamber. This insulator is always required, as the sputtering chambers are grounded and the metal cathodes are powered at -300 to -500 V DC. The ground shield may also be used to prevent metal deposition onto this insulator ring. This function may also be designed into the insulator mounting with a reentrant design such that sputtered metal must hit one or two other surfaces before it can get to the insulator.

The spacing criteria for shields within the chamber is such that the openings be on the order of a Debye length or two. From Eq. (2.5), the Debye length is (in cm) 743 $(T_e/n_e)^{1/2}$, where T_e is the electron temperature in eV and n_e is the electron (plasma) density in electrons/cm^3. The low-density plasmas in the magnetron chamber might have densities of at most 10^9/cm^3 and an electron temperature of 1 eV, which leads to a Debye length of 0.25 mm. This is a fairly tight tolerance (1/2 mm), particularly given that most production cathodes are 12–14 in. (300–350 mm) in diameter and that they are mounted on deformable o-rings. In addition, this small spacing must hold off 500 V in the presence of a weak plasma as well as small particulates and flakes. Therefore, shielding tolerances are generally on the order of a few mm, and the small stray currents that might be observed by the very weak plasma across the insulator are ignored.

RF plasmas are less well confined than DC magnetron discharges. In an RF plasma, the cathode functions as the cathode for most of the RF cycle, but then switches and becomes the anode for a small fraction of the cycle. At this point, the rest of the chamber becomes the defacto cathode even though the bombardment energy is low. The result is that the plasma spreads farther around the chamber and substrate biases are increased. There is also the possibility of the formation of secondary discharges in tight, confined areas of the chamber. This is due to a hollow-cathode trapping effect [3.17] in which secondary electrons are geometrically trapped in places such as pump ports, tubes, gas lines, and so on. Since virtually all production semiconductor PVD tools are operated DC, this is not an issue. However, RF tools will be needed for the complex dielectrics (k = very high, or $k < 4$), and this extraneous discharge may become a problem.

The shields in a magnetron tool are almost always grounded by design: They are made of metal and screwed or clamped to the grounded chamber walls. As such, they function as anodes in the plasma circuit and may actually carry a reasonable current to ground from the plasma. If a shield were oxidized at its contact point, this situation could change as the shield became negatively charged and attained the floating potential. At this point, the shield is like a large capacitor: It contains a stored negative charge that may be released rapidly in the form of an arc, which may result in material emission and deposition onto a sample. This same effect could occur for a heavily deposited shield that develops cracks or breaks in its thick sputter-deposited coating. Upon air exposure, these cracks may allow moisture or air penetration under the film, which may result in flaking or eventual arcing.

The arcs that occur on the cathode surface (see Chapter 11 for a discussion of target issues) are called unipolar arcs, in that they appear to have only one point of contact. In reality, since an arc implies current flow, the arc is between the cathode surface and the plasma, hence the term *unipolar.* As will be described in Chapter 11, arcs may result in droplet emission from the cathode as well as physical damage to the cathode. Present-day magnetron power supplies (1998) are solid state, switching supplies, which are less susceptible to arcing than older style, RC-filtered supplies. However, unless specific precautions are taken — such as arc-supressing circuits — miniarcs of a few amperes may occur during sputtering and it will be very hard to detect these arcs. This is an ongoing field of interest, and one that is only recently been recognized as a contributor to defect formation during deposition.

3.11 Plasma Diagnostics and Optical Emission in Magnetrons

Commercial magnetron sputtering tools are generally not equipped with any sort of plasma, film, or optical diagnostics. For laboratory-scale tools, several authors have described the use of Langmuir probes, which are small wires inserted into the plasma. By biasing this wire tip in the presence of the plasma, it is possible to determine the electron temperature, electron density, and plasma potential [3.17–3.20]. This type of experiment is greatly complicated by the large fluxes of sputtered metal atoms that coat the insulators holding the Langmuir probe and make determination of the functional area of the tip difficult.

Laboratory-scale tools are also occasionally outfitted with deposition rate monitors that can be operated during deposition. These rate monitors are typically small quartz disks driven at 5 Mhz or so, whose resonant frequency changes in a known way depending on the mass of the deposited film. Unfortunately, the geometry of cluster tools (Chapter 5) is such that there is no room for this sort of film deposition diagnostic.

One laboratory-scale technique that could be applied to magnetron sputter deposition is optical emission spectroscopy (OES). This is a passive diagnostic, in that it only measures light emitted from the discharge. However, the amount of light emission turns out to be a rather complex function of both the species being observed and the operating conditions of the tool. For example, if a wavelength of light is observed that is due to neutral Ar, it would be expected that the intensity of that light would increase as the discharge power is increased. (The excitation probability for an atom depends on the plasma density, which should scale linearly with the discharge power.) And this is indeed what is observed. However, the same observation for light from Ar ions is more complex. Since the number of Ar ions is dependent on the discharge power, as is the probability of light emission, the light intensity from the Ar ions will scale roughly as the square of the discharge power.

The same argument can be extended to the metal species from the cathode. Since the rate of metal sputtering scales with the discharge power, the light intensity will also scale with the square of the discharge power. For metal ions, since the number of metal atoms scales with the power, the probability of ionization scales with the power, as does the emission probability, the light intensity from the metal ion species scales as the cube of the discharge power (Fig. 3.21). This is all explained in more detail in [3.21]. The end result, though, is that unless the emission wavelengths from the discharge are well identified, measurements of their intensity to diagnose or monitor the discharge will be very hard to decipher.

Needless to say, production PVD tools are rarely, if ever, outfitted with optical emission detectors. In fact, most tools have virtually no windows

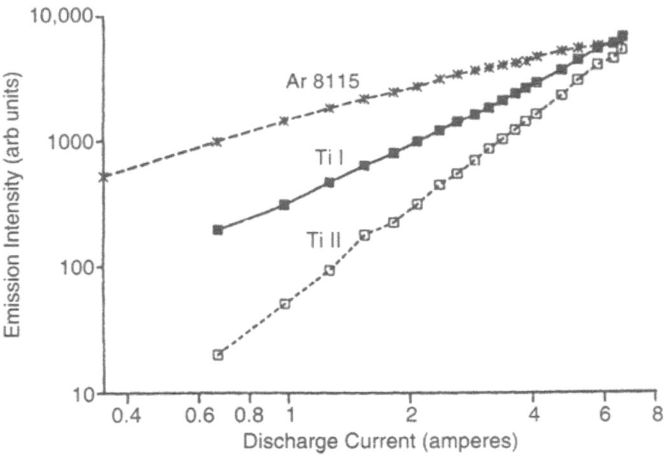

FIG. 3.21 Optical emission spectroscopy (OES) of emission from Ar neutrals, Ti neutrals, and Ti ions as a function of discharge current during magnetron sputtering of Ti in Ar [3.21].

and it is quite hard to see the plasma. This is because the tooling is generally quite tight to capture all of the deposition for subsequent tool cleaning: If a window to see the plasma were there, it would eventually become deposited with metal, and eventually that metal would flake off and form particulates. However, since OES is widely used for diagnosis of reactive etching technology, its use in PVD tools cannot be precluded. Any such use, though, should be carefully tailored to observe only the desired species.

A very common diagnostic now being outfitted on most production PVD tools is a mass spectrometer or residual gas analyzer (RGA). The role of this tool is obvious: to observe and quantify the composition of the background gas inside the chamber both during sputtering and at the nominal base pressure. Since most sputtering runs occur at pressures of a few mTorr, it has often been necessary to differentially pump the RGA, which tends to not function well at pressures above 0.5 mTorr. Differential pumping requires a small turbo pump and backing pump and brings up the possibility of oil back-contamination to the chamber. It also provides only a moderately accurate look at the internal gas composition of the chamber, as the plasma region is generally sampled through a small aperture and with some distance to the RGA (typically tens of cm). However, the RGA can be extrememly useful for leak detection as well as for observation of residual water vapor and impurities in the gas lines. Since newer-generation RGAs have become avaiable that are compact and may not even require differential pumping, they have become nearly standard equipment on the manufacturing production-scale tools. Simply detecting a problem with the RGA that would compromise a single wafer run could easily justify the added expense of the RGA ($10k), as wafers late in their production cycle can easily be worth $5–10k per wafer.

References

3.1. H. R. Kaufman and R. S. Robinson, *Operation of Broad Beam Ion Sources*, Commonwealth Scientific, Alexandria, VA, 1987.

3.2. D. B. Meveded, P. Mahadevan, and J. K. Layton, "Potential and kinetic electron ejection from Mo by argon ions and neutral atoms," *Phys. Rev.* **129**: 2086 (1963).

3.3. C. D. Child, "Discharge from hot CaO," *Phys. Rev.* **2**: 492–511 (1911).

3.4. B. Chapman, *Glow Discharge Processes*, Wiley, New York, 1980.

3.5. D. Bohm, "Minimum Ionic Kinetic Energy for a Stable Sheath," in *The Characteristics of Electrical Discharges in Magnetic Fields*, pp. 77–86, A. Guthrie and R. K. Wakerling, Eds., McGraw-Hill, New York, 1949.

3.6. F. Chen, *Introduction to Plasma Physics*, Plenum Press, New York, 1974.

3.7. J. S. Logan, "RF Diode Sputter Deposition and Etching," in *Handbook of Plasma Processing Technology*, pp. 140–150, S. M. Rossnagel, J. J. Cuomo, and W. D. Westwood, Eds., Noyes Publications, Park Ridge, NJ, 1990.

3.8. Applied Materials, Santa Clara, CA.

3.9. S. M. Rossnagel, "Induced drift currents in circular planar magnetrons," *J. Vac. Sci. & Tech.* **A5**: 88–91 (1987).

3.10. P. Clarke, U.S. Patent No. 3,616,450 (1971).

3.11. D. W. Hoffman, "Design and capabilities of a novel cylindrical post magnetron sputtering source," *Thin Solid Films* **96**: 217 (1982).

3.12. W. D. Sproul, "New routes in the preparation of mechanically hard films," *Science* **273**: 889–892 (1996).

3.13. J. M. Schneider, W. D. Sproul, M. S. Wong, and A. Matthews, "Scalable process for pulsed DC magnetron sputtering of nonconducting oxides, submitted (1996) to *Surface and Coatings Tech.*

3.14. S. Schiller, U. Heisig, Chr. Korndorfer, G. Beister, J. Reschke, K. Steinfelder, and J. Strumpfel, "Reactive DC high rate sputtering as a production technology," *Surface and Coatings Tech.* **33**: 405–423 (1987) also **61**: 331 (1993).

3.15. S. Schiller, V. Kirchhoff, K. Goedicke, and N. Schiller, "Pulsed plasma deposition creates new era for PVD," *Solid State Tech.*, S12–S14 (Dec. 1996).

3.16. B. Windows and N. Savvides, "Charged particle fluxes from planar magnetron sputtering sources," *J. Vac. Sci. & Tech.* **A4**: 196–202 (1986). N. Savvides and B. Windows, "Unbalanced magnetron ion-assisted deposition and property modification of thin films," *J. Vac. Sci & Tech.* **A4**: 504–508 (1986). B. Window and N. Savvides, "Unbalanced DC magnetrons as sources of high fluxes," *J. Vac. Sci. & Tech.* **A4**: 453–456 (1986).

3.17. C. M. Horwotz, "Radio frequency sputtering, the significance of power input," *J. Vac. Sci. & Tech.* **A1**: 1795–1800 (1983).

3.18. S. M. Rossnagel and H. R. Kaufman, "Langmuir probe characterization of magnetron operation," *J. Vac. Sci. & Tech.* **A4**: 1822–1825 (1986).

3.19. T. E. Sheridan, M. J. Goeckner, and J. Goree, "Electron and ion transport in magnetron plasmas," *J. Vac. Sci. & Tech.* **A8**: 1623–1626 (1990).

3.20. M. Dickson, F. Qian, and J. Hopwood, "Quenching of electron temperature and electron density in ionized physical vapor deposition," *J. Vac. Sci. & Tech.* **A15**: 340–344 (1997).

3.21. S. M. Rossnagel and K. L. Saenger, "Optical emission in magnetrons: nonlinear aspects," *J. Vac. Sci. & Tech.* **A7**: 968–971 (1989).

3.22. L. McCaig and R. Sacks, "Current sensitivity of plasma voltage and emission line intensities on a planar magnetron glow discharge device," *Appl. Spectr.* **46**: 18–24 (1992).

3.23. A. Okamota and T. Serikawa, "Reactive sputtering characteristics of Si in an argon-nitrogen mixture," *Thin Solid Films* **137**: 143 (1986).

3.24. W. D. Sproul, P. J. Rudnick, C. A. Gogol, and R. A. Mueller, "Advances in partial pressure control applied to reactive sputtering," *Surface and Coatings Tech.* **39/40**: 499–506 (1989).

Chapter 4 The Planar Magnetron

4.1 The DC Magnetron

The introduction of commercial DC magnetrons in the 1970s provided the IC industry with a production-worthy alternative to both evaporation and diode sputtering. Magnetron-based sputter tools deposited thin films at much higher rates than diodes and operated at lower pressure, where gas-phase scattering and gas-phase impurities were minimal. As discussed in detail in Chapter 3, a DC magnetron is basically a magnetically enhanced diode in which the spatial relationship of electric (**E**) and magnetic (**B**) fields is engineered to confine secondary electrons produced by Ar^+ bombardment of the target. Restricting these electrons to remain close to the target surface increases their probability of ionizing the Ar working gas, which in turn results in a more intense plasma discharge that can be sustained at a lower pressure. Since the Ar ions are much heavier than electrons (mass ratio $\approx 7 \times 10^4$), they are not affected to first order by the confining magnetic fields, and sputter bombardment proceeds much as in a diode.

The underlying physics of electron confinement is based on the Lorentz force **F** given by the vector cross product of the two fields ($\mathbf{F} = e\mathbf{E} \times \mathbf{B}$). Since **E** is perpendicular to the target surface, application of a **B** field tangential to the surface gives the electron a component of velocity parallel to the target. Keeping the electrons confined is then achieved by creating a situation where the locus of these tangential **B** fields form a closed path. In this way, an electron whose initial velocity would have launched it away from the target surface will remain in "low orbit" above it, undergoing a cycloidal hopping motion along the orbit path. Each electron can then ionize many neutral gas atoms before being collisionally scattered out of the plasma region. A similar approach, but without the closed-path confinement, has been used to increase the sensitivity of high vacuum pressure gauges based on electron impact ionization of residual gas molecules. In this case, thermionically emitted electrons undergo helical motion in the presence of an applied magnetic field, which increases their path length in the vicinity of a biased collector grid.

As described in Section 3.7, the basic physics of the crossed-field DC magnetron has been implemented in many different hardware designs over the past 20 years, with the cathode shape and the spatial arrangement of magnetic fields limited only by the creativity of the designer (the reader is referred to refs. 4.1– 4.7 for excellent reviews of magnetron sputter source development). While a variety of magnetron designs are still in active use for IC production, the planar magnetron cathode with circular target is by far the most widely deployed in VLSI and ULSI device manufacturing.

4.2 The Planar Magnetron

Planar magnetrons have been designed for a variety of target shapes — e.g., square, rectangular, wedge, and circular. However, the circular shape is widely used in IC production because it matches the circular geometry of the wafer and lends itself to rotating the magnet arrays that are employed to improve target utilization and film uniformity. Figure 4.1 schematically presents a DC planar magnetron with a circular target (Fig. 9.10 in ref. 4.7). In the design shown, the surface-parallel **B** field needed to produce closed electron paths is provided by a ring of permanent bar magnets arranged around a central magnet, the entire array being located behind a nonmagnetic target backing plate. The magnets are connected to each other by a *field return* plate made of a highly permeable material such as iron. This iron rail completes the magnetic circuit and prevents magnetic flux from spreading into regions other than the desired one in front of the target surface. The field lines pass through the target and arch over the target surface, which for thick targets at start of life can be 1 inch or more away from the backing plate. Therefore, to obtain the desired field strength at the target surface, strong permanent magnets having an energy product on the order of 30 megagauss-oersted (MGO) are often used. The value of the energy product is determined from the point on an experimental plot of B versus H, where the product of $B \times H$ is maximum — analogous to determining the maximum power point on a loadline plot of current versus voltage for an electrical device. The square of magnetic field strength has units of energy density that would normally be given as ergs/cm^3 in the CGS system or joules/m^3 in the MKS system. However, since the convention is to give B in gauss and H in oersted, the value of the energy product is given in mixed units of gauss-oersted.

A variety of high-energy-product permanent magnets have been used in planer magnetrons, including ferrites and rare earth alloys such as samarium-cobalt (Sm-Co) and neodymium-boron-iron (Nd-B-Fe). Using dense arrays of compact bar magnets with high coercive force (e.g., $\approx$ 30 MGO and volume $<$ 30 cm^3), one can produce tangential fields at the target surface on the order of 500 G, or 0.05 Tesla (T) in SI units. A modern planer magnetron design might incorporate perhaps 30–50 such magnets. Electromagnets have also been used in some modern designs and offer the ability

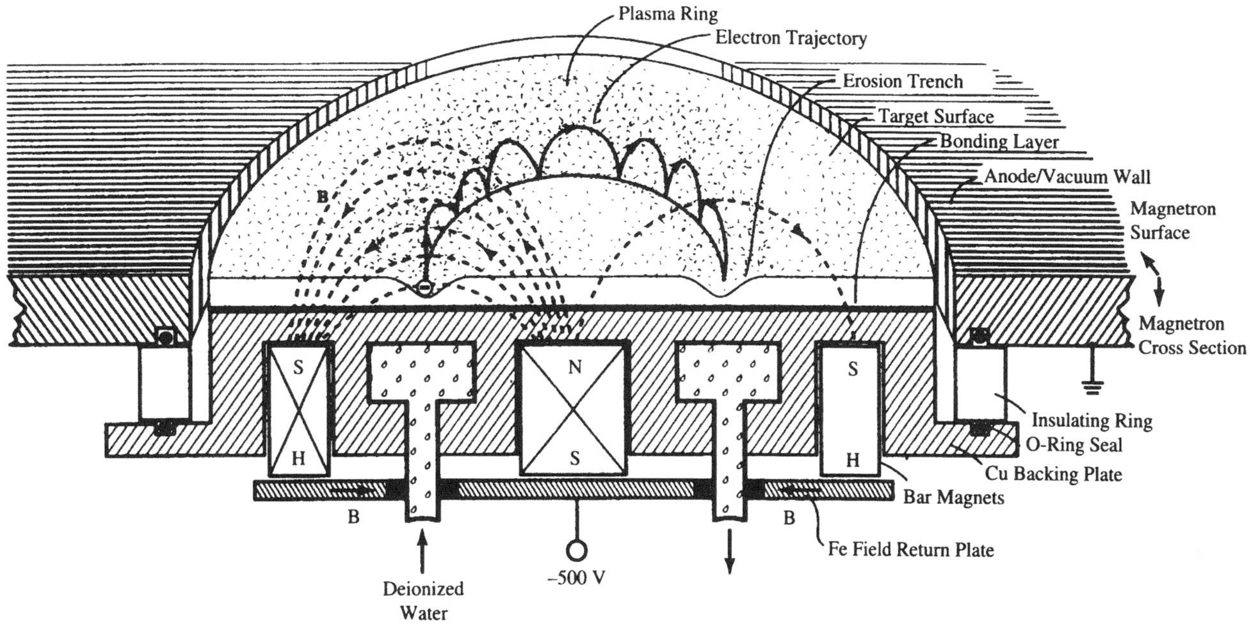

FIG. 4.1 Schematic of a planar magnetron with circular target and stationary magnet array. (Reproduced with permission of the McGraw-Hill Companies from D. L. Smith, *Thin Film Deposition: Principles & Practice*, McGraw-Hill, NY, 1995.)

to electrically vary field strength (e.g., to jump-start the plasma at a low pressure by temporarily using a high field) or to optimize field distribution for improved film uniformity. However, permanent magnets offer a number of advantages over electromagnets. For example, there is no DC power required; no additional heat to dissipate; no danger of magnet insulation failure or field interruption from a power failure; and far less weight, volume, and complexity for the equivalent magnetic field. Figure 4.1 shows how the ring of magnets confines the electrons into an annular closed path, which in turn produces an annulus of intensified sputtering plasma and a corresponding "race-track" sputter erosion groove in the target surface. The narrowness of the groove results from radial compression of the plasma by a magnetic-mirror effect, which is common to many magnetic plasma systems [4.5]. The boundaries of the track correspond to the location of the pole pieces of the magnetic array where the **E** and **B** fields are nearly parallel ($\mathbf{E} \times \mathbf{B} \approx \mathbf{0}$). Since the crossed-field confinement effect is minimal here, electrons can escape from the cathode without causing any localized enhancement of plasma density.

The planar magnetron shown in Fig. 4.1 with fixed permanent magnet elements suffers from poor utilization of the target material, which is preferentially etched in a narrow annular region. If the target is too close to the wafer, the deposited thin film thickness profile will mirror the race track and be highly nonuniform. By using a source-to-substrate distance that is very large compared to the mean free path of the sputtered neutrals, isotropic gas-phase scattering can be used to smooth out this nonuniformity; however, the same scattering will reduce the deposition rate by directing atoms away from the wafer and toward the chamber walls. Also, as shown in Fig. 4.2, the angular distribution of atoms leaving the target will change as the groove deepens since the local surface normal is no longer in the same direction as when the target was planar. This will affect both the step coverage and thickness uniformity of thin film deposition as the target ages. Target erosion also brings the target surface physically closer to the source magnets, with the most strongly eroded regions being closest to the magnets below. This tends to increase the magnetic field in grooved regions and serves to accelerate the effect with target life. Finally, due to the finite geometric size of the source as seen by the wafer and the nondirectional nature of the sputtered flux, a uniformly eroded target will not produce a uniform thin film. In batch PVD systems, one can scan the wafer beneath the target to address these concerns; however, in single-wafer PVD cluster tools this has not been practical. Therefore, the job of the magnetron designer is to tailor the target erosion profile over the target to

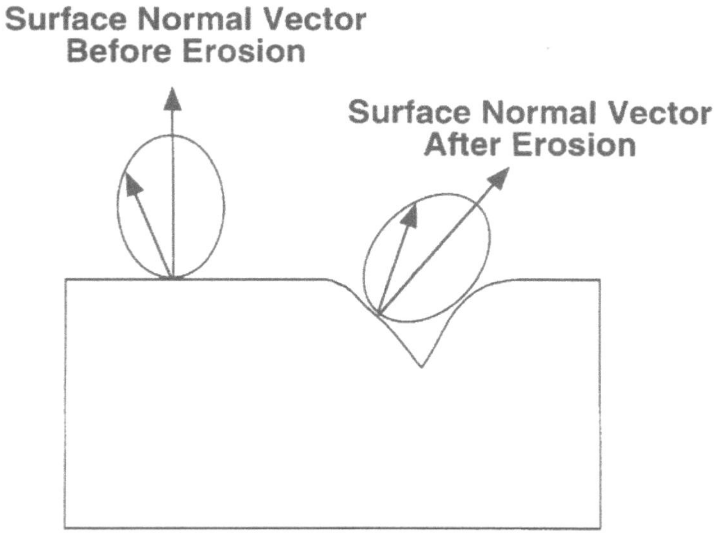

FIG. 4.2 Development of a narrow sputter erosion groove during target life can change the distribution of sputtered flux.

produce an acceptably uniform thin film, while achieving good utilization (e.g., full-face erosion) of the expensive target material.

4.3 The Swept-Field Magnetron

The interrelated problems of target grooving, target utilization, and thin film uniformity have led to the use of so-called swept-field magnetrons in which the permanent magnet array, rather than being stationary, is mechanically rotated behind the wafer in the plane of the target at relatively low speed ($\approx$ 40–60 rpm). By tailoring the shape of the magnet array and its center of rotation, designers have been able produce a more uniformly eroded annular region or uniform erosion over virtually the full face of the target. Since the engineering design rules and manufacturing details of advanced source designs are highly proprietary (or disclosed only in the patent literature), we only indicate the basic idea.

Figure 4.3 shows a circular ring of bar magnets that upon off-axis rotation produces an annular race track with two broad erosion grooves. To

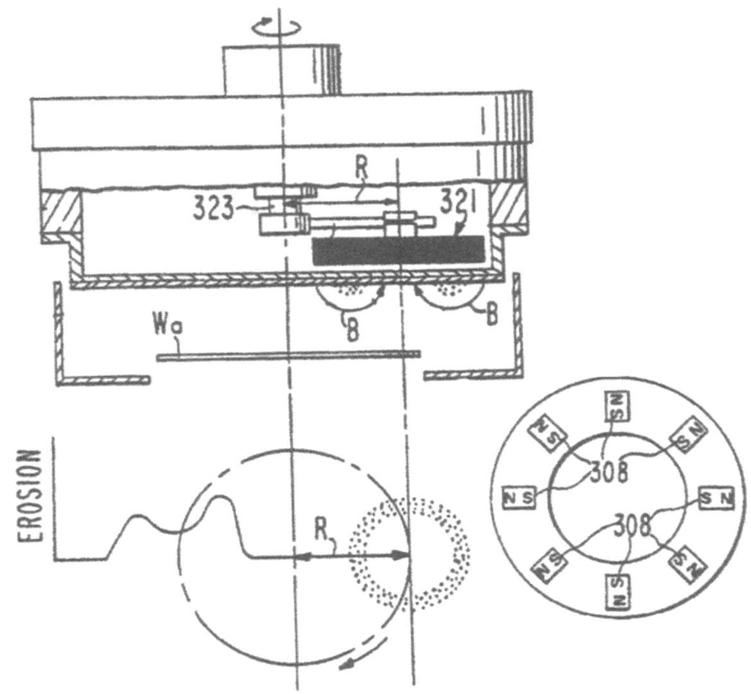

FIG. 4.3 Planar magnetron with a circular target and a rotating ring-shaped magnet array (after Figs. 3C and 3D in ref. 4.8).

erode the central region of the target, more advanced designs have been developed. Figure 4.4 shows a heart-shaped magnet array (Quantum™ source from Varian Associates), which can be precisely tailored both in shape and magnet configuration to produce extremely uniform, full-face erosion. The permanent magnets (M_1 through M_{14}) are sandwiched between two parallel, iron keepers (labeled K_1 and K_2), which distribute the magnetic field uniformly along the magnets and serve to define and hold the contour of the array during rotation. If this heart-shaped magnet is substituted for the ring-type magnet in Fig. 12.3 and rotated about point C, a broad and uniform erosion profile can be obtained, as indicated in Fig. 4.4.

Just as there are many different ways of implementing the basic DC magnetron concept, the swept-field magnetron design has been modified and optimized to deal with specific target materials and PVD process conditions. For example, because the angular emission of sputtered flux from

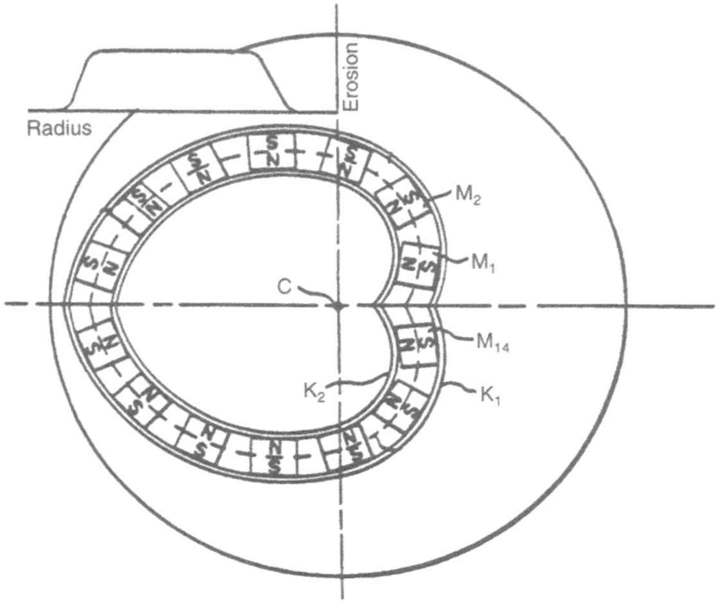

FIG. 4.4 Planar magnetron with a circular target and a rotating, heart-shaped magnet array (after Fig. 3L in ref. 4.8).

an Al target is different from that of Ti, one would not expect the same target erosion profile (i.e., the same magnetron design) to produce the same thin film uniformity. Hence, it is not unusual to have separate sources for Al, Ti, or TiN, with a source for PVD Ti/TiN further differentiated depending on whether a collimated (e.g., TiN barrier) or noncollimated process (e.g., TiN ARC layer) is being used. Differences in magnetron design also relate to differences in uniformity and step coverage needs at the wafer. For example, a thick, blanket Al alloy film intended for an interconnect line requires very high target utilization with little need for step coverage. This might be accomplished with a magnet design producing a broad, annular erosion groove in the target or a design producing a number of concentric erosion grooves. On the other hand, the step coverage for Ti/TiN liners is critical to their application, so target utilization could be reduced to achieve the higher priority. In this case, preferential erosion near the edge might be used to compensate for the fact that outwardly facing via sidewalls at the wafer edge "see" a reduced flux of sputtered material due to the finite size of the target.

In addition the influence of materials and processes on magnetron design, the spacing of the magnetron from the wafer can have a strong effect on uniformity. In fact, as the target erodes and changes both its thickness and surface contour, the source-to-substrate spacing may need to be changed for optimum uniformity. Figure 4.5 shows representative curves of film uniformity for a 200-mm wafer versus source-to-substrate spacing calculated for two different designs of swept-field planar magnetrons with a 12-inch-diameter target. The corresponding film thickness profiles are also shown (a diameter scan of sheet resistance R_s would give the inverse

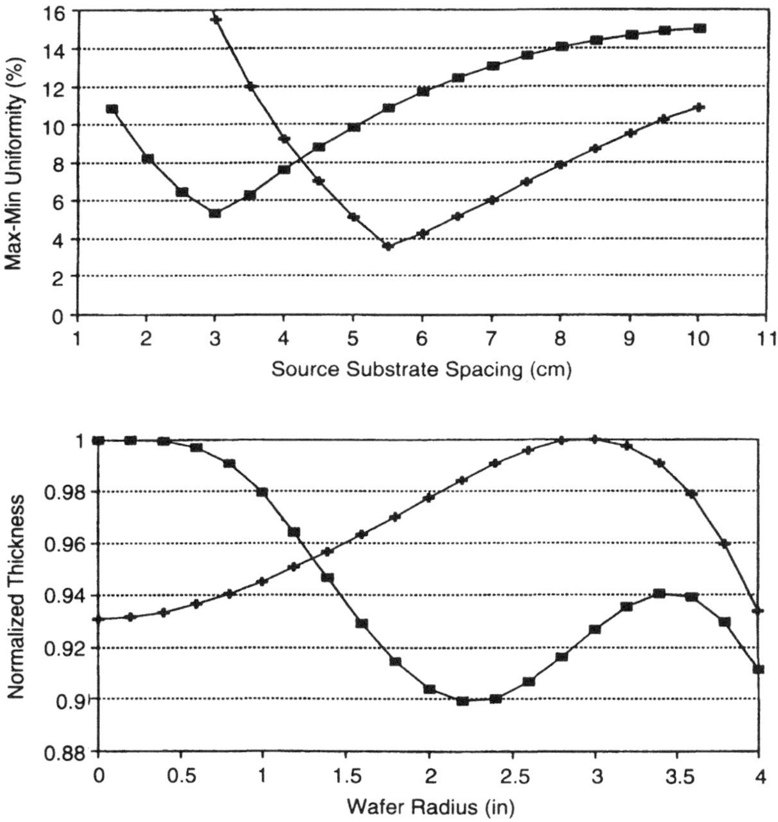

FIG. 4.5 (Upper) Film uniformity over a 200-mm wafer versus source-to-substrate separation calculated for two different designs of 12-inch diameter magnetron source. (Lower) Calculated profile of normalized film thickness for the two sputter sources.

of this profile since $R_s = \rho/t$ is inversely proportional to film thickness t). Such curves can be directly measured or calculated to first order by using the physically measured target erosion profile as a source function and then assuming a simple angular emission from the target (e.g., cosine) without gas scattering. One can also approximate a symmetric target erosion profile as a finite sum of ring sources and then sum their flux at the wafer location to estimate film uniformity.

We see from Fig. 4.5 that either of the two magnetron designs represented can provide excellent uniformity (max-min of $\approx \pm 5\%$), although the optimum uniformity is obtained at quite different source-to-substrate spacing and can be degraded significantly if the spacing changes by as little as 1 cm. Fortunately, by choosing a proper separation (e.g., using a wafer table with vertical z-axis motion) and modifying this separation either manually or automatically as the target erodes, good uniformity can generally be obtained throughout target life. In some cases, it is possible to design a target erosion profile that will keep film uniformity from drifting outside of a specified upper limit (e.g., $3\sigma < 5\%$) for a fixed source-to-substrate separation throughout target life. Alternatively, one can sometimes modify operating conditions (power, pressure, etc.) enough to compensate for changes in film thickness uniformity associated with target erosion.

Historically, the diameter of planar magnetrons has increased along with the diameter of the wafers they were intended to coat. For example, targets 8 inches in diameter are commonly used for PVD of 150-mm wafers, while targets 12 inches in diameter are used for 200-mm wafers. Advanced swept-field magnetron designs are scalable to 300-mm wafers and, based on the historic trend, we would expect that magnetrons with 18-19 inch diameter will be used for this application. It is possible, however, that a different geometry may be required to obtain acceptable uniformity and cost-of-ownership for 300-mm wafers. For example, a rectangular magnetron source could be used and the wafer moved through the beam — which is similar to how PVD is used to coat large-area substrates such as flat panel displays and architectural glass.

4.4 Source Arcing

An arc is the general term used for any low-impedance condition created during the PVD process that then appears at the output of the magnetron power supply. As a result of the abrupt drop in impedance, the arc can dump the energy stored in the output of the power supply into the target.

Arcs are constantly occurring during magnetron sputtering and, if one is not careful, sufficient energy can be delivered to them to create fine particulates or even melt microscopic amounts of target material that "splat" onto the wafer. Hence, increasing effort by source designers has been directed at preventing arcs, suppressing them once they begin to form, and/or limiting the delivered energy to extremely low levels (< 1 mJ for a 10-kW process). As a result, modern PVD sources often include arc suppression circuitry within the source assembly housing or as an auxiliary piece of hardware associated with the DC power supply. In spite of the importance of arc formation and suppression during magnetron sputtering, many thin film scientists and engineers are unfamiliar with the topic. Also, much useful literature on this topic is found in product brochures, patents, and proceedings of industrial coating conferences that PVD workers are not likely to have. With this in mind, the reader is directed to refs. 4.11–4.17 for an introduction to the field.

Arcs can occur in a number of ways. One possibility is a bipolar arc from the cathode to some other part of the chamber that acts as an anode. For example, a low-impedance path can form between the target and a grounded surface in close proximity, such as a dark-space shield. This can occur by electrical breakdown of the low-pressure gas in the intervening space or by direct electrical conduction through a metallic flake that bridges the target and shield. This kind of arc can generally be prevented by proper spacing of the shield and periodic cleaning of the chamber. Similar bipolar arcs can form between the PVD shields and the substrate.

Far more common is the unipolar arc, or *microarc*, whose onset and termination occur on one and the same electrode. It has been estimated that microarcs account for 99.9% of the arc events in a properly designed PVD system. A microarc is initiated whenever a small region of target is able to supply a sufficiently high current of electrons into the plasma, such that the resulting fields cause the plasma to locally collapse into a thread of ions and electrons. Microarcs can be caused by such things as inclusions or irregularities in the target surface that can act as field emission sources, local hot spots that lead to thermionic emission, and microbursts of trapped gas released from the target during sputtering. Microarcs can also occur when insulating material on the target surface charges up to voltages that exceed the material's dielectric breakdown strength. This electrical breakdown can often initiate an arc and also produce negatively charged particles that can travel quite long distances, being accelerated by the electric fields at the target. This could be the case if insulating hydrocarbon contamination or native oxides were on the target surface, e.g., as the result of an improper target burn-in procedure (see Section 11.3). Alternatively, reactive gases could form insulating compounds on the target surface, such as a high partial pressure of O_2 or H_2O in Ar forming Al_2O_3 on lightly eroded parts of an Al target. It is instructive to consider the time to initiate such an arc under typical PVD conditions using elementary electrical arguments.

Consider a small insulating region at the target surface of area A and thickness d that is undergoing Ar^+ bombardment with ion flux density J. This area acts like a microcapacitor with one plate being the metal target and the other plate being the top surface of the insulator. Assuming that the dielectric constant of the insulator is ϵ, its capacitance is then given by $C = \epsilon\epsilon_0 A/d$, where $\epsilon_0 = 8.8 \times 10^{-12}\ C^2/\text{N} - \text{m}^2$. As ions bombard the capacitor, it builds up a positive surface charge after time t, given by $Q = JAt$. Since the voltage across a capacitor is $V = Q/C$, it is then easy to show that an electric field of strength $E = V/d$ develops after time t such that

$$t = \frac{\epsilon\epsilon_0 E}{J} \tag{4.1}$$

The time to break down the insulating region is then given by Eq. (4.1) but with the breakdown field strength substituted for E. Using parameters appropriate to Al_2O_3 ($\epsilon = 10$, $E = 10^8$ V/m) and an ion current density of 40 mA/cm^2 = 400 A/m^2, we calculate $t = 2$ msec. This is a very short time and is comparable to the time it takes for the plasma to extinguish from electron-ion recombination after shutting off the power supply.

There are a number of hardware fixes that have been provided to address microarcing. These generally involve modifying the circuitry of the DC power supply to actively sense and then suppress arcs as soon as they form. Arc formation is signaled by an abrupt drop in voltage across the cathode dark space. Once the arc is sensed, it is possible to suppress it by rapidly reversing the polarity of the target voltage and biasing it so that it is $\approx$ 10–20 V more positive than the plasma potential. This serves to attract electrons from the plasma to the target and quench the arc. Advanced arc suppression circuits are capable of reacting to cathode arcs in a few 100 nsec and can suppress more than 2000 microarcs per second. As a result, the energy delivered to a given arc can be sufficiently low ($<$ 1 mJ) to avoid particulate generation and other target-related damage. For information on these and other methods of controlling arcs such as low-frequency AC techniques, the use of RF alone or in combination with DC power, dual magnetron sources, and unipolar pulsed DC magnetron sputtering, the reader is referred to refs. 4.11– 4.17.

4.5 Low-Pressure Sputtering

The efficient use of electrons in the DC magnetron allowed sputter deposition to be carried out at much lower pressure than had been possible with diode sputtering. In fact, proper magnetron action requires that pressure be low enough so that the electron mean free path associated with gas scattering is not significantly less than the electron gyratron radius. For representative DC voltages and magnetic field strengths (e.g., 500 V and 300 G), the gyration radius is calculated to be ≈ 2 mm, with the result that operating pressure must be less than ≈ 50 mTorr for efficient magnetron sputtering — this is 10–20 times lower than had been possible with diode sputtering. The reduced gas-phase scattering at these pressures provided benefits such as improved sputtered flux directionality, a greater fraction of emitted material reaching the wafer, and retention of enough energy in the incident adatoms to influence and control film morphology.

In IC production, PVD films are typically deposited at operating pressure ≈ 2–5 mTorr; however, there is a desire to reduce magnetron operating pressure to well below 1 mTorr to further reduce the effects of gas-phase scattering and thereby improve directionality. For example, long-throw sputtering must be carried out at source-to-substrate distances that are several times greater than in conventional PVD (e.g., 300 mm versus 100 mm), which requires a comparable increase in mean free path to prevent atom scattering. Hence, operating pressure in long-throw PVD is several times lower than in conventional PVD. In this regard, it should be noted that highly directional PVD methods based on ionized metal species have been developed that use much higher than conventional pressure (≈ 20 mTorr) to thermalize sputtered neutrals so they can be more efficiently ionized and directed at a biased wafer (see Chapter 9). In this case, the ion acceleration occurs over such a short distance (the dark space) that scattering is not an issue even though the chamber pressure is very high. That is, it is the product of distance and mean free path of the atom or ion that is important and not the value of either quantity alone.

Lower PVD operating pressures have already benefited conventional processes such as collimation since this reduces the possibility that atoms exiting the collimator will scatter away from near-normal incidence before reaching the wafer. Also, since the transmission of a collimator decreases strongly with its cell aspect ratio, reduced pressure has allowed lower aspect ratio collimators to be used to achieve the same directionality, with a corresponding increase in deposition rate. Whether reactive PVD processes such as collimated TiN will be possible at very low pressure is problematic since the flux of the reactive component may be insufficient to produce a stoichiometric compound in the allotted time. For example, sputter deposition of a 200Å TiN film in 60 sec is equivalent to ≈ 1 monolayer per second of TiN, or $\approx 2 \times 10^{15}$ Ti atoms/cm^2. Assuming the working gas is a 50/50 mixture of Ar/N_2 at 0.1 mTorr, the molecular incidence rate of N_2 at the wafer is about 10^{16} cm^2/sec, which is not much greater than the flux of Ti. In addition, the shields and collimators that become coated with Ti act as a getter-pump of chamber N_2. This serves to further reduce the amount of nitrogen available at the wafer surface to form TiN.

Initiating and sustaining magnetron discharges at pressures << 1 mTorr is difficult due to the reduced number of Ar atoms available for ionization. This leads to reduced Ar^+ bombardment of the target and consequently decreased production of discharge-sustaining secondary electrons. On the other hand, the average electron energy may be somewhat greater at reduced pressure, which increases their cross section for Ar ionization and mitigates the reduction in Ar gas density. Greater magnetic field strength can also be applied to increase the electron mean free path and therefore the probability of ionizing the working gas at low pressure; however, this has its limits. As described in Chapter 9, one way to reduce magnetron operating pressure is to remove the working gas entirely and utilize self-sputtering of the metal to sustain the discharge. This method is well-suited to materials such as Cu that have a high self-sputtering yield; however, it is not generally applicable to all materials of interest to IC fabrication.

More generally, magnetron discharges can be initiated at quite low pressures and sustained at even lower pressure (≈ 0.1 mTorr) if a sufficient supply of low-energy electrons is provided. This can be done by using an additional source of electrons (e.g., injection of electrons from a hollow cathode electron source) or by preventing secondary electrons that were not trapped in $\mathbf{E} \times \mathbf{B}$ drift orbits from leaving the plasma volume. In the latter case, one method that has been successfully used is the so-called bucking magnet in which a secondary, fixed ring of magnets is arranged at the edge of the magnetron, at or behind the target plane. The magnetic poles in the fixed ring are aligned opposite to that of the primary magnets in the rotating array (see Fig. 4.6), which creates a net magnetic field acting to redirect secondary electrons emitted at the edge of the target back toward the plasma region.

Other approaches to low-pressure magnetron operation are discussed in refs. 4.18 and 4.19 and citations therein. Considering that many crossed-field devices already operate at pressures much less than 1 mTorr (e.g., sputter ion pumps and UHV ion gauges) and that high-vacuum planar magnetrons have been demonstrated [4.19], there is reason to expect that magnetrons designed for sub-0.1 mTorr operation will be applied to future IC production.

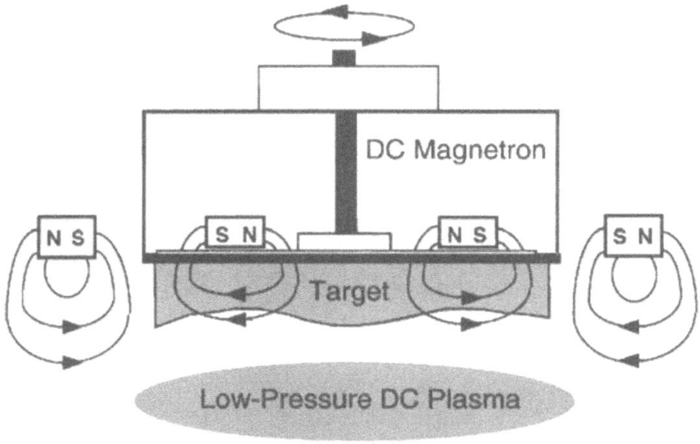

FIG. 4.6 Bucking magnet hardware used to further confine secondary electrons and reduce operating pressure of a DC magnetron (after ref. 4.9).

References

4.1. R. W. Wilson and L. E. Terry, "Application of high-rate **E** × **B** or magnetron sputtering in the metallization of semiconductor devices," *J. Vac. Sci. & Tech.* **13**(1): 157–164 (1976).

4.2. J. A. Thornton and A. S. Penfold, "Cylindrical Magnetron Sputtering," in *Thin Film Processes*, pp. 75–113, John L. Vossen and Werner Kern, Eds., Academic Press, New York, 1978.

4.3. David B. Fraser, "The Sputter and S-Gun Magnetrons," in *Thin Film Processes*, pp. 115–129, John L. Vossen and Werner Kern, Eds., Academic Press, New York, 1978.

4.4. Robert K. Waits, "Planar Magnetron Sputtering," in *Thin Film Processes*, pp. 131–173, John L. Vossen and Werner Kern, Eds., Academic Press, New York, 1978.

4.5. Stephen M. Rossnagel, "Glow Discharge Plasma and Sources for Etching and Deposition," in *Thin Film Processes II*, pp. 11–77, John L. Vossen and Werner Kern, Eds., Academic Press, New York, 1991.

4.6. Robert Parsons, "Sputter Deposition Processes," in *Thin Film Processes II*, pp. 177–208, John L. Vossen and Werner Kern, Eds., Academic Press, New York, 1991.

4.7. Donald L. Smith, *Thin Film Deposition: Principles & Practice*, McGraw-Hill, NY, 1995.

4.8. R. E. Demaray, J. C. Helmer, R. L. Anderson, Y. H. Park, R. R. Cochran, and V. E. Hoffman, "Rotating Sputtering Apparatus for Selected Erosion," U.S. Patent No. 5,252,194 (Oct. 12, 1993).

4.9. K. Lai, "Design of Magnetron Sputtering Source for Low Pressure Operation," U.S. Patent No. 5,593,551 (Jan. 14, 1997).

4.10. R. Scholl, "Process improvements for sputtering carbon and other difficult materials using combined AC and DC process power," in *Proc. 35th Ann. Soc. of Vacuum Coaters (SVC) Tech. Conf.*, pp. 391–394 (1992).

4.11. R. A. Scholl, "Advances in arc handling in reactive and other difficult processes," technical report of Advanced Energy Industries, Fort Collins, CO (1994).

4.12. L. Anderson, "A new technique of arc control in DC sputtering," in *Proc. 35th Ann. Soc. of Vacuum Coaters (SVC) Tech. Conf.*, pp. 325–329 (1992).

4.13. S. Beisswenger et al., "Economic considerations on modern web sputtering technology," in *Proc. 35th Ann. Soc. of Vacuum Coaters (SVC) Tech. Conf.*, pp. 128–134 (1992).

4.14. R. L. Cormia et al., "Method for Coating a Substrate," U.S. Patent No. 4,046,659 (Sept. 6, 1977).

4.15. S. Schiller, K. Goedicke, J. Reschke, V. Kirchhoff, S. Schneider, and F. Milde, "Pulsed magnetron sputter technology," *Int. Conf. on Metallurgical Coatings and Thin Films* (ICMCTF93), San Diego, CA, April 19–23, 1993.

4.16. S. Schiller, K. Goedicke and Ch. Metzner, "Advances in pulsed magnetron sputtering (PMS process)," *Int. Conf. on Metallurgical Coatings and Thin Films* (ICMCTF94), San Diego, CA, April 25–29, 1994.

4.17. G. Este, "A quasi-direct current sputtering technique for deposition of dielectrics at enhanced rates," *J. Vac. Sci. & Tech.* **A**: 1845ff (1988).

4.18. T. Asamaki, T. Miura, G. Nakamura, K. Hotate, and S. Yonaiyama, "High-vacuum planar magnetron discharge," *J. Vac. & Sci. Tech.* **A10**(6): 3430–3433 (1992).

4.19. T. Asamaki, T. Miura, K. Hotate, S. Yonaiyama, G. Nakamura, K. Ishibashi and N. Hosokawa, "High-vacuum planar magnetron sputtering," *Jpn. J. Appl. Phys.* **32** (part I, no. 2): 902–906 (1993).

Chapter 5 Sputtering Tools

> CAVEAT: The purpose of this book is educational and not commercial, and the specific equipment used to illustrate pedagogical points should not be construed as a product endorsement. Furthermore, while we have tried to depict essential equipment features, neither the dimensions nor the details of a given piece of hardware are intended to be an engineering drawing. Finally, PVD tools are continuing to evolve with new and/or improved models introduced every few years. Therefore, the reader is encouraged to consult the actual suppliers' product brochures or contact the suppliers directly to obtain accurate, up-to-date details on tool construction and product specifications.

5.1 Evolution of PVD Tools for Microelectronics

PVD equipment for microelectronic fabrication has evolved greatly since the early 1970s both in terms of technical performance and productivity. This evolution is inexorably linked to the phenomenal growth in complexity of Si devices over the same period (e.g., 1975 = 4K DRAM on 3-inch wafers; 1998 = 256 Mb DRAM on 8-inch wafers) and to the increasingly stringent requirements placed on both the PVD process and the PVD process tool to allow these devices to be manufactured in a cost-effective way. Fortunately, PVD tool designers were able to take advantage of concurrent improvements in such supporting technology as vacuum pumping, gas delivery, robotic wafer handling, microprocessor-based control, and process automation. As a result, PVD equipment was able to progress relatively rapidly from manual-loaded, stand-alone batch tools to fully automated, vacuum-integrated, single-wafer cluster tools. In this regard, the development of deposition tools based on PVD was similar to the development of dry etching tools based on plasma technology. Pattern definition by dry etching began to replace wet chemistry in the late 1970s and, not surprisingly, today's state-of-the-art cluster tools for PVD and plasma etching look quite similar.

Figure 5.1 summarizes the evolution of microelectronic manufacturing from 1975–1998, and the following sections present a brief discussion of PVD hardware evolution during this time frame, with representative illustrations of equipment.

Microelectronic Manufacturing 1975-1998

	1975	1985	1995	1998
Cost of Fab	$10M	$100M	$1B	$1.5B
Automation	Semiautomatic	Cassette-to-Cassette	Cassette-to-Cassette	Cassette-to-Cassette
Leading Wafer Size(s)	4 inch	5 inch/6 inch	8 inch	8 inch/12 inch
Production Linewidth	3-6 μm	1.5-3 μm	0.35-1 μm	0.25-1 μm
Clean Room Class	1000	10	1	1
Product Class	LSI	VLSI	ULSI	ULSI
Number of Components	10^3 (4K DRAM)	10^6 (1M DRAM)	10^8 (64M DRAM)	5×10^8 (256M DRAM)
Workers Garments	Smocks	Bunny Suits	Ultra-clean Suits and/or Robots	Ultra-clean Suits and/or Robots
Where Located	Silicon Valley	Worldwide	Worldwide	Worldwide
Engineering Tools	Electronic Calculator	Workstation	Local Area Network	Local Area Network
PVD Tool	Standalone 1975-1979: one chamber 1980-1984: multi-chamber	Cluster Tool 1st Generation	Cluster Tool 2nd Generation	Cluster Tool 200 mm: 3rd Gen. 300 mm: 1st Gen.

FIG. 5.1 The evolution of microelectronic manufacturing and PVD manufacturing tools.

5.1.1 Pre-1975: The Stage Is Set

Prior to 1975, the preferred method of metallization for semiconductor production was vacuum evaporation with heating provided by e-beam bombardment or resistance-heated filaments. Although sputtering hardware was available (e.g., Western Electric used sputtered Ta and TaN thin film resistors in hybrid circuits of the first Touch-Tone phones), PVD was not deployed in mainstream Al metallization. Instead, sputtering was utilized to deposit insulators and refractory metals that were difficult to evaporate (see Fig. 5.2 for a typical RF diode configuration of the time).

As device size decreased, the increased aspect ratio of features challenged the limited step coverage of e-beam evaporation, which is basically a line-of-sight deposition. Also, the move away from pure Al and toward

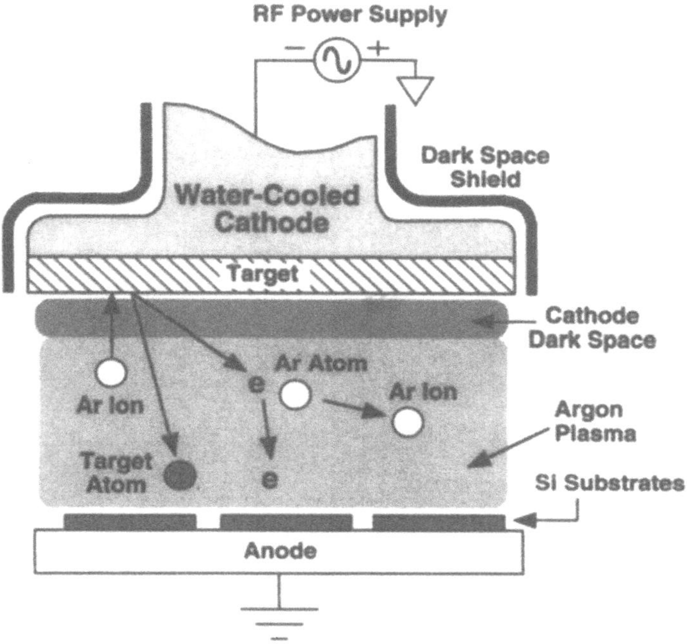

FIG. 5.2 Early RF diode configuration of PVD showing batch substrates.

Al-Si alloys to deal with junction spiking and toward Al-Cu to deal with electromigration was a complication for evaporation technology due to the different vapor pressures of the alloy constituents. Although sputtering offered a solution to both these problems, it could not be implemented into manufacturing because RF diode sputtering of Al had relatively low deposition rates and poor electrical and optical film quality. The solution turned out to be the development of DC magnetron sputtering based on the pioneering invention of the 3-inch sputter gun by Peter Clarke in 1968 [5.1]. By confining secondary electrons near the surface of the sputter target, argon ion generation could be increased considerably with the magnetron configuration over the diode with resulting higher deposition rate, and electron bombardment heating of the substrate was reduced [5.2]. And though RF magnetron sputtering was also possible, DC power supplies were much less expensive than RF and film quality was better. By 1975, the DC magnetron had been refined from a small-scale laboratory curiosity to a production-worthy source (Fig. 5.3).

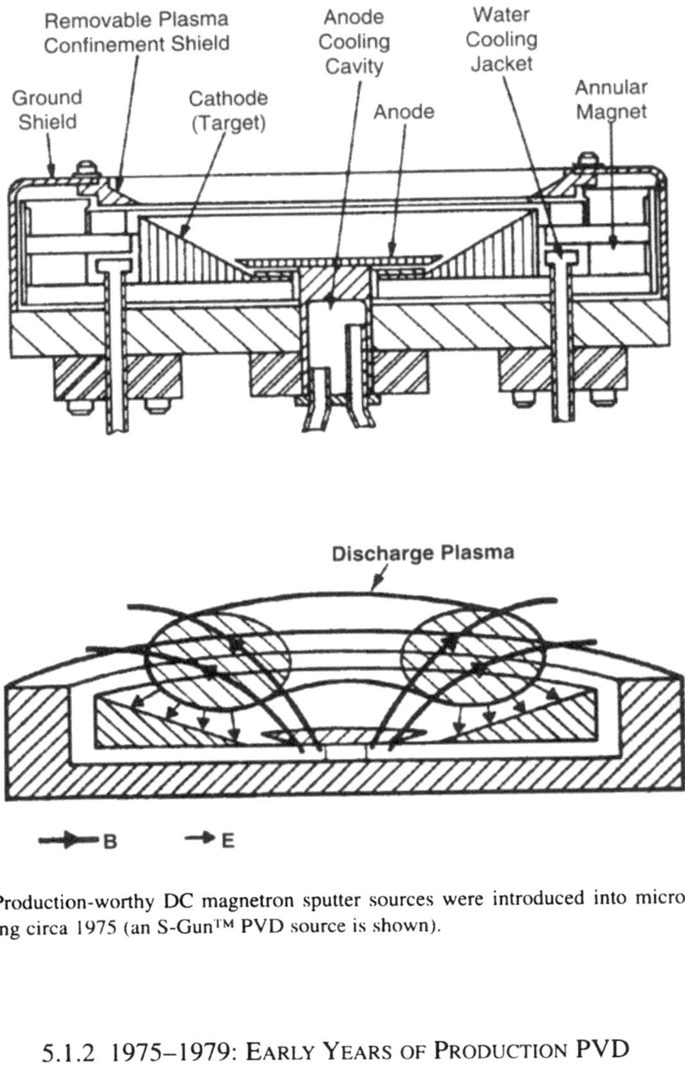

FIG. 5.3 Production-worthy DC magnetron sputter sources were introduced into microelectronic manufacturing circa 1975 (an S-Gun™ PVD source is shown).

5.1.2 1975–1979: Early Years of Production PVD

This period saw the implementation of DC magnetron source technology into a variety of PVD system configurations from about a dozen suppliers, including Airco Temescal, Leybold-Heraeus, MRC, Perkin-Elmer Ultek, Sputtered Films, and Varian [5.3]. However, it was not until the late 1970s that PVD began to displace e-beam evaporation. For example, the chapter on IC metallization in a well-known "how-to" book published by Fairchild

Semiconductor in 1979 [5.4] refers only to the diode configuration of sputtering, noting that "[sputtering] can be accomplished using both DC and RF voltages, and it can be used to deposit almost any material, although the deposition rate is often extremely low."

Hardware development was characterized by a strong effort to lower the cost per wafer through source designs that gave more uniform target erosion and thus more effective percent utilization of target material (25–50% was typical). A variety of target shapes were utilized (rectangular, circular, wedge-shaped) with sputtering being done with the wafer above (face-down), below (face-up), or at right angles (side sputtering) to the source.

Representative of hardware from this period are the Varian batch sputtering system shown in Fig. 5.4 in which wafers were manually loaded onto a planetary assembly that was rotated during deposition to improve uniformity, and the Perkin-Elmer system shown in Fig. 5.5 that used a rotating wafer table to improve uniformity. The latter system is noteworthy in that it had a vacuum loadlock and an optional heater for outgassing wafers before deposition, anticipating future trends in vacuum cleanliness. Given the large installed base of evaporation equipment, an upgrade market also developed whereby existing e-beam and filament evaporation systems could be retrofit with a suitable PVD source (see Fig. 5.6).

FIG. 5.4 Batch sputtering system c. 1978 (Varian model 3125 coater is shown).

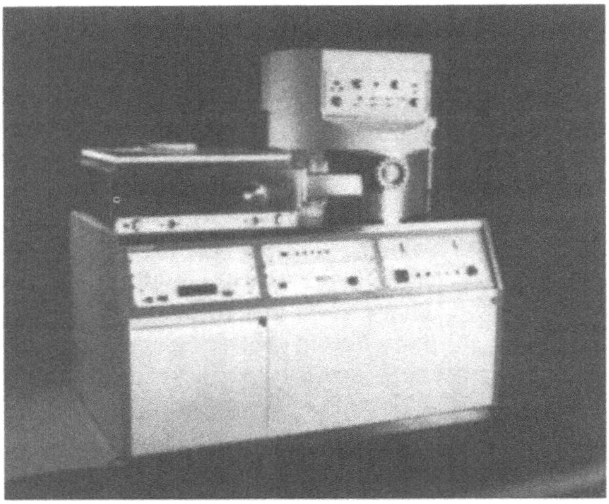

FIG. 5.5 Batch sputtering system c. 1978 (Perkin-Elmer model 4410 coater is shown; see ref. 5.3).

5.1.3 1980–1984: PVD Stand-Alone Tools

Semiconductor International could report in 1980 that "In many semiconductor metallization applications, high vacuum evaporators are being challenged by DC magnetron sputtering systems. However, in no danger of extinction, high vacuum evaporators are a competitive alternative, featuring cost effective operation and high deposition rates" [5.5]. By 1985, however, DC magnetron sputtering — both planar and conical magnetron configurations — was regarded as the state-of-the-art method of depositing thin films for interconnect applications, including Al-Si and Al-Si-Cu from alloy targets and refractory metal silicides such as $MoSi_2$, WSi_2, and $TiSi_2$ from composite targets [5.6, 5.7]. The reason for this paradigm shift in metallization technology was the development of production-oriented PVD systems with high vacuum quality (for improved film purity), better temperature control and uniformity (for improved repeatability and uniformity of film properties), reduced particle levels (for higher device yield), and automated substrate transport that eliminated wafer breakage due to operator mishandling and that supported production-worthy throughputs of 60 wafer per hour.

FIG. 5.6 Users of evaporation equipment in the late 1970s could upgrade to PVD by replacing evaporation sources with one or more sputter sources (retrofit of an e-beam system for PVD operation c. 1979 is shown).

Introduced in mid-1980, the 3180 coater from Varian Associates was the first of these production-oriented PVD systems and offered fully automated, cassette-to-cassette, single-wafer PVD processing (Fig. 5.7). An inline architecture was used in which individual wafers were moved sequentially from one vacuum process station to the next and which anticipated the coming of truly modular PVD cluster tools in the 1985–1990 time frame. Vacuum loadlocks and cryopumped PVD process chambers replaced the manual loading and diffusion pumping of earlier generation tools, with the result that base pressure was now in the low 10^{-7} Torr range. The 3180 coater was also significant in that it utilized backside-gas-assisted heat transfer that allowed one to rapidly change and/or control wafer temperature while retaining the mTorr-level vacuum ambient needed for PVD. In addition, the method produced improved temperature uniformity, which was of particular interest since the industry was now facing the transition from 4-inch-diameter substrates to 6-inch ones. It is interesting to note that this backside-gas-assisted method of wafer temperature control has become pervasive in the equipment industry and is still used 20 years later in today's most advanced PVD cluster tools (see Section 5.3.4).

FIG. 5.7 The Varian model 3180 coater illustrates the move toward automated, cassette-to-cassette single-wafer PVD processing in the 1980–1985 time frame.

5.1.4 1985–1989: PVD Cluster Tools — I

This period saw the development of single-wafer, multichamber standalone sputtering systems and the introduction of PVD-specific cluster tools in which modular, vacuum-isolated, and independently operated process chambers were interfaced to a central robotic wafer-handling platform. The cluster tool concept provided a vacuum-integrated processing capability that allowed complex process flows for selected applications such as barrier/liner (e.g., degas + preclean + PVD Ti + PVD TiN) and slab Al interconnect (e.g., degas + preclean + PVD Ti/TiN+ PVD Al-Cu + PVD TiN ARC layer). The control of film properties, particles, and interface quality needed to carry out such process sequences in a production environment could not have been accomplished without the process isolation and vacuum integrity of these PVD tools. In addition, it was believed that a flexible, cluster-tool approach to semiconductor processing would allow so-called best-of-breed modules to be integrated by the primary equipment supplier onto their tool — e.g., addition of a third party's rapid thermal annealing module. This direction was encouraged in 1990 by the Semiconductor Equipment and Materials Institute (SEMI) through the formation of a Modular Equipment Standards Committee (MESC) to establish "MESC standards" for such things as the mechanical and electrical interface between process module and cluster tool backbone.

In other developments, the planar magnetron was beginning to replace the conical magnetron in high-end applications where uniformity of step coverage over 8-inch wafers was a concern, and the need to reduce yield-killing fine particles was forcing suppliers to pay increased attention to such hardware details as materials of construction; sputter shield design; wafer clamping; and gas delivery, pumping, and venting.

Representative tools from this period include the ClusterLine™ model from Balzers (Fig. 5.8a), the Sigma™ model from Electrotech (Fig. 5.8b), the Eclipse™ model from MRC (Fig. 5.8c), the Loadlok™ model from CVC (Fig. 5.8d), and the M2000™ model from Varian (Fig. 5.8e). The M2000 was introduced in 1987 as the first open-architecture PVD cluster tool with totally independent and interchangeable process modules.

5.1.5 1990–1997: PVD Cluster Tools — II

Since much of the tool design strategy and PVD technology embodied in these more recent PVD tools will be discussed in the remaining sections of this chapter, only a few comments will be made here. Probably the most significant change in design philosophy was the recognition that technology and productivity (i.e., cost-of-ownership [CoO]) had now become equally important in an advanced semiconductor fabrication tool and that the tool provider (equipment supplier) and the tool user (IC manufacturer) must work together in the spirit of the quality improvement process (QIP) to develop and improve such hardware. This view, quantified and promoted by organizations such as SEMATECH, meant that mean-time-to-failure (MTTF) and mean-time-to-repair (MTTR) considerations were now being given equal attention to film uniformity and step coverage. Also, much more attention was being given to the reduction of process-added particles, which generally have a dominant impact on CoO calculations. In addition, it began to be recognized that PVD was far from a mature technology and that the technology could be extended to deal with advanced devices through improved directional PVD source concepts and process

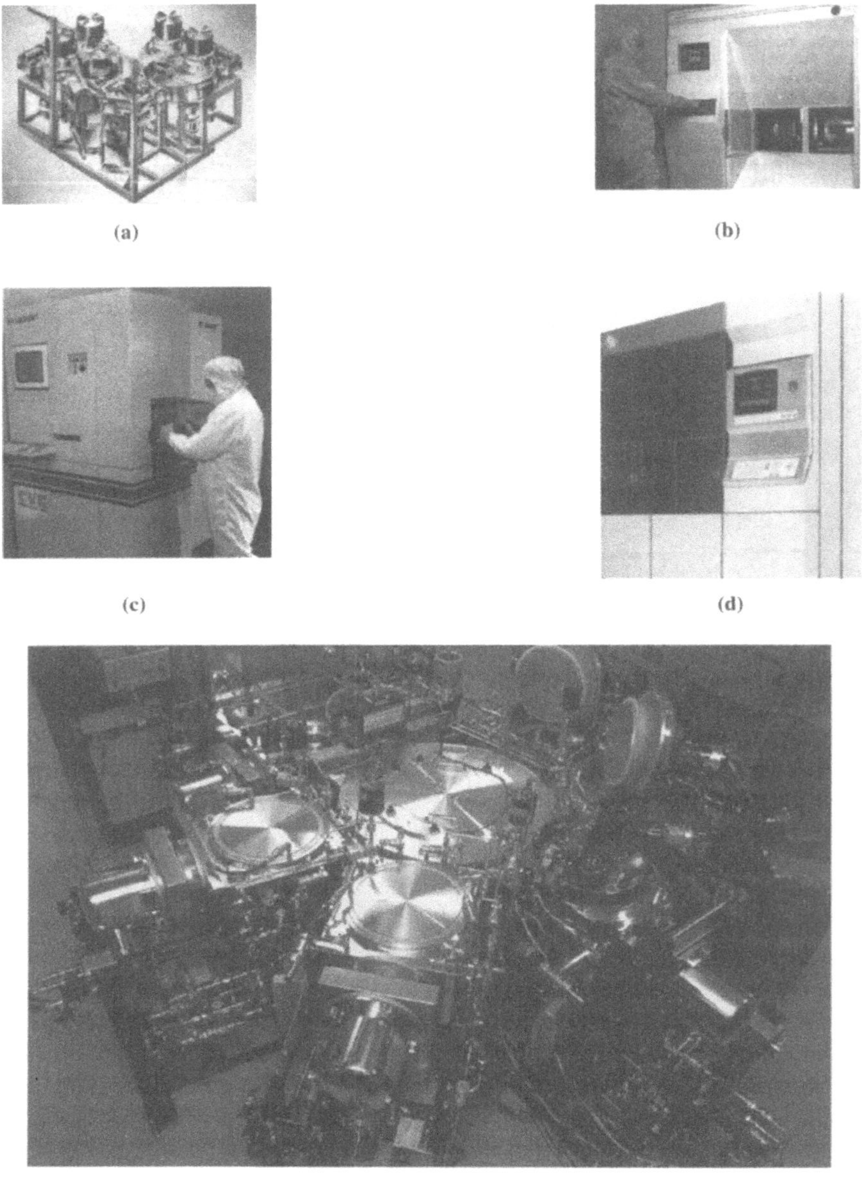

FIG. 5.8 PVD systems of the 1985–1990 period were strongly influenced by the move toward vacuum-integrated cluster tools for semiconductor processing.

recipes enabled by high-vacuum compatible cluster tools — e.g., Al reflow-type planarization, collimated barriers and liners, long-throw directional sputtering, and ionized metal plasma PVD. Cluster tools capable of true ultrahigh vacuum (UHV) base pressure in the process chamber ($< 10^{-9}$ Torr) were also of interest based on data that this level of vacuum quality would result in Al alloy interconnect films with improved electromigration resistance. On the other hand, whether or not UHV base pressure is actually required to obtain a given film property, UHV-compatible *practices* (e.g., polished chamber surfaces with low outgassing rates) are desired to meet the stringent requirements of ULSI device fabrication. A strong motivation to extend PVD also stemmed from the enormous installed base of sputter metallization tools. Finally, since sputtering uses inert gases and nontoxic targets, a PVD push was provided by environmental awareness and increasingly stringent legislation regarding the handling, use, and disposal of hazardous materials.

Addressing the above noted customer needs for lower CoO and higher vacuum technology, Applied Materials introduced its first PVD tool in mid-1990 — the Endura™ 5500 system. The Endura system was designed to be a robust, high-throughput ultrahigh vacuum PVD system [5.8] directed to such high-volume applications as slab-Al interconnect (Fig. 5.9). Other new and/or improved PVD systems introduced during 1990–1997 included the CVC Connexion™ model and Sputtered Films Endeavor™ model for general purpose, integrated processing; the MRC Solarus™ model for advanced Al applications; and the applications-specific Varian mb2™ model for barriers and liners. In 1997, Novellus Systems bought the Thin Film Systems business from Varian Associates and introduced the INOVA™ PVD cluster tool shown in Fig. 5.10.

5.1.6 1998–2001: PVD Cluster Tools — III

Based on the established success of the cluster tool concept, changes in PVD tools in the 1998–2001 are expected to be evolutionary and driven primarily by the need to further reduce cost-of-ownership while meeting the process and thin film requirements of 0.18-μm (1 Gb DRAM) devices on the MLM Interconnect Roadmap. For maximum productivity, any overhead time associated with wafer transfer will need to be minimized — e.g., by use of rapid acceleration-deceleration robotic handling with dual arms for extremely high mechanical throughput ($>$ 60 wafers per hour). Software on PVD tools will be accessed via operator-friendly, highly intuitive graphical user interfaces (GUI), and factory automation of the wafer

FIG. 5.9 Endura™ PVD cluster tool (courtesy of Applied Materials, Inc., Santa Clara, CA).

fab line will require PVD tools to provide direct loadlock access for wafer-carrying robots (such as the automated guided vehicle [AGV] or less automated rail guided vehicle [RGV]) and cassette containers or "pods" having an environmentally controlled or vacuum ambient (e.g., the standard mechanical interface [SMIF] box). Dealing with the transition to 300-mm wafer fabrication in the most advanced fabs (see Section 5.4) will probably result in scaling up proven 200-mm PVD cluster tool designs and using correspondingly larger process modules capable of uniformly degassing, precleaning, and coating 300-mm substrates.

With regard to process modules, just as collimation was used to improve the directionality of PVD in the 1990–1997 time frame, even more directional PVD methods based on ionized metal plasmas are expected to be used for barrier/liner and even fill applications after 1998. Also, it is likely that PVD tool architecture will permit a mix-and-match approach to PVD and CVD in which both methods can be used to advantage on a common backbone (see Section 9.9). For example, a vacuum-integrated PVD Ti

FIG. 5.10 INOVA™ PVD cluster tool (courtesy of Novellus Systems, Inc., San Jose, CA).

wetting layer + CVD Al + PVD Al sequence might be used to fill a high aspect ratio structure. Low-damage, in-situ cleaning of such steep features prior to PVD will probably require the use of reactive gas chemistry, and we can expect future PVD tools to incorporate modules for reactive plasma precleaning similar to the technology used on high-density plasma etching tools.

Finally, the transition from Al alloys to Cu interconnects — at least at the higher levels of metallization — will require PVD tools to deposit suitable barriers such as Ta and TaN and to interface with potentially exotic wet deposition methods such as electroless plating or electroplating. For example, a PVD tool might provide a vacuum-integrated stack of PVD Ta + CVD Cu or PVD Ta + PVD Cu on which to plate Cu.

5.2 Generic PVD Cluster Tool

All PVD equipment for advanced microelectronic device metallization is currently based on the single-wafer, vacuum-integrated cluster tool design. There are as many ways of implementing the basic design concept as there are suppliers, and Fig. 5.11 shows schematically how several have configured their tools. As shown, both the maximum number of process chambers and their placement around and/or within the central vacuum handlers vary greatly from tool to tool. Also, the flexible design allows one to use fewer process modules than the maximum number (e.g., removing a module for preventive maintenance without taking the entire tool off-line) and to double up on modules for optimum throughput (e.g., devoting two modules to the same low-deposition-rate process).

The basic architecture of an application-flexible PVD cluster tool is illustrated in Fig. 5.12 (based on an Applied Materials design). This sort of tool might be used to perform a multistep integrated process sequence such as Al-slab deposition (e.g., preclean/Ti/TiN/Al-Cu/TiN). Application-

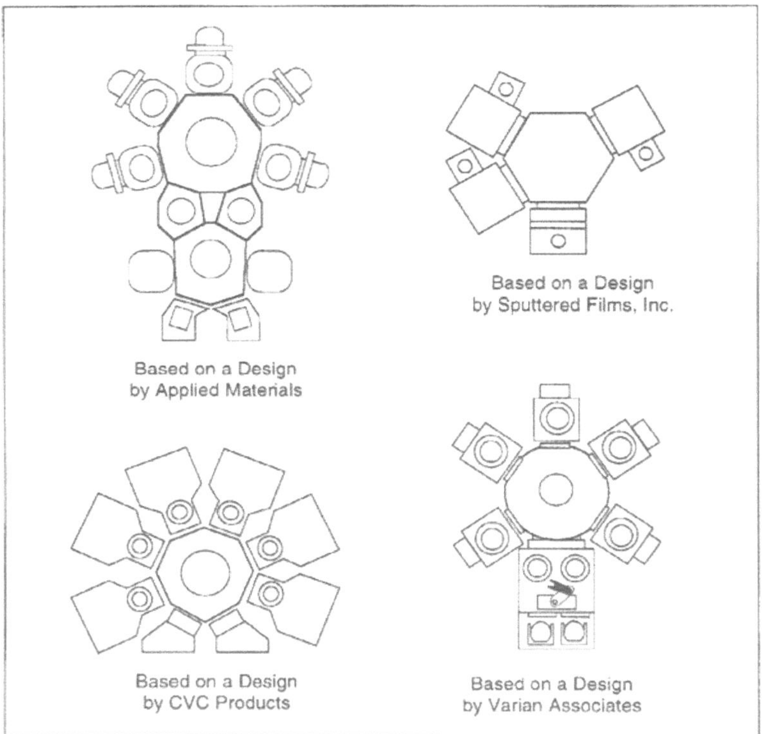

FIG. 5.11 PVD cluster tools can be configured in a variety of ways, including both the number of modules and their positioning on the wafer handling backbone (tools are not drawn to scale).

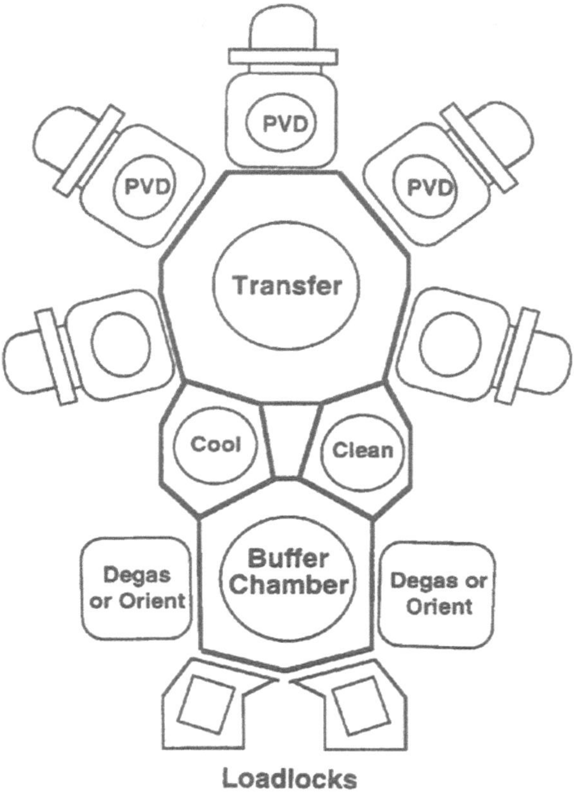

FIG. 5.12 Illustration of the basic architecture of an applications-flexible PVD cluster tool (based on an Applied Materials design).

specific PVD cluster tools having a smaller number of process chambers have also been developed (e.g., the mb2™ model from Varian Associates, which targeted barrier/liner applications such as preclean/Ti/TiN). However, in spite of important supplier-specific differences in tool design and construction, PVD cluster tools generally exhibit four basic building blocks:

1. *Front-end* for cassette-to-cassette, wafer loading/unloading into/out-of the tool. Load and unload stations can be vacuum isolated from each other, and sometimes the load station includes a wafer flat alignment as well. Cassettes can be handled manually by an operator or robotically by

an automated guided vehicle (AGV). In some tool designs, an incoming cassette of wafers is stored at atmospheric pressure under a flow of dry, filtered air or nitrogen to prevent additional exposure to water vapor. Individual wafers are then loaded into the tool from the cassette via a vacuum loadlock. An alternative design is to place the entire cassette in a vacuum loadlock that is then pumped down so that subsequent removal of individual wafers for processing occurs with the cassette under clean, vacuum conditions.

2. *Degas/cool station* to heat-treat the wafer for subsequent process steps or to cool the wafer sufficiently to allow placement in a plasticized cassette. In some cases, flat alignment is done in this station as well.

3. *Transfer module* (sometimes called a central wafer handler) in which the wafer is robotically moved with high positional accuracy ("pick-and-place") between vacuum-isolated process modules. In the early years of PVD, wafer orientation during handling and processing was a topic of much discussion. Today the generally accepted practice is having the wafer horizontal and face-up during both handling and PVD.

4. *Process modules* in which sputter deposition and other process steps such as a preclean before PVD or a rapid thermal anneal after PVD are carried out. The modules are vacuum-isolated from each other, which permits parallel processing to be done at different vacuum levels or incompatible gas chemistries to be used without cross-contamination (e.g., PVD of TiN using Ar/N_2 in module 1, Ar^+ sputter etch removal of native SiO_2 in module 2, and PVD of Al-Cu using Ar in module 3 without the formation of insulating AlN or Al_2O_3).

5.3 The Technology of PVD Cluster Tools

We now consider in detail a number of interrelated hardware and process issues that are common to all PVD cluster tools, such as vacuum pumping and gas delivery, wafer handling and holding, wafer thermal management, contamination, and particles. Recognizing that tool productivity is today as important as technology, topics related to cost-of-ownership (capital and consumables cost, throughput, maintenance, etc.) will also be discussed.

5.3.1 Vacuum Considerations

The vacuum range encountered during PVD processing ($\approx$ 12 orders of magnitude) is arguably the largest in microelectronic production since wafers enter the tool from a clean room at atmospheric pressure (760 Torr) and are ultimately processed at a few mTorr in PVD modules that can have ultrahigh vacuum (UHV) base pressure of $< 10^{-9}$ Torr. Figure 5.13 shows one possible range of vacuum levels (both base and operating pressures) for a PVD cluster tool.

A PVD cluster tool must be designed to simultaneously deliver high vacuum base pressure in the PVD process module and high wafer throughput ($\approx$ 20–50 wafers per hour or more, depending on the overall process complexity and process time per module). The vacuum requirement is related primarily to the fact that metals such as Al and Ti are highly reactive to water vapor and other oxidizing ambients (forming insulating Al_2O_3 and TiO_2) and that film microstructure and electrical properties can be adversely affected by very small amounts of hydrocarbons, O or N [5.9]. The throughput requirement means that pump-down and vent-up times must be as short as possible.

These requirements have led designers of PVD systems to adopt vacuum system design techniques previously used in UHV molecular beam epitaxy (MBE) deposition systems in which the level of vacuum seen by the wafer improves in a stepwise way from the loadlock (low vacuum) to transfer

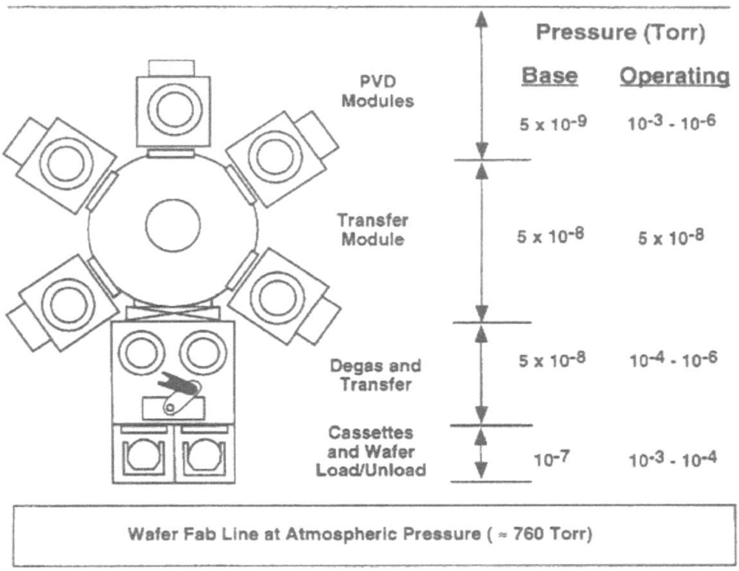

FIG. 5.13 Representative vacuum levels encountered in PVD cluster tool processing.

chambers and precleaning chambers (medium vacuum) to process chambers (high or ultrahigh vacuum). For example, a wafer might be taken from a cassette under atmospheric pressure (high purity N_2 gas at $\approx$ 760 Torr), into a vacuum loadlocked degas station ($\approx 10^{-4}$ Torr), into the central transfer module ($\approx 10^{-7}$ Torr), and finally into a PVD process module ($\approx 10^{-8}$ Torr). In some PVD tools, the volume containing the wafer cassettes is pumped down and maintained under low mTorr-level vacuum [5.8, 5.10], thereby reducing the pressure change to the next vacuum-isolated stage. In other designs (see Fig. 5.12 and ref. 5.8), an intermediate level of vacuum can be added to further control interstage pressure gradients and/or to achieve true UHV PVD conditions ($\approx 10^{-9}$ Torr). The so-called *vacuum buffering* [5.10] inherent in all these designs reduces the chance of process chamber contamination from atmospheric gases. It should be remarked that obtaining ultralow pressure in a PVD Ti module is facilitated by the fact that Ti is an excellent getter of oxygen and water vapor, and the Ti that invariably coats the relatively large surface area of shields, collimator, etc. provides an intrinsic vacuum-pumping capability. Also, even though a module achieves ultrahigh vacuum base pressure, the pressure during PVD is about 10^6 times greater so that even trace levels of impurities in the process ambient can compromise the expected benefits of higher quality vacuum.

With regard to the vacuum levels in a PVD cluster tool, it is important to note that operating pressure and base pressure are sometimes confused and that these pressures can be very different in practice. Operating pressure is primarily determined by process conditions and how long one is willing to pump on the chamber before it is opened to another part of the tool. Base pressure is the ultimate, lowest pressure that can be achieved in a particular vacuum chamber after, in theory, an infinite pump-down time (see Eq. 5.1). Base pressure depends on such things as outgassing from the chamber walls, real and virtual leak rates of gas into the chamber, and the type of vacuum pumping being used. For example, an unbaked PVD chamber pumped down from atmosphere with a mechanical roughing pump might "base out" at 10^{-4} Torr. The same chamber after vacuum baking and pumping with a cryopump might base out at 10^{-9} Torr. During PVD the chamber is backfilled with Ar so that its operating pressure at this point in the wafer process sequence might be $\approx$ 1 mTorr. After deposition, but before the vacuum valve between chamber and transfer module were opened, the PVD chamber might be quickly pumped down to a pressure of, say, 10^{-6} Torr — which is much higher than the base pressure of the module but sufficiently low to prevent an unacceptable gas load from getting into the transfer module.

In other words, even though the base pressure of the module is 10^{-9} Torr, it might actually vacuum cycle between 10^{-3} Torr and 10^{-6} Torr during operation. In a similar vein, the wafer degas station at the front end of a PVD cluster tool may in fact base out in the high vacuum range, but would effectively operate in the medium vacuum range as a result of the water vapor outgassing of wafers being continually processed and the throughput-driven need to keep pump-down times as short as possible. Finally, while base and operating pressure are certainly relevant parameters in PVD cluster tool design, the residual gas composition of the vacuum ambient — e.g., the partial pressure of water vapor and oxygen — can be equally relevant with regard to film properties.

The scope of this book does not permit a tutorial on vacuum science and technology; the reader is referred to several excellent monographs on the subject [5.11–5.13] and articles with a focus on vacuum technology for semiconductor processing [5.14, 5.15]. On the other hand, state-of-the-art PVD cluster tools share the same vacuum considerations, leading to identifiable trends in the types of pumps, seals, and materials of construction [5.16].

Vacuum Pumping

Given the great variation in vacuum level and chamber volume in a PVD cluster tool (5×10^{-6} Torr in a 10-liter degas/cool module, 5×10^{-8} Torr in a 40-liter transfer module, etc.), it is not surprising that a wide variety of vacuum pumps must be used. Figure 5.14 shows the general types of vacuum pumps available for semiconductor processing under atmospheric to UHV conditions. Fortunately for PVD, the pumping of hazardous and toxic gases is not an issue (as it can be with CVD and plasma etching) since the process gas is typically inert Ar or a mixture of Ar/N_2 for reactive deposition of TiN. This greatly simplifies pump selection and also allows certain pumps to serve dual purposes. For example, the same dry pump could be used as the primary roughing pump of a transfer module and the backing pump for a turbomolecular roughing pump of a UHV PVD module.

Thus, pump selection for a PVD cluster tool is determined primarily by the pressure and gas throughput requirements of the process; the time to pump out the chamber volume; and the usual considerations of cost, reliability, and cleanliness common to all semiconductor process tools. With regard to cleanliness, the major contaminant of interest to PVD continues to be water vapor, and the ability of pumps to maintain a low partial pressure of water vapor during deposition ($< 1 \times 10^{-8}$ Torr) is required to produce film quality suitable for ULSI devices.

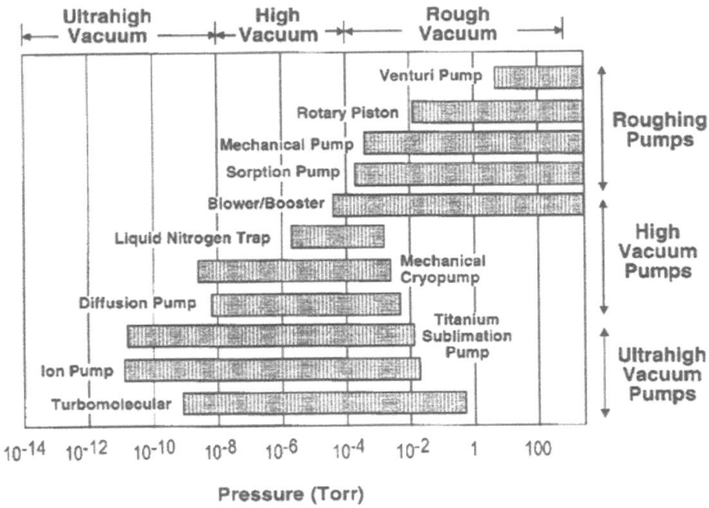

FIG. 5.14 General types of vacuum pumps available for semiconductor processing.

Figure 5.15 presents one possible pumping scheme for a PVD cluster tool, and Figs. 5.16–5.18 illustrate the major pump types used on PVD tools for low, medium, and high vacuum, respectively: dry pump (Fig. 5.16), turbopump or "turbo" (Fig. 5.17), and cryopump or "cryo" (Fig. 5.18). In the system configuration shown in Fig. 5.15, a common, dry roughing pump is used to rough out the process chambers to a pressure low enough for use of a dedicated cryopump. The turbo/drag pump on the chamber can be used to further rough out the process chamber and/or pump gases released by the cryopump during its *regeneration* cycle (discussed in more detail later in this section). The industry trend is away from oil-sealed mechanical pumps and toward the use of oil-free, dry, mechanical pumps for roughing and backing purposes, using either established multi-stage pumps or more recent orbital scroll pumps. Dry pumps virtually eliminate the possibility of oil backstreaming and offer reduced maintenance. Pumping is provided by trapping and removing small pockets of gases in several stages from the inlet to the exhaust, with each stage compressing the gases more.

Small turbomolecular or turbopumps are sometimes used to evacuate wafer load stations from atmospheric pressure as well as being used on degas modules. The turbopump is a clean, compression pump that basi-

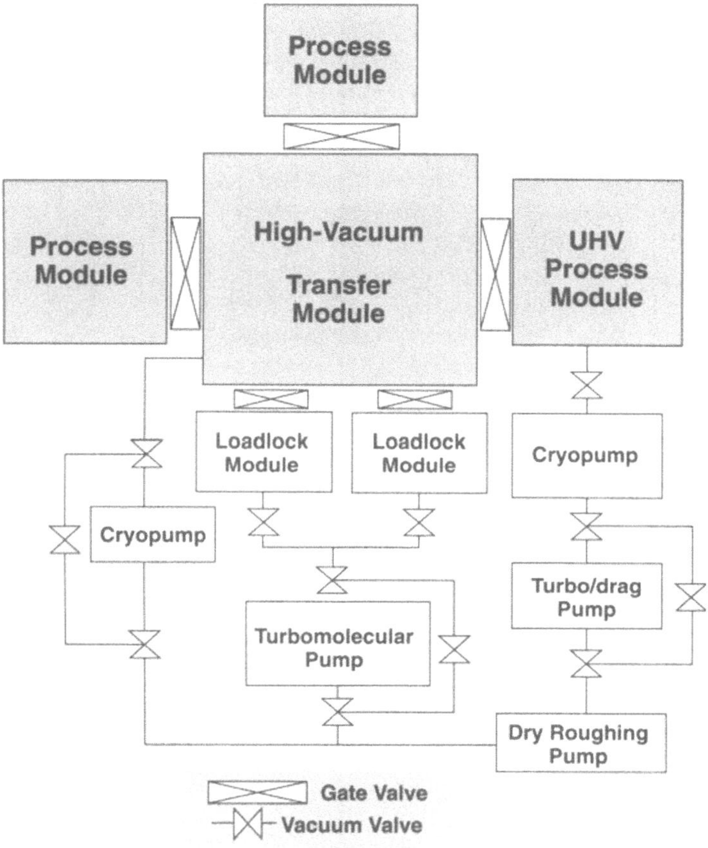

FIG. 5.15 Pumping scheme for a PVD cluster tool, illustrating that a mix of pump types is used depending on the level of vacuum and process conditions required. For simplicity, the pumping on only one of the three process chambers is shown.

cally consists of an alternating stack of rotating and fixed disks into which have been machined a large number of angled blades (see Fig. 5.17). Pumping action occurs when gas molecules bounce off of the rapidly moving rotors (e.g., 70,000 rpm), which are angled so as to increase the molecules' momentum in the direction of the pump exhaust. Turbopumps offer very high pumping speed and constant throughput at moderate pressures, and are sometimes used to rough out UHV chambers. On the other hand,

FIG. 5.16 Representative dry pump used for low-vacuum application in a PVD tool (dry iQ-series pump shown, courtesy of Edwards High Vacuum International, Wilmington, MA).

turbopumps are slow at pumping light gases and water vapor. Also, a backing pump must be used to prevent the turbopump from being overloaded by the gas load it is compressing into its foreline. For example, the turbopump shown in Fig. 5.17 has N_2 pumping speed of 70 liter/sec and might be backed with a 4-liter/min mechanical pump. The trend in turbos is toward compound pumps that combine a regular turbopump with a so-called molecular drag pump that is like a turbopump (it uses a rotating drum or disk) and allows the compound pump to be exhausted at pressures high enough to use simple, low-cost backing pumps.

FIG. 5.17 Representative turbopump used for medium- to high-vacuum application in a PVD tool with shaft removed to reveal alternating disks of rotors and stators (turbopump model V70 shown, courtesy of Varian Vacuum Products, Lexington, MA).

PVD processing requires a contamination-free, high-vacuum pump with high pumping speeds for both process gases and residual gases. As a result, PVD chambers are primarily pumped using cryopumps due to their cleanliness (no pump oil means no hydrocarbons) and high pumping speed for water. In a two-stage cryopump, water vapor and other condensible gases are pumped in the first stage via physisorption on a cryogenically cooled surface ($T \approx 77$ K), while the second stage is used to trap gases with high vapor pressure — such as Ar, He, and H — in the molecular-scale pores of a charcoal array ($T \approx 15$ K). Given typical PVD chamber volumes (20–50 liters) and pump gate valves (e.g., 20-cm internal diameter with a 10-inch conflat flange), a cryopump equivalent to the one shown in Fig. 5.18 might be used. It should be noted that modern PVD tools (since 1990) are generally operated without throttling of the pump — i.e., at full pumping speed. This means that the base pressure of the tool (e.g., 1×10^{-8} Torr) is also effectively the base pressure during operation. Also, during high-temperature deposition, radiant heat from the sample holder ($\approx 500°C$) may be sufficient to warm the front of the cryopump and "dump" the pump. In this case, it is necessary to provide some protective

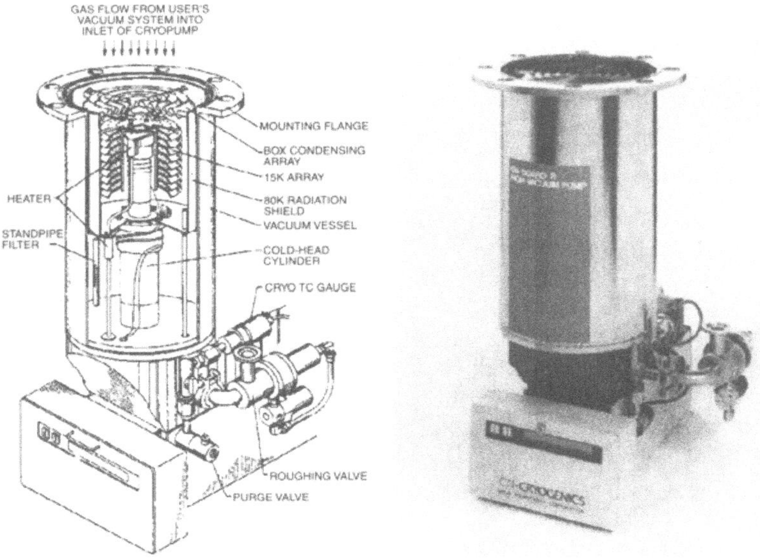

FIG. 5.18 Representative cyropump used for high-vacuum application in a PVD tool (ON-BOARD® high-vacuum pump, courtesy of CTI-Cryogenics, Mansfield, MA). For use on an ultrahigh vacuum PVD process module, the elastomer-sealed mounting flange would be replaced by a metal-sealed conflat flange.

radiation shielding for the cryopump. Cryopumps have a high pumping speed for argon, but they cannot handle very high Ar flow rates for long, due to their limited absorption capacity. Under these conditions, the cryopump would be throttled back. Unfortunately, this also reduces the pumping speed for water vapor, which can have an adverse impact on oxidation-sensitive hot PVD processes such as reflow Al and the two-step "cold-hot" Al process (see Chapter 7). Hence, some PVD modules utilize an unthrottled turbopump with a relatively small pumping speed for argon (e.g., 200 l/sec) in tandem with a cold trap having high capacity and pumping speed for water (e.g., > 1000 l/sec).

There is a slight reliability advantage to cryos over turbos in that cryos tend to fail slowly, making loud noises and generally breaking down within a day or two. Hence, there is usually enough warning to schedule a routine pump replacement without losing any wafers. On the other hand, turbos tend to fail quickly and without much external notice. Conversely, there is a slight maintenance disadvantage with cryos associated with the downtime associated with regeneration of the pump — namely, the cryogenic surfaces have a finite gas capacity that requires a periodic bakeout (heated N_2 gas is flowed through the pump) to remove the adsorbed species. Fast regeneration cryopumps are therefore of great interest. These pumps work by only heating up the colder, second stage. This can greatly reduce regeneration times (from $\approx$ 2.5 hr at 500°C to 0.5 hr); however, since the adsorbed water is not baked off, such pumps are best used where the absolute volume of residual water vapor is low — e.g., the internal chambers of a loadlocked cluster tool.

Finally, we note the recent application of nonevaporable getter (NEG) technology for PVD applications, which has been used to reduce the time for a PVD chamber to recover to high-vacuum base pressure after chamber venting [5.17, 5.18]. Gettering materials remove residual active gases such as O_2, N_2, CO_2, H_2O, and H_2 from a vacuum chamber by forming stable chemical bonds or compounds. Semiconductor processing has long taken advantage of high-vacuum pumping based on the gettering action of a thin film of reactive titanium metal — either through evaporation as in a titanium sublimation pump (TSP) or by sputter deposition as in a sputter-ion pump, such as the Varian Vac-Ion™ design. Less familiar to IC manufacturing are bulk getters (or nonevaporable getters) that have been used for many years in other industries — e.g., to produce high vacuum in linear accelerators for high-energy physics research. NEGs are alloys of metals from Group IV-A of the periodic table — such as Ti, Th, and Zr — that are capable of dissolving their own oxides in the solid state at elevated but moderate temperature (e.g., 350–500°C for certain Zr-alloys). Therefore, even though the NEG surface eventually saturates during use, it can be renewed by a vacuum anneal that produces a clean and highly reactive metal surface. As a practical matter, NEG activation annealing might be done as part of a preventive maintenance cycle during chamber bakeout.

Mounted internally within the PVD chamber in conjunction with an external cryopump, the NEG boosts pumping speed in the high-vacuum regime — particularly for H_2, which is difficult to cryopump (see Fig. 5.19). This greatly reduces the time for the module to reach base pressure after preventive maintenance (PM), which results in improved tool productivity. Figure 5.20 shows the time associated with a PM process, which consists of venting a PVD chamber to atmosphere, changing sputter shields and target, and pumping down to the 10^{-9}-Torr range base pressure. When the cryopump was augmented by in-situ NEG pumping, the chamber based out in 30 minutes. On the other hand, 5 hours were required using a cryopump

FIG. 5.19 Nonevaporable getter (NEG) pumping package using high-surface-area disks of zirconium-alloy (InsiTorr™ fast pump module, courtesy of SAES Pure Gas, Inc., San Luis Obispo, CA).

Event	Time (hr) without NEG	Time (hr) with NEG
Chamber Vent	0.05	0.05
Shield Change	0.08	0.08
Target Change	0.58	0.58
Pump to 10^{-6} Torr	0.42	0.42
Leak Check	0.17	0.17
Cycle Purge	0.25	0.25
Bakeout	2.00	2.00
Chamber Cool	2.00	2.00
25 Wafer Burn-In	0.83	0.83
Pump to Base (3×10^{-8} Torr)	5.00	0.33
Total Time	11.38 hrs	6.71 hrs

FIG. 5.20 Time associated with a preventive maintenance step, with and without NEG-assisted pumping in the high vacuum regime (after table 1 in ref. 5.17). Reprinted from the August 1996 edition of Solid State Technology (copyright 1996 by PennWell).

alone. Since pump-down time represents a significant fraction of total module downtime, the net effect of using NEG technology was to reduce total maintenance cycle time by 35%.

Vacuum Practices

Elementary vacuum theory gives the relation for the pressure $P(t)$ of a chamber of volume V being evacuated from an initial pressure P_0 by a pump of speed S as

$$P(t) = \left(P_0 - \frac{Q}{S}\right)\exp\left(\frac{-tS}{V}\right) + \frac{Q}{S} \tag{5.1}$$

where Q is the leakage of gas into the chamber, either intentional (e.g., Ar gas for PVD) or unintentional from outgassing and vacuum leaks. Eq. (5.1) shows that the base pressure ($t = \infty$) is given by the ratio Q/S, so that reducing base pressure requires either increasing pump speed or reducing leaks and outgassing. Increasing the vacuum conductance (C) of connecting tubing and orifices between pump and chamber (e.g., using a large-diameter, close-coupled pump line) is desirable to bring the effective pump speed (S_{eff}) as close to the theoretical maximum speed S as possible ($1/S_{\text{eff}} = 1/C + 1/S$). However, using much larger pumps is not cost-effective since the price of UHV pumps increases greatly with size (i.e., pump speed). Therefore, great attention is paid in PVD cluster tools to the integrity of vacuum flanges, welds, etc. and to surface preparation to reduce outgassing. Since the speed of cryopumps is usually $S < 5000$ l/sec, we estimate that leak-up rates $Q < 5 \times 10^{-7}$ Torr-l/sec are required to achieve $Q/S = 10^{-9}$ Torr UHV base pressures (1 Torr-l/sec corresponds to $\approx$ 79 sccm flow of gas at atmospheric pressure). A similar order-of-magnitude calculation shows why the use of UHV practices are desired for PVD. The gas throughput of Ar when sputtering at 5 mTorr is $\approx$ 1 Torr-l/sec. Also, it is known that partial pressures of water in Ar process gas as small as $\approx$ 1 ppb can adversely affect PVD Al film properties. Therefore, taking the surface area of a PVD chamber as $\approx 10^4$ cm^2, we estimate that 1 ppb of water will be introduced into the process gas by an outgassing rate of only 10^{-13} Torr-l/sec per cm^2.

Preparing surfaces to achieve this level of cleanliness in a PVD cluster tool requires great attention to such details as (1) surface finish of chambers and fixtures (e.g., an electropolished or a mirror finish is preferred to reduce surface area), (2) materials of construction (e.g., stainless 316L is preferred over 304 to minimize hydrogen outgassing), (3) sealing surfaces (e.g., metal bonnet-sealed gate valves are preferred over more gas-permeable elastomers), and (4) bakeout (e.g., internal bakeout lamps can be used to outgas both chamber walls and sputter shields). High-vacuum and UHV design practices are discussed in a number of texts and articles on vacuum technology to which the reader is referred (e.g., see refs. 5.11–5.14).

5.3.2 Wafer Degas

Proper bakeout of chamber and shields is required to minimize outgassing of water vapor and other unwanted residual gases during PVD. Similarly, proper in-situ degas of oxide-patterned wafers is needed prior to both precleaning and PVD to prevent outgassing that could impede contact formation or give rise to "poisoned" vias with poor via chain resistance. In extreme cases, Al via plugs can actually be forced out of the via opening by outgassing from below. While a combination of CVD oxide and spin-on glass (SOG) are used in interlevel dielectric stacks, the SOG is generally of greater concern with respect to PVD degas since it is very porous on a nanometer scale and prone to adsorption/desorption of relatively large amounts of water. This water exists as adsorbed surface water on the glass and within its network of interconnected micropores, and in the form of silanol groups (Si-OH) in the bulk of the film. Even though much of this water can be driven out of the film by a high-temperature furnace cure after spin coating, the film can readily reabsorb moisture when left in humid room air. For example, a cured 2000-Å SOG film might absorb several weight percent of water during only a few minutes of storage or transport in humid air before being loaded into a PVD tool.

Water evolved from the SOG during a subsequent precleaning or deposition step can then, if the partial pressure is great enough, lead to unwanted oxidization of a contact (SiO_2 formation) or via (e.g., Al_2O_3 formation) [5.19, 5.20]. Even though the SOG might be protectively sandwiched between two CVD glass layers, the area of the SOG that is directly exposed on the sidewall of the contact or via provides a ready source of unwanted water vapor. Since the out-diffusion rate of water from SOG tends to increase greatly with temperature, the problem is most severe for elevated temperature steps. Also, water vapor is readily dissociated in the sputtering plasma to form atomic H, and this hydrogen can accumulate along the grain boundaries of many sputtered films, (e.g., for Ti-W alloys) increasing their intrinsic film stress.

To deal with this problem, an in-situ vacuum anneal ($\approx 10^{-5}$ Torr) is given to wafers prior to both precleaning and PVD deposition. The empirical rule seems to be as follows: Degas the wafer for as long as possible consistent with high wafer throughput (typically ≈ 60 sec) and at least 50°C higher than the hottest process step. For example, if an Al two-step flow process (see Chapter 7) requires wafer temperature of 380°C, degas should be above 430°C. It is worth pointing out that although loosely bound surface H_2O can be removed by a moderate vacuum bake at ≈ 200°C, more tightly bound H_2O in the bulk (e.g., hydrogen-bonded silanol groups) may require outgassing temperatures of ≈ 400°C. Extremely stable water, such as that associated with isolated silanol groups in the bulk of the glass, may not be completely released until temperatures above 700°C are reached. On the other hand, since PVD process temperature rarely exceeds 500°C, such water is not likely to be mobile during deposition and is not usually of concern.

Regarding hardware for degas, lamp heating of wafers has been employed as well as hot plates with backside-gas-assisted heat transfer. While the ramp-up time to degas temperature can be much less with lamps, consistent with high wafer throughput, this must be considered against the added hardware complexity and cost.

5.3.3 Wafer Preclean

The presence of interfacial contamination and oxides can inhibit desired thin film solid state reactions and lead to repeatability problems. For example, the presence of native silicon oxide at a contact before PVD Ti deposition can inhibit the Ti silicidation process and lead to increased and/or variable contact resistance. This has led to the widespread use of precleaning, which refers to the in-situ removal of native oxides and possible dry etch residues (e.g., Teflon-like polymers) prior to PVD metal deposition (although in the case of contact metallization with Ti, the high-purity PVD Ti also serves to clean the interface by reducing native silicon oxides). PVD precleaning is sometimes referred to as "etching," which while strictly true, can cause confusion since etching is also used to describe the wet chemical and dry plasma-assisted methods that are widely used for photolithographic pattern transfer in IC fabrication. For example, plasma etching of photoresist-patterned PVD AlCu is used to define separate interconnect lines out of the blanket film.

Precleaning is typically done using inert Ar^+ sputter etching (inert ion milling) with ion energy of ≈ 50–500 eV — in effect treating the wafer as a sputter target. This approach works because physical sputtering is a kind of universal solvent, the sputter yield of most elements of interest in microelectronics being greater than ≈ 0.01 for ion energy above ≈ 50 eV (see Chapter 2). Ion bombardment can also create dangling bonds that promote the adhesion of a subsequent sputtered overlayer. On the other hand, unlike reactive ion etching (RIE) and other plasma-assisted methods used for pattern definition etching in microelectronics, the selectivity of physical sputtering is not particularly high. Since Si and SiO_2 sputter at similar rates, overetching when cleaning native SiO_2 at a Si contact will remove the Si as well. For the ion energies typically used in precleaning, the sputter yield of Al_2O_3 is approximately 5 times lower than SiO_2, making it more time consuming to clean equivalent oxide thickness from the via level than at the contact.

Another challenge for conventional sputter precleaning is removing unwanted oxides and contamination from the bottom of steep features without simultaneously sputtering the interlayer dielectric from the sidewalls onto the bottom. Tanaka et al. [5.21] have found that the yield of cleaned submicron vias was a strong function of Ar^+ energy, which they attributed to the lower gas scattering and increased directionality of higher-energy Ar ions. It seems likely that some chemical component will need to be added to sputter cleaning in the future to deal with very high aspect ratio features (e.g., the use of Ar/H_2 or other reactive gas mixtures) or that a more traditional RIE or even a vapor-phase clean will be used.

Under conditions of high-rate sputter etching (e.g., 500 Å/min removal of SiO_2 with 500 eV Ar^+), the incident power density can be > 0.5 W/cm^2, which can heat the wafer well above 100°C. The temperature reached by the wafer during preclean is important for several reasons. First of all, if the wafer gets too hot during sputter cleaning, sufficient water vapor can be released from the exposed SOG sidewalls to poison contacts and vias during PVD deposition. This same effect must be considered when precleaning, as pointed out by Wolters and Heesters [5.22]. For example, consider the 1.0-μm-diameter via shown in Fig. 5.21, which is patterned in a sandwich of 0.2 μm of SOG over 0.8 μm of CVD oxide. In this case, the goal of the sputter preclean is to remove Al_2O_3 from Al at the bottom of the via. Water vapor that is evolved from the via sidewalls is assumed to come from the exposed "ring" of SOG, whose surface area is ≈ 10^{-8} cm^2. Studies have shown that the desorption rate of H_2O from SOG is ≈ 5×10^{-6} Torr-l/sec-cm^2 (≈ 2×10^{14} H_2O molecules/sec-cm^2) for glass temperatures in the range of ≈ 100–400°C. Therefore, at elevated temperatures, ≈ $2 \times 10^{14} \times 10^{-8} = 2 \times 10^{6}$ H_2O molecules/sec are released. Given the area of the via bottom (8×10^{-9} cm^2) and a typical sputter etch removal rate of Al_2O_3 (50 Å/min = 2×10^{13} Al_2O_3 molecules/sec-cm^2), it is easy to see that the amount of water vapor entering the via from its side-

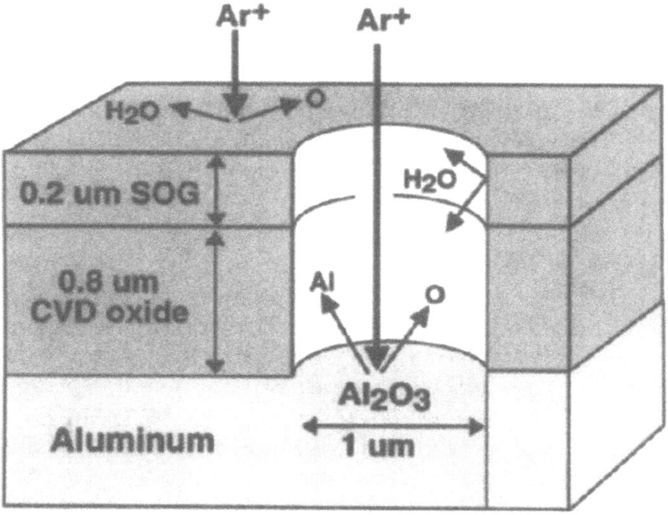

FIG. 5.21 Illustration of how outgassing of SOG can reoxidize the bottom of a via during sputter cleaning.

wall could be 10 times greater than the amount of material being removed from the via bottom (2×10^5 Al_2O_3 molecules/sec). This can lead to the aluminum being oxidized at a faster rate than it is removed, which underscores the need for a proper degas.

A related oxygen-contamination issue is the fact that the blanket nature of sputter etching means that the field oxide (SiO_2) over the entire wafer surface is also being removed, sputter ejecting O from a surface area that vastly exceeds that of the exposed vias to be cleaned of Al_2O_3. Under conditions of high-rate via etching, SiO_2 etch rates can be 600 Å/min, and one can calculate that undesirably high O partial pressures ($> 7 \times 10^{-4}$ Torr) can result unless effective pumping speed for O is very high ($\gg$ 40 l/sec). Therefore, vacuum pumps used on a preclean module should have relatively high pumping speed for both oxygen and water vapor.

Another concern with elevated temperature during preclean of contacts relates to device damage — primarily wafer charging that damages thin gate oxides. More specifically, charging induces Fowler-Nordheim tunneling of electrons through thin oxides that generate traps, resulting in degraded electrical properties such as leakage. Device damage during reactive plasma or inert sputter etching is a complex topic, dependent on such things as oxide thickness, plasma uniformity, and device structure [5.23, 5.24]. One measure of an oxide's susceptibility to charging damage is its charge-to-breakdown (Q_{bd}), with damage starting to appear when the electron fluence (electron current × time) passing through the oxide exceeds a critical threshold value — typically ≈ 1% of Q_{bd}. Since Q_{bd} decreases with temperature, lower-temperature precleaning allows thin oxides to be exposed to a greater total charge before the onset of damage. In addition to traps generated by electrons tunneling through the oxide toward the Si below, holes injected from the Si into the oxide can also generate traps. Since the mobility of the holes is temperature dependent, higher temperatures lead to more damage. As a practical matter, keeping wafer temperature less than ≈ 100°C during contact preclean is probably acceptable for 0.25-μm device fabrication, although more advanced ULSI devices may require preclean temperatures of room temperature or below. Concerns about temperature are relaxed for via cleaning, and some users employ relatively high temperatures (≈ 400°C) during via preclean because the additional wafer outgassing (the wafer has already been vacuum-annealed in the degas module) is found to improve the reliability of a step such as reflow Al, which is influenced by even trace amounts of water.

Figure 5.22 lists representative process attributes of an advanced precleaning module. Throughput requirements for the overall PVD process require etch rates sufficiently high to remove native oxides of Si and Al at the bottom of high aspect ratio features in times < 60 sec. On the other hand, customer concerns about ion-induced damage have required the use of lower Ar^+ ion energy for both contact- and via-level cleaning (e.g., ≪ 100 eV for contacts). Because the sputter yield is reduced by going to lower energy, these two requirements have led suppliers to develop higher density plasma sources with high ion flux at low energy. It is instructive to estimate the required ion flux density to remove 100 Å of SiO_2 in 1 minute or less. At relatively low ion energy (≪ 1 keV), the sputter yield Y is found to have a square-root dependence on ion energy E of the form $Y = A\,(E^{1/2} - E_0^{1/2})$, where both the sputtering threshold E_0 and the constant A depend on the ion-substrate combination. For Ar^+ on SiO_2, $E_0 \approx 30$ eV and $A \approx 3.5 \times 10^{-2}$ molecules/ion-eV$^{1/2}$, giving a sputter yield at 100 eV of $Y \approx 0.15$ molecules/ion. Since the atomic density of silicon oxide is 2.3×10^{22} molecules/cm^3, sputter etching away 100 Å of SiO_2 in 1 minute or less will then require an incident Ar^+ flux of $> 2.5 \times 10^{15}$ ions/cm^2-sec. Advanced plasma sources can provide such ion flux densities, and the power density at 100 eV is small enough (≈ 40 mW/cm^2) that appreciate wafer heating does not occur during the short preclean time.

Desired Attributes for an Advanced Sputter Preclean	Representative Values
High-rate removal of SiO_2 and Al_2O_3 over a range of bias voltages	200-800 Å/min (SiO_2) at bias = 50-500 V
Ability to clean the bottom of high aspect ratio holes	AR > 4:1
Low operating pressure High-vacuum base pressure	≈ 1 mTorr $< 5 \times 10^{-7}$ Torr
Near-ambient temperature	T < 100°C
Excellent uniformity	3σ < 10% over 200 mm with no edge exclusion
No damage to thin gate oxide devices (gate oxide < 80 Å)	100% yield
Ultralow levels of mechanical and process-added particles	< 0.01 cm^2 @ > 0.3 μm

FIG. 5.22 Process attributes desired for an advanced precleaning module for PVD.

Concerns about contamination and directionality (from gas-phase scattering) are also driving preclean process pressure below 1 mTorr, while the need to reduce particles is promoting a move away from mechanical edge clamps or rings (which can be sputter deposited onto the wafer) and toward electrostatic chucks (e-chucks) or clampless processes. A further advantage of an electrostatic chuck is that it allows one to retain the use of backside-gas-assisted heat transfer between the wafer and chuck, facilitating control of temperature during the process. Even using an e-chuck, however, material sputtered from the wafer surface can redeposit onto the walls of the precleaning chamber. Although 200 Å or less of oxide is typically removed per wafer, after a large number of wafers are precleaned (e.g., 1000), a rather thick film deposit can build up on the shields (e.g., 20 μm), which itself becomes a source of contamination and/or particles. A number of methods have been developed to increase the amount of deposit that the interior shielding can bear before they have to be replaced and/or cleaned. For example, shield surfaces can be perforated, which allows the contaminants to pass through and coat the outside of the shields and other interior parts of the chamber. This effectively increases the surface area of the shields and the time between shield cleaning. Another approach is to texture the shield surfaces by bead or sand blasting or by sputter etching in an effort to increase the adhesion of the deposit. Suitably treated, the shields can then retain a greater mass of deposited material before flaking or exfoliating particles.

Figure 5.23 shows several hardware configurations that have been used for preclean: diode and magnetically enhanced diode, and two high-density, low-pressure plasma approaches — electron cyclotron resonance (ECR) and inductively coupled plasma (ICP). Both the ECR and ICP are shown with an independent RF bias applied to the wafer. The flexibility provided by a dual-frequency etch permits incident ion energy and ion flux

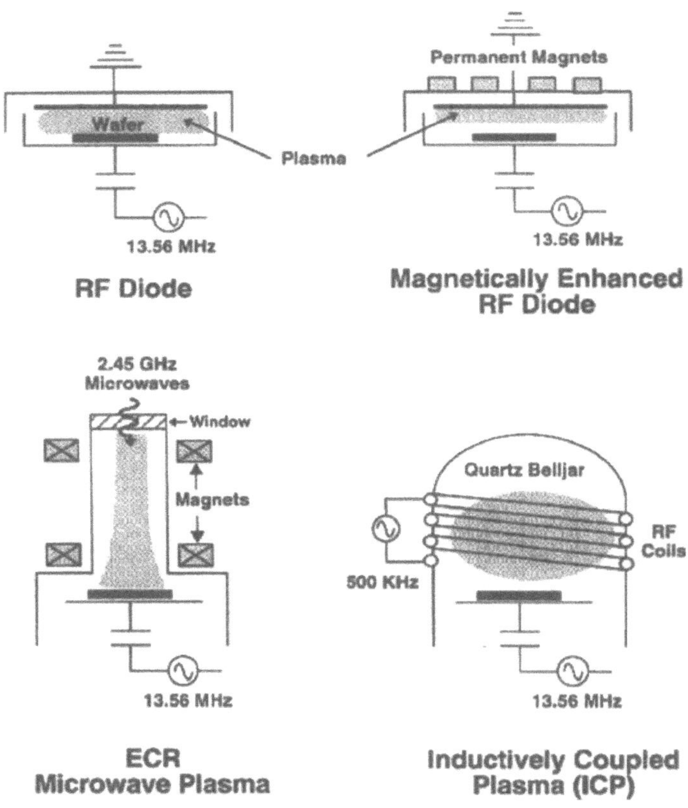

FIG. 5.23 Several hardware configurations have been used for precleaning of wafer prior to PVD, including diode, magnetically enhanced diode, electron cyclotron resonance (ECR), and inductively coupled plasma (ICP).

to be varied more or less independently. Also note that there are many similarities between hardware for dual-frequency ICP preclean and hardware being developed for directional, RF-ionized PVD (see Chapter 8) — although the application (sputter preclean vs directional deposition) and the process pressure (< 1 mTorr vs ≈ 20–50 mTorr), among other things, are quite different.

5.3.4 Wafer Temperature

Even though wafers enter and exit a PVD cluster tool at near room temperature (≈ 20°C), their thermal history in passing through the tool can be extremely complicated, exhibiting a temperature vs time profile that is dependent on the specific process and hardware used. In general, wafer temperature varies from ≈ 20°C to 450°C in a cluster tool during PVD processing, in contrast to front-end-of-line processes such as thermal oxidation or ion implantation activation annealing, which can involve temperatures as high as 900–1000°C. As we have seen, during degas the wafer is brought rapidly up to a high temperature (≈ 450°C) and held for nearly 60 sec for water vapor outgassing of thick interlayer oxides. However, during a similar 60 sec of sputter precleaning, it is desirable to limit wafer-temperature rise to below ≈ 100°C to minimize damage to thin gate oxides. During PVD itself, the deposition temperature can be used to control such properties as film morphology, grain size and crystal orientation, and stress and step coverage (i.e., surface adatom mobility). Also, many advanced PVD processes — such as the two-step of Al planarization process (so-called cold-hot process) — depend critically on control of wafer temperature both to deposit a suitable PVD Ti or TiN wetting layer prior to PVD Al and to provide a time–temperature profile during PVD Al that minimizes void formation during the "cold" Al step and maximizes void annihilation during the "hot" Al step.

Thermal Budget

Although there are many process-driven reasons to control and/or reduce wafer temperature in PVD, it is usually stated that PVD temperature must come down to meet a device-driven thermal budget that quantifies the total time at elevated temperature allowed to manufacture an advanced device. Actually, this statement can be misleading. The thermal budget concept was introduced to address the thermal diffusion of dopants that are precisely introduced in semiconductor devices to delineate such things as the position and depth of p-n junctions. The ability to form a junction that is both shallow and precisely controlled is at odds with excessive dopant diffusion. Simple one-dimensional diffusion theory shows that the distance diffused by a dopant in a time t is given by $x \approx 2(Dt)^{1/2}$, where the diffusion coefficient D depends exponentially on temperature T (viz., $D = D_o \exp(E_a/kT)$). This argues for keeping the temperature of a given process step low or greatly reducing the time at elevated temperature as with a rapid thermal anneal. Also, since the total diffused distance is statistically additive ($x_{total} = \sqrt{x_1^2 + x_2^2 + \cdots}$), successive elevated temperature steps can quickly use up the thermal budget set by the dimensions and dimensional tolerances of the device design.

The problem in applying this dopant-related concept to PVD is that PVD process steps rarely exceed 500°C and the total time at elevated temperature is usually less than a few minutes, so that dopant diffusion is negligible (e.g., the fast-diffusing p-type dopant B diffuses about 10 Å in crystalline Si after 400 hours at 600°C). On the other hand, there are other compelling reasons to reduce PVD time and/or temperature. Clearly, even a small reduction in time of a lengthy degas or reflow step could result in measurably improved throughput. With regard to temperature, low-dielectric-constant (low-k) interlayer dielectrics are under active development to reduce on-chip RC time-constant signal delays in multilevel metallization. Since many of the candidate materials are polymer-based and are degraded or destroyed by elevated temperature (> 400°C), there is a need to reduce maximum process temperatures during PVD to compatible levels and to reduce the time spent close to the temperature of degradation. Also, when a metal interconnect line is subjected to high-temperature excursions, a phenomenon known as *stress voiding* can occur. In this situation, voids are formed in the metal as the result of stress imposed by the thermal expansion mismatch with the underlying and overlying dielectrics and by the intrinsic stress and microstructure of the interconnect itself. Stress voiding can also accelerate failure due to electromigration. Therefore the thermal budget for PVD really translates into a rather complex materials-dependent limit on process temperature and a wafer throughput-dependent limit on process time.

Temperature during PVD

Control of wafer temperature has allowed PVD to be successfully applied in advanced microelectronic applications and is often considered an enabling technology for PVD. For convenience we will focus on sputter deposition, although the basic physical arguments also will apply to other vacuum-processing steps associated with PVD such as degas and preclean.

Also, the mathematical treatment of wafer heating/cooling is intentionally simplified to illustrate key aspects of the topic. For a more rigorous treatment, the reader is directed to refs. 5.25–5.27.

Figure 5.24 illustrates the thermal environment of a wafer during PVD. The wafer is assumed to be held in close proximity to a substrate holder and is surrounded by grounded sputter shields. Although a generalized substrate heater is shown in Fig. 5.24, in the following discussion we do not consider active substrate heating. The principal thermal input to the wafer are then associated with irradiation, particle bombardment, and thermodynamic effects. Regarding the kinetic energy associated with particle bombardment, one should consider (1) incident-sputtered metal atoms, (2) bombardment by Ar^+ due to the wafer being more negative than the plasma, (3) bombardment by energetic neutralized Ar, and (4) bombardment by energetic electrons from the magnetron plasma that escape along non-reentrant field lines. Regarding thermodynamic effects, one should consider (5) the heat of condensation, which is released when the depositing sputtered atoms go from the gaseous phase to the solid phase as they adsorb onto the wafer surface and (6) the heat of neutralization, which is given up as Ar^+ ions combine with electrons at the wafer surface. (7) Blackbody radiation emitted from either the plasma or from potentially hot

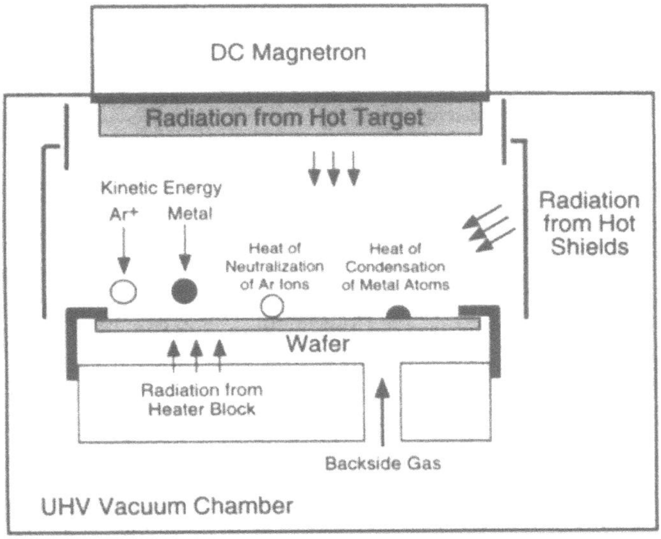

FIG. 5.24 Wafer thermal environment during PVD.

objects — such as the target and sputter shields — can all be absorbed to some extent by the wafer. (8) Finally, there is also the hot gas within the chamber, which can be at quite high temperatures (> 400°C). While the thermal situation during DC magnetron sputtering is complex, it turns out for many microelectronic applications that energetic atom and ion bombardment make up the majority of the heat input (≈ 60–70%), with the heat of condensation accounting for the remainder.

Regarding thermal output from the wafer, the possible mechanisms are conduction, convection, and/or radiation. At the low pressures used in PVD (≈ 1–10 mTorr), convection is not relevant. The thermal conductivity of gases at low pressure is so small (e.g., the heat transfer coefficient of Ar at 1 mTorr is ≈ 10^{-5} W/cm^2-K) that gas-phase conductive heat loss is also negligible. Also, even though the wafer and substrate holder may be in physical contact, the point nature of the contact on a microscopic scale greatly reduces the effective cross-sectional area through which heat can flow, ruling out effective solid-phase conductive heat loss. It is often believed that clamping a wafer more tightly against the substrate holder will improve heat conduction, but this does little to increase the actual microscopic contact area. The result is that, unless intentional backside-gas-assisted heat transfer is employed, radiation is the dominant heat loss mechanism.

With these assumptions, the one-dimensional heat equation for the wafer in Fig. 5.24 can be written as

$$\rho C_p z \frac{dT}{dt} = \Phi_{\text{in}} - \Phi_{\text{out}} \tag{5.2}$$

where ρ, C_p, and z are the density, specific heat, and thickness of Si, respectively. The density of Si is 2.33 gm/cm^3, and while C_p is a function of temperature, a value of ≈ 0.7 J/gm-K is a typical PVD temperature. Φ_{in} is the incident energy flux, and Φ_{out} is the sum of the separate radiative heat losses from the front and back faces of the wafer.

We now consider the various mechanisms for heat gain and loss at the wafer, assuming for the moment that neither intentional wafer heating nor cooling are being employed.

1. *Kinetic Energy of Sputtered Atoms.* Sputtered-atom incident kinetic energy flux can be calculated using the Thompson energy distribution presented in Chapter 2. However, as a first approximation we take the average atom energy <E> for materials such as Al and Ti to be ≈ 12 eV. Knowing the incident deposition rate (e.g., 1 μm/min of Al) and atomic density of

the metal ($\approx 6 \times 10^{22}$ Al atoms/cm^3) allows one to calculate the incident sputter flux density J ($\approx 10^{17}$ Al atoms/sec-cm^2), leading to an incident power density given by the product of $J \times \langle E \rangle \approx 1.2 \times 10^{18}$ eV/sec-cm^2 = 0.2 W/cm^2.

2. *Kinetic Energy of Incident Ar Ions.* For an electrically nonbiased wafer, the plasma potential is in general $\approx$ 10 V more positive than the wafer surface and Ar^+ bombardment of the wafer occurs with energy not too different than that of the incident metal atoms. Experimental data for high-rate PVD Al suggests that the flux of Ar^+ at the wafer is $\approx$ 20–30% of the flux of Al atoms [5.26].

3. *Kinetic Energy of Energetic Neutrals.* An energetic, or "fast," neutral is an Ar ion that has been neutralized at the target surface and reflected back toward the wafer. These atoms can have considerably higher energy (e.g., 50 eV = 600,000 K) than the Ar atoms in the process gas (e.g., 0.05 eV $\approx$ 600 K). Both the number and energy of these fast neutrals depends strongly on whether the mass of the incident Ar ion (m_i = 40 amu) is greater than or less than the mass of the target atoms (m_t). Figure 5.25 shows a Monte Carlo calculation of fraction R of reflected, neutralized Ar^o and maximum reflected energy (E_r) for 1-keV Ar^+ bombardment of various mass targets. The two most commonly sputtered microelectronic materials are Al alloys and Ti or TiN. Since the mass of Al (m_t = 27 amu) is less than that of Ar, we see from Fig. 5.25 that energetic neutral production is insignificant. Also, even though the mass of Ti (m_t = 48 amu) is greater than that of Ar, both the fraction of reflected Ar (< 0.1%) and its maximum energy (< 60 eV) suggest that they are not a significant source of wafer heating. Whereas energetic neutrals should be considered for specific targets with ultrahigh mass such as Ti-W, we ignore their contribution in this general discussion of wafer heating.

4. *Electron Bombardment.* For a DC magnetron with grounded shields, electron bombardment of the substrate is not significant.

5. *Heat of Condensation.* The latent heat of condensation ΔH is the change in enthalpy of the sputtered material associated with its going from the gas phase to the solid state. The energy given up upon surface adsorption is on the order of 10 eV, so that the energy flux associated with adsorption is given by $J \times \Delta H$ ($\approx$ 0.2 W/cm^2 for 1 μm/min of Al).

6. *Heat of Neutralization.* Each Ar ion that is neutralized at the wafer gives up the Ar ionization potential of 15 eV, with the total power determined by the flux of Ar^+ at the wafer. Since the ion current to the wafer in conventional PVD is $\ll$ 1 mA/cm^2, the power associated with the heat of Ar neutralization is $\ll$ 0.01 W/cm^2. This is not the case for ionized

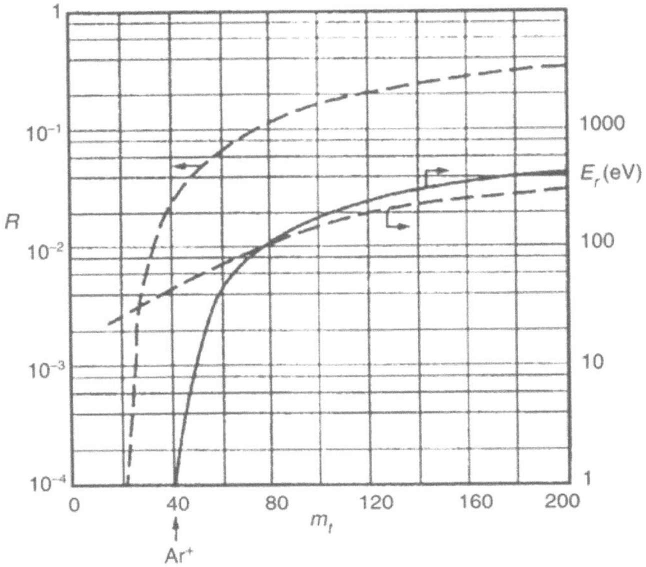

FIG. 5.25 Reflected fraction (R) and maximum reflected energy (E_r) of reflected neutralized Ar^0 during 1 keV Ar^+ bombardment of various targets. (Reproduced with permission of The McGraw-Hill Companies from D. L. Smith, *Thin Film Deposition: Principles & Practice*, McGraw-Hill, NY, 1995.)

magnetron PVD, in which an appreciable flux of Ar ions can be incident at the wafer (see Chapter 8).

7. *Blackbody Irradiation.* In general, the temperature of a properly water-cooled target and the sputter shields will be below 400°C. Since the emissivity of metals is very low, the radiated power is in general not a significant contribution to wafer heating, with outgassing of these surfaces during PVD usually of greater concern.

8. *Radiative Heat Transfer.* Radiative loss is governed by Stefan's Law and involves the fourth power of wafer temperature (in K):

$$\Phi_{out} = \epsilon\sigma(T^4 - T_0^4) \tag{5.3}$$

where σ is the Stefan-Boltzmann constant (5.67×10^{-12} W/cm^2-K^4), ϵ is the effective emissivity of the wafer front or back surface — which takes into account both the emissivity of the emitting surface and the environment into which it radiates (e.g., the hot shields) — and T_0 is the temperature of the environment. In general, the emissivity of a PVD metal is wavelength dependent and a complicated function of surface roughness, temperature, and film thickness. As a practical matter, the emissivity of thick (> 1000 Å) Al, Ti, or TiN films used in PVD is relatively low (< 0.2).

Figure 5.26 presents calculated time–temperature profiles for a Si substrate under conditions representative of PVD Ti deposition (1000 Å/min; 25 mW/cm²) and PVD Al deposition (1 μm/min; 250 mW/cm²). Given the maximum thickness of Ti/TiN (< 1000 Å) and Al (< 1 μm) encountered in advanced devices, deposition time is < 60 sec. The important points to recognize from Fig. 5.26 are the following: (1) The initial temperature rise of the wafer is linear with deposition time and independent of emissivity. Radiative losses are not significant at low temperature, and Eq. (5.2) shows that the slope of the temperature rise is $1/(\rho C_p z)$. (2) Wafer temperature saturates at a radiation-limited temperature that can be quite high under conditions of high deposition rate. (3) However, for realistic process times below ≈ 60 sec, neither process leads to excessively large wafer temperature increase ($\Delta T \approx$ 50–100°C). On the other hand, Fig. 5.24 does not take into account such things as radiation from a very hot shield or a collimator, or the possibility of a heated wafer chuck. Also, variations in wafer-to-wafer temperature can have a harmful effect on process repeatability. Finally, the incident power density during high-rate, sputter preclean etching (≈ 1 W/cm²) can be several times larger than during sputter deposition, resulting in a proportionately larger temperature increase.

While this discussion has focused on temperature rise during processing, wafer cooling is equally important. Figure 5.27 shows the cooldown of a Si wafer after deposition of 1 μm of Al that brought its temperature to 225°C [5.27]. Depending on your point of view, the cooldown is either unacceptably rapid so that starting wafer temperature at the next process module depends strongly on intramodule robotic handling time or unacceptably slow so that waiting for the wafer to stabilize or cool to near room temperature will compromise throughput. In view of all the above considerations, an improved method of controlling wafer temperature during PVD and related vacuum processing is required.

Temperature Control

The most widely used method of controlling wafer temperature during PVD is by backside gas conduction [5.29–5.31] in which a few Torr of gas (vs the few mTorr of process gas used) are introduced into a volume between the wafer and the platen (see Fig. 5.28). The gas provides a thermally conducting path between the wafer and the platen, which could be actively heated or cooled if desired. Typically a wafer clamp

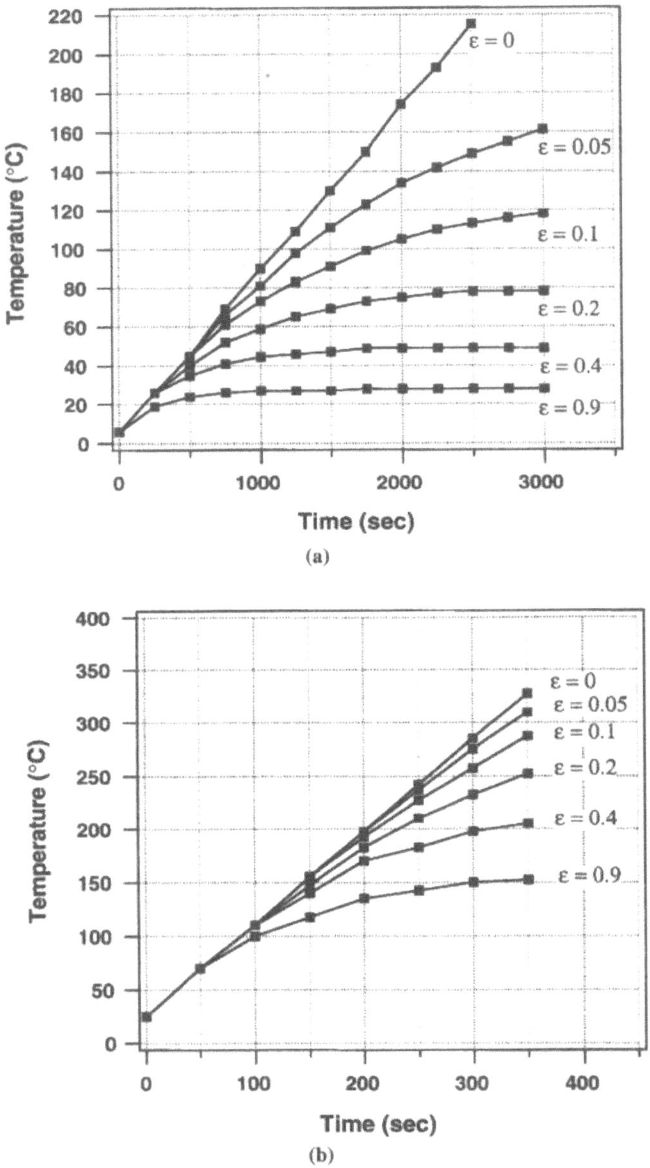

FIG. 5.26 Calculated time–temperature profiles from ref. 5.27 for a Si substrate under deposition rate and thermal loading conditions representative of (a) PVD Ti (1000 Å/min; 25 mW/cm^2), and (b) PVD Al (1 μm/min; 250 mW/cm^2).

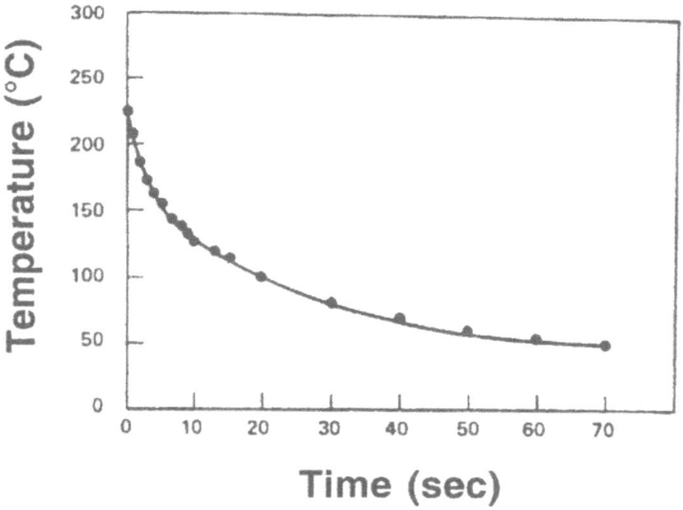

FIG. 5.27 Experimental cooldown of a Si wafer after deposition of 1 μm of Al (data from ref. 5.27).

ring assembly is used to provide sufficient holding force to keep the wafer in place against the backside gas pressure (4 Torr results in nearly 4 lb of force over a 200-mm wafer) with sufficiently low leakage of gas to prevent contamination or variations in process pressure. Because some leakage of gas into the process chamber is inevitable, Ar gas is typically used

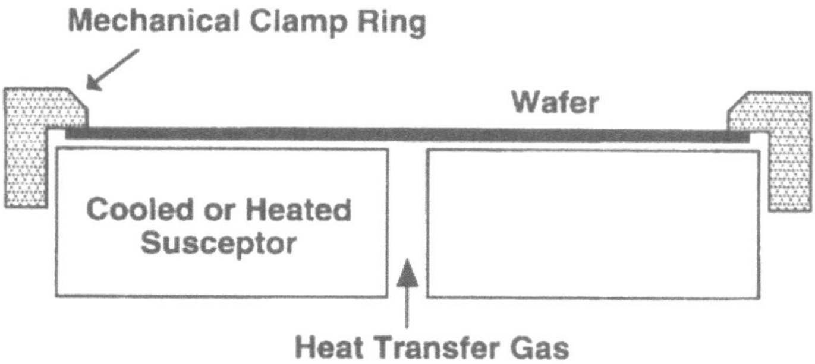

FIG. 5.28 Illustration of the use of backside-gas-assisted heat transfer to control wafer temperature during vacuum processing.

as the backside gas (e.g., as opposed to inert gases like He, which is costly and difficult to pump with a cryopump, or H_2, which is reactive with Ti). As a result, the acronyms BSA (backside argon) and BSG (backside gas) are often used interchangeably in referring to backside-gas-assisted heat transfer.

Well-designed wafer chucks with gas-assisted heat transfer can achieve heat transfer coefficients in the range of h = 10–30 mW/cm^2-K, which is ≈ 100 times greater than without BSA. The heat transfer coefficient, h, is a measure of the efficiency of the process, and using heat theory we can estimate how well temperature can be controlled. At relatively low wafer temperature, radiative losses can be ignored and the wafer will cool more or less exponentially ($\exp(-t/\tau)$ with a time constant $\tau = \rho C_p z/h$. Using z = 0.05 cm for a 200-mm wafer and h = 20 mW/cm^2-K gives $\tau \approx 5$ sec. So wafers can very quickly be brought to equilibrium. Figure 5.29 shows data for the rise time of a wafer with and without backside gas admitted between the heater and wafer. Since h is finite, in general there will be a

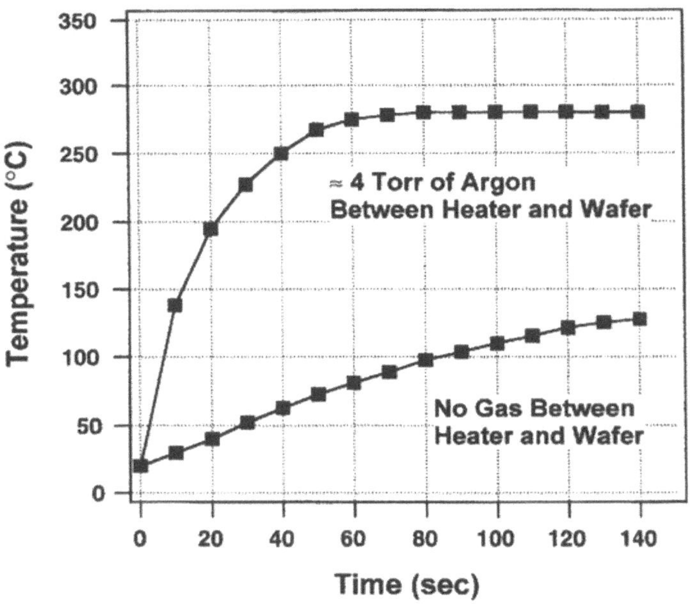

FIG. 5.29 Temperature rise time for a wafer with and without backside gas between heater block and wafer.

difference or offset between the equilibrium temperature of the wafer and the heated chuck, which is on the order of $\Delta T = \Phi/h$, where Φ is the energy flux from the heater. For a blackbody heater at 500°C (773 K) with emissivity of 0.5, $\Phi = \epsilon\sigma T^4 \approx 1$ W/cm^2.

Assuming all this is absorbed by the wafer, a value of $h = 20$ mW/cm^2-K leads to the wafer being 50°C cooler than the chuck. Therefore, care must be taken when using reported process temperatures, since what is often reported is the heater block set point temperature (e.g., measured with a thermocouple embedded in the block) and not the actual wafer surface temperature, which is extremely difficult to measure accurately without physical wafer contact in the temperature range of 300–500°C where many PVD processes are carried out. For example, conventional optical pyrometry is difficult to apply since the spectral emissivity of Si depends on emitted wavelength and is a complex function of wafer temperature, doping, roughness, and the presence of thin dielectric films. Also, the T^4-dependence of the radiated power quickly reduces optical signal strength as wafer temperature falls below 500°C.

Although noncontact methods of wafer temperature based on optical effects such as absorption, interference, scattering, and luminescence have been developed, they are not widely deployed in production and not always cost-effective to retrofit onto the installed base of PVD tools. Therefore, a tool supplier will often make controlled measurements using an instrumented test wafer whose surface is in intimate thermal contact with an array of thermocouples of thermal mass sufficiently small compared to the wafer to provide an accurate reading and rapid response time (e.g., the wafers provided by Sensarray Corporation, Santa Clara, CA). In this way, the supplier can provide data to a customer — albeit not obtained on the customer's actual wafers or module during processing — of the temperature- and time-dependent correlation between heater block temperature and wafer temperature for a PVD module, preclean module, degas/cooldown station, etc.

The physics of backside gas conduction heat transfer is discussed in refs. 5.30 and 5.31. Depending on the value of the dimensionless Knudson number Kn given by the ratio of the mean free path in the gas (λ) to the gap (z) between the wafer and wafer platen, two regimes need to be considered: the viscous flow or continuum regime where $\mathrm{Kn} = \lambda/z < 0.01$, and the molecular regime, where $\mathrm{Kn} \geq 1$. In the viscous regime, the heat conductivity is independent of gas pressure and varies inversely with z. However, the molecular regime is the one relevant to PVD processing. In this case, the efficiency of heat transfer is independent of the exact gap spacing, and the heat conductivity is proportional to the gas pressure.

Figure 5.30 presents calculated heat transfer parameters for a variety of gases. Even though lighter gases such as H_2 and He conduct heat better than Ar, they are not typically used for the practical reasons stated earlier. In any case, as long as the gap is small compared to the gas mean free path, the heat transfer coefficient h at pressure P is given by $h = \alpha\Lambda P$, where Λ is the free molecular conductivity and the accommodation coefficient α measures the extent to which molecules reflecting from a wall will adjust their kinetic energy to the temperature of the wall. For completely roughened surfaces, $\alpha \approx 1$; for highly polished surfaces, $\alpha \ll 1$. Taking $\alpha = 0.5$ and using the data for Ar in Fig. 5.30, we see that at 4 Torr, a value of $h = \alpha\Lambda P = 0.5 \times 9.3$ mW/cm^2-K-Torr $\times$ 4 Torr $\approx$ 20 mW/cm^2-K is obtained — provided that the gap spacing is less than the mean free path, which, from Fig. 5.30, is $\lambda = (52\ \mu$m-Torr)/(4 Torr) $= 13\ \mu$m. Maintaining a gap this small is challenging since the pressure behind the wafer tends to bow its center outward.

The theory of mechanics allows one to calculate the central deflection, y, of an edge-clamped Si wafer of radius r and thickness t subjected to a BSA gas pressure P:

$$y = \frac{3}{16}P\left(\frac{r^4}{t^3}\right)\left[\frac{1}{E}\right]\left[\frac{(m-1)(5m+1)}{(m^2)}\right] \tag{5.4}$$

where the material-dependent quantities in brackets are the elastic modulus, E, and m is the inverse of the Poisson ratio ν, i.e., $m = 1/\nu$. When $y = 0$, the wafer is assumed to be directly in contact with the planar wafer chuck. For Si, $E = 1.5 \times 10^7$ psi $\approx 7.8 \times 10^8$ Torr and $m = 3.3$ ($\nu = 0.3$).

Gas	Mean Free Path (μm-Torr)	Molecular Heat Conductivity Λ (mW/cm^2-K-Torr)	High-Pressure Thermal Conductivity k (mW/cm-K)
H_2	93	60.7	1.62
He	147	29.4	1.35
N_2	46	16.6	0.22
O_2	53	15.6	0.26
H_2O	30	26.5	0.14
CO_2	30	16.7	0.14
Ar	52	9.3	0.16

FIG. 5.30 Gas cooling coefficients for a variety of gases encountered in PVD processing, with Ar highlighted because of its use in backside-gas-assisted heat transfer. (Adapted from ref. 5.31 with kind permission from Elsevier Science-NL, Sara Burgerhartstraat 25, 1055 KV Amsterdam, The Netherlands.

For a 200-mm Si wafer ($r = 100$ mm; $t \approx 0.675$ mm), Eq. (5.4) then reduces to

$$y\,(\text{mm}) = 0.29P\,(\text{Torr}) \tag{5.5}$$

We see that an increase of 1 Torr of BSA widens the gap by 0.29 mm $\approx$ 10 mil. Hence, as one increases backside gas pressure to improve heat transfer efficiency, the added displacement due to wafer bowing can result in $z \gg \lambda$ at the center. This results in nonuniform heat transfer. In addition, even though the thermal conductivity of the backside gas is greater at the higher pressure, the added gap spacing translates into an increased thermal transfer path length and can actually increase the thermal resistance between the wafer and chuck. A widely used method to deal with this situation is to clamp the wafer over a domed platen whose shape is machined in advance to anticipate the bowed wafer shape taken during BSA operation.

Although the backside-gas-coupled chuck is widely used to control wafer temperature during PVD, other methods can be used. For example, radiant heating of wafers by a tungsten-halogen lamp array located outside of the process chamber has been reported by Clarke [5.32] as a way of improving wafer thermal response time and reducing contamination from chuck outgassing and mechanical clamps. In this case, radiation directly couples into the backside of the Si wafer through a quartz window and no backside gas is used during PVD — an approach similar to the method used in lamp-based rapid thermal process (RTP) systems for annealing and for rapid thermal CVD.

5.3.5 PVD Module

Figure 5.31 shows schematically the basic elements of a generic PVD module. Sputtering is provided by a planar DC magnetron with a large-diameter circular target (e.g., $\approx$ 12–14-inch diameter for an 8-inch wafer) and relatively slow mechanical rotation ($\approx$ 40 rpm) of a permanent magnet array behind the target is used to improve and/or to tailor the erosion profile of the target. This type of sputter source is not the only one used in PVD for microelectronics (e.g., an S-Gun™ sputter source with separately biasable dual cathodes is used by Sputtered Films); however, the planar magnetron is quite common, and the technology of this source was discussed in more detail in Chapter 4. Source cooling is provided by flowing water over the magnets within the source housing — the magnets being suitably potted in a water-resistant epoxy to prevent corrosion. The entire

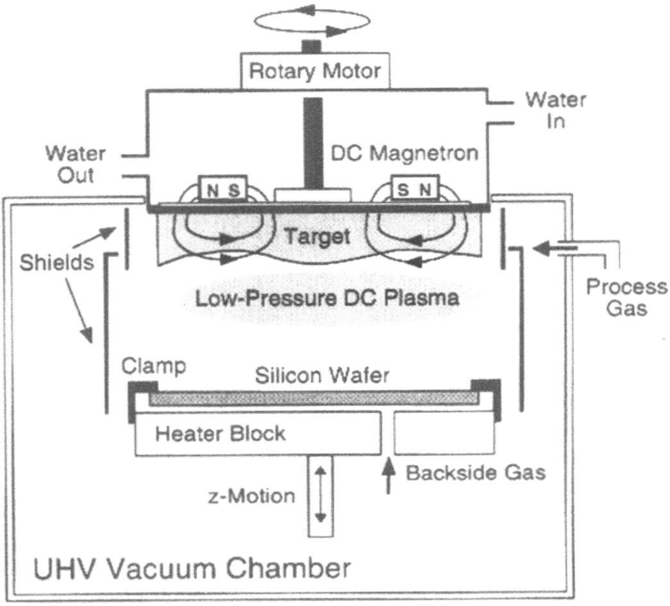

FIG. 5.31 Basic elements of a generic PVD module.

source housing containing the source–target assembly can be pivoted into the face-up position to allow for target changes or chamber maintenance. Due to the weight of such assemblies ($>$ 100 lb is not unusual), the mechanical leverage of a source manipulator is often used to assist the operator. Figure 5.32a illustrates how a PVD source on a vented module can be rotated from the face-down to the face-up position for target change or maintenance. A new and eroded target on two such adjacent modules are shown in Fig. 5.32b with an eroded target (right) and newly replaced one (left).

To confine the plasma discharge to the vicinity of the target face and to prevent unwanted sputtering of other parts of the target assembly, a metal ground shield is contoured around these surfaces at a distance less than that of the cathode dark space. Metals shields are also used to prevent sputtered material from directly coating the chamber walls, although the coated shields themselves must be periodically removed for cleaning. Process gases such as Ar and N_2 are introduced into the chamber by single-point

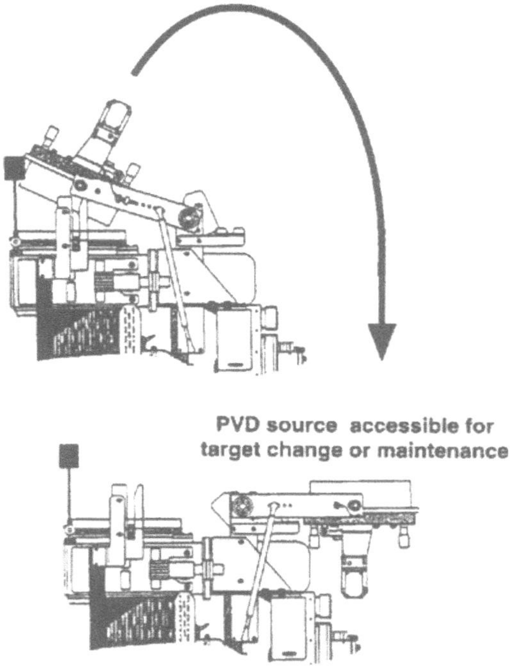

FIG. 5.32 **(a)**

injection from a simple tube or by multiple-point injection — e.g., from a perforated tube or "spider" manifold.

For reactive sputtering of films such as TiN, a uniform distribution of the reactive gas is desired over the wafer surface, and equipment suppliers carefully engineer the location of both gas injection and pumping ports (often guided by computer modeling of gas flow around the sputter shields and other chamber fixtures). A mechanical shutter is also sometimes provided between the source and wafer to allow a Ti target that was nitrided during PVD TiN deposition to be cleaned before a subsequent PVD Ti step. The wafer is located relatively close to the target in most systems ($\approx$ 5 cm) and is positioned on a substrate holder that usually is configured for active wafer heating via a heater block with backside-gas-assisted heat transfer.

Mechanical clamping of the wafer to the wafer holder is generally accomplished through mechanical pressure at the wafer edge using a continuous clamp ring or finger clamps. More recent designs employ electrostatic

(b)

FIG. 5.32 Example of a manipulator used to assist the operator in removal of PVD source assembly (Quantum™ PVD source shown courtesy of Varian Associates).

chucks (ESCs), which permit uniform holding over virtually the entire wafer area and avoid such things as edge exclusion and particles associated with mechanical clamp rings. However, it should be noted that many PVD steps can successfully be carried out clampless — only gravity holds the wafer down. Some chuck assemblies also allow for *z*-axis motion to vary source-to-substrate spacing up or down by a few cm. This *z*-axis motion can be used as a "tweak" to optimize film uniformity during target life or to use the same chamber to perform processes such as PVD Ti and PVD TiN that in general require different source-to-substrate spacing.

As we have seen, RF bias can be applied to the wafer to enhance and/or control ion bombardment during precleaning. Wafer bias is also used to improve the directionality of PVD, as in the case of RF-ionized PVD (see Chapter 8), and to improve film properties during film growth — e.g., to increase film density by removing loosely bound gas atoms or to improve step coverage by resputtering metal atoms and increasing their surface mobility. As a result, some wafer chucks for PVD also provide an RF bias capability. We note that designing a high-temperature electrostatic wafer chuck for PVD (> 500°C wafer temperature) that is also capable of RF bias and z-motion is a nontrivial engineering exercise.

Argon Gas for PVD

Argon gas makes up less than 1% of the composition of room air in a PVD module before pump-down, but can make up 20–100% of the process gas (e.g., Ar/N_2 or Ar) during PVD. Being the primary sputtering medium, Ar is deserving of some discussion. There are several reasons why Ar is used so widely in PVD. Physical sputtering involves a direct momentum transfer, which depends on the relative masses of incident ion (m) and target atom (M) through the reduced mass $\mu = 2(mM)/(m + M)$. $^{40}Ar^+$ is relatively heavy ($m = 39.948$ amu) so that its sputter yield will in general be much larger than that of a typical residual gas that is ionized in the same DC magnetron plasma ($^{16}O^+$, $^{32}O_2^+$, $^{14}N^+$, $^{28}N_2^+$, $^{1}H^+$, and $^{2}H_2^+$). Mass 40 is also a reasonable compromise given the mass of common sputtered metal atoms such as ^{27}Al and ^{48}Ti.

In addition, Ar is inert and will not react chemically with the wafer, the material being sputtered, or the residual gas atoms. It is true that Ar can be incorporated into the growing PVD film and thereby affect film density and growth morphology. However, at typical incorporation levels (< 0.5% due to very low sticking coefficient of Ar) these effects are generally not severe, and the levels are reduced further at elevated wafer temperature. In any event, the situation would be far more complicated if the Ar were chemically reactive. Although other inert gases can be used (e.g., ^{4}He, ^{84}Kr, ^{131}Xe), Ar is readily available even in ultrahigh purity grades such as "6 nines" grade = 6N = 99.9999% and is relatively inexpensive (e.g., Kr and Xe are ≈ 20 and 75 times more expensive than Ar, respectively). Also, He is difficult to pump with cryopumps and turbopumps and could interfere with vacuum leak checking with He leak detectors.

The important vacuum concept with regard to Ar gas purity is that of relative arrival rate. Namely, to ensure high-purity sputtered films, the arrival rate of sputtered atoms should be much greater than the arrival rate of residual gas atoms/molecules. From the kinetic theory of gases, the surface arrival rate R of gas-phase molecules at temperature T (K) and pressure P

is related to the average molecular velocity $<v>$ and the gas density n by the expression

$$R\ (\text{molecules/cm}^2\text{-sec}) = n\,\frac{<v>}{4} = 3.51 \times 10^{22}\,\frac{P}{(MT)^{1/2}} \tag{5.6}$$

where P is in Torr and M is the mass of the molecule in amu. For the case of Ar at room temperature, this reduces to $R = 3.2 \times 10^{20}\ P$. Therefore at typical PVD pressure of 3 mTorr, the Ar arrival rate is $\approx 10^{18}$ atoms/cm^2-sec. Given the extremely low sticking coefficient of Ar for typical PVD films and process temperatures (< 0.001), the effective Ar incorporation rate at the film surface can then be below 10^{15} atoms/cm^2-sec, which is about 1 monolayer per sec ($\approx$ 1 Å/sec). Since film deposition rates for Al are $\approx$ 1 μm/min (160 Å/sec), Ar incorporation of 0.5% is understandable.

While Ar incorporation is generally not a problem, small levels of contamination in the Ar can have a significant impact on film quality. Assuming a sticking coefficient near unity (not a bad assumption for residual species such as O and metals such as Ti and Al), $\approx$ 1–2 Å/sec of contamination will "deposit" at a partial pressure of 10^{-6} Torr — which represents only about 0.03% of the total 3-mTorr Ar pressure. Viewed in another way, incorporation of contamination at a rate of $\approx$ 2 Å/sec during Al deposition at 167 Å/sec (1 μm/min) leads to a 1% level of contamination in the film. Figure 5.33 shows the effect of residual gas on film resistivity and reflectivity of a sputtered 1.2-μm Al-1%Si film [5.33, 5.34] and indicates the need to keep reactive gas partial pressures very low ($< 10^{-7}$ Torr) for critical applications. The use of ultrahigh-purity Ar gas (6N grade or better is recommended) with short Ar delivery lines to reduce outgasing from interior line surfaces is also necessary. In addition, point-of-use purifiers are available for low-flow-rate applications such as PVD that can cost-effectively purify 5N grade Ar to 7N grade; remove H_2O, O_2, CO, CO_2, H_2, and N_2 to below 10 ppb; and thereby cost-effectively purify 5N grade Ar up to 7N at the point of entry to the PVD chamber without the tenfold cost premium usually associated with a 7N vs 5N purity cylinder of Ar.

Even if the Ar gas is 100% pure, outgassing from the walls of the chamber and/or from the wafer itself can introduce contamination into the sputtering ambient. For a given pumping speed on a vacuum module, the ultimate base pressure is determined by the outgassing rate of the chamber (see Eq. (5.1)), and this in turn is strongly influenced by such things as choice of materials and seals, surface finish, and bakeout procedures. Figure 5.34 plots process pressure versus level of contamination introduced from outgassing. Assuming that base pressure is 10^{-9} Torr, we see

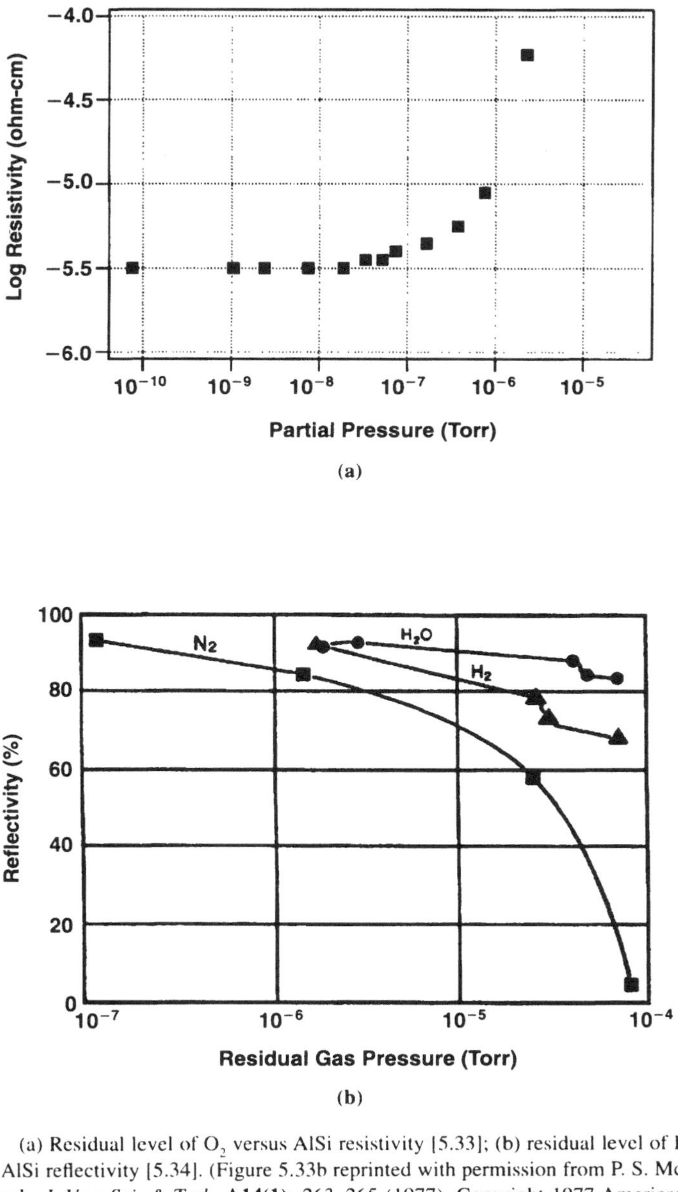

FIG. 5.33 (a) Residual level of O_2 versus AlSi resistivity [5.33]; (b) residual level of H_2, H_2O, and N_2 versus AlSi reflectivity [5.34]. (Figure 5.33b reprinted with permission from P. S. McLeod and L. D. Hartsough, *J. Vac. Sci. & Tech.* **A14(1):** 263–265 (1977). Copyright 1977 American Institute of Physics.)

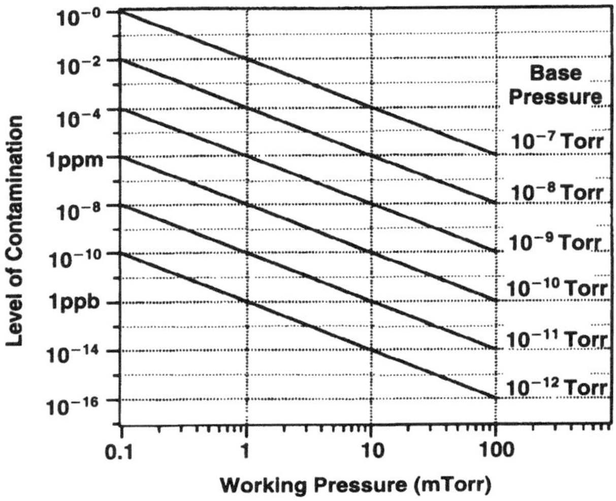

FIG. 5.34 Calculated level of contamination introduced due to outgassing versus process pressure, for several base pressures. Contamination-free PVD manufacturing favors UHV base pressure.

that contamination levels < 1 ppm can be obtained with process pressure > 2 mTorr — assuming the Ar process gas is 100% pure. Since the trend in conventional PVD is toward lower process pressure (< 0.5 mTorr) to reduce gas-phase scattering, this will increase the need for UHV-type vacuum practices. One exception to this is ionized PVD, which is carried out at much higher than conventional (15–20 mTorr) process pressure (see Chapter 8). In this case, it is necessary to throttle the chamber pump, which degrades the base pressure.

Wafer Holding

There are basically only two ways of holding a wafer during PVD: mechanical and electrical. As mentioned earlier, some PVD processes can be done clampless; however, this rules out the possibility of backside gas, which requires several Torr for effective heat transfer, and would blow the wafer off the platen (the weight of a 200-mm wafer corresponds to a pressure of ≈ 0.08 Torr). Clampless processing also makes knowledge and control of deposition temperature more difficult. Holding at the wafer edge with a spring-loaded mechanical clamp ring or using only the weight of the clamp (a gravity clamp) are commonly practiced, although provision for the physical flat or notch at the wafer edge must be considered. Since the PVD film also deposits on the clamp ring, the ring can be a source of particles as the film builds up and flakes off. In addition, the ring could "stick" to the wafer during some elevated-temperature processes (e.g., a Si ring could stick to a CVD W film by forming WSi_2), requiring careful design and process-specific choice of materials. Finally, the edge exclusion of the clamp ring (typically $\approx$ 6 mm) prevents coating of valuable "Si real estate." For a 200-mm wafer this means that $\approx$ 12% of the wafer area is covered by the clamp and cannot be occupied by a die. For these reasons, the trend is away from frontside, mechanical holding and toward backside, electrical holding of wafers using electrostatic chucks (ESCs), sometimes called e-chucks.

The physics and technology of ESCs are dealt with in a number of excellent reviews (e.g., see refs. 5.35 and 5.36). ESCs operate via the principle of coulombic charge attraction, in which charges on the chuck electrode attract real or image charges in the Si wafer being clamped — which is analogous to the attractive force between the plates of a parallel-plate capacitor, only in this case one of the plates is the wafer.

A simple e-chuck is illustrated in Figure 5.35 and compared with a mechanical edge clamp. For simplicity, the lift pins that would raise and lower the wafer off of the chuck and onto the end effector of a robotic arm are not shown. An upper insulator of thickness h_{diel} and relative dielectric constant k coats the metal electrode of the chuck and is separated from the wafer by an effective vacuum gap h_{gap}. A vacuum gap is inevitable over some fraction of the wafer surface when using a nondeformable dielectric (e.g., a material such as alumina and not an elastomer) due to the micro-roughness of the wafer backside and possible wafer bow or warp. Also, gaps are intentionally created at the dielectric surface by the grooves or channels that are used as a distributed path for backside gas.

The expression for the electrostatic pressure P_{ESC} holding the wafer down (force per unit area of clamping) can then be shown to be

$$P_{ESC} = \frac{\frac{1}{2}\epsilon_0 V^2}{\left(\frac{h_{diel}}{k} + h_{gap}\right)^2} \tag{5.7}$$

where the dielectric constant of free space $\epsilon_0 = 8.85 \times 10^{-10}$ farad/meter. Assuming perfect contact ($h_{gap} = 0$), Eq. (5.7) reduces to

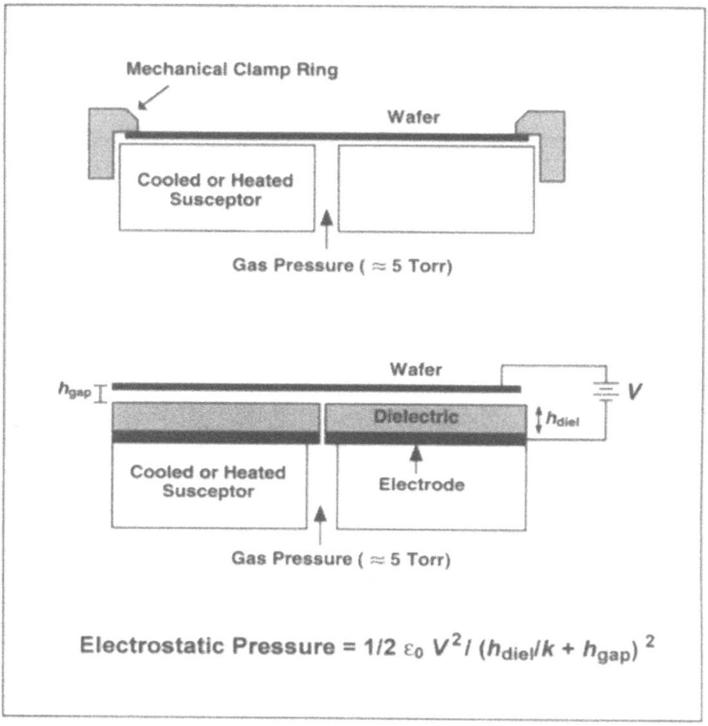

FIG. 5.35 Illustration of mechanical chuck and electrostatic chuck (ESC) for wafer holding during PVD.

$$P_{ESC} = \frac{1}{2}\,\epsilon_0 k^2 \left(\frac{V}{h_{diel}}\right)^2 = \frac{1}{2}\,\epsilon_0 (kE_{diel})^2 \tag{5.8}$$

where E_{diel} is the electric field across the dielectric in volts/meter and P_{ESC} is the pressure in pascals (1 Pa = 7.5 mTorr). Therefore, the maximum possible holding pressure is determined by the electric field strength of the dielectric at breakdown.

A popular ceramic insulator for e-chucks is alumina (Al_2O_3 with $k \approx 10$), whose breakdown strength depends on the method of preparation (e.g., bulk, plasma sprayed, anodized) but is typically in the range of 10–15 V/μm. Using these values in Eq. (5.8) gives a maximum pressure of ≈ 350–500 Torr. To provide a margin of safety, one would operate at a voltage several times lower than the breakdown point of the dielectric, and the ESC pressure would be much lower than the calculated maximum (P_{ESC} goes as V^2). In addition, as noted earlier, perfect contact with $h_{gap} = 0$ is never possible over the entire wafer area given a wafer's backside roughness and the presence of gas conduction grooves machined into the chuck surface. Therefore, a realistic holding pressure of the chuck might be 10 times lower than given by Eq. (5.8), but in any case is more than sufficient to hold a wafer in place against backside gas pressures of $\approx$ 5 Torr. Also, since thin films of alumina and related ceramics can be deposited with good quality (e.g., low leakage current at elevated temperature), relatively low applied voltages can be used (e.g., 500 V for a 200-μm thick insulator) to provide suitable holding force for PVD applications.

An important potential benefit of an ESC is that backside gas can be used without frontside holding. However, to prevent wafer bowing the wafer should be pulled down uniformly over its entire area (e.g., not just electrostatically clamped at the edge) and the distribution of gas should be such that it provides uniform heat transfer. Figure 5.36 shows several designs of open gas channels, or grooves, that have been created at the dielectric surface for this purpose. In general, a combination of concentric

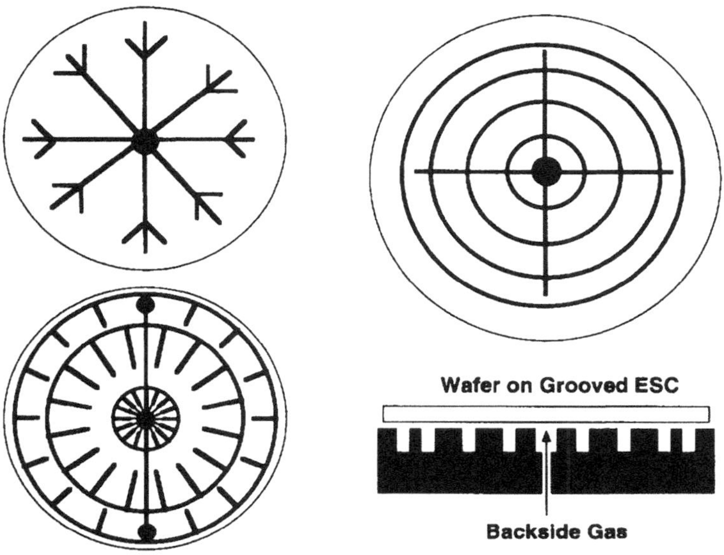

FIG. 5.36 Illustration of grooving in the surface of an ESC to obtain good surface contact and distribution of backside gas over the wafer area. Gas enters at the large dots and is distributed through a network of radial and linear troughs.

rings and radial lines are used. Making the grooves too numerous and/or too deep can reduce the attractive force by increasing h_{gap} in Eq. (5.7). Also, heat transfer efficiency within the groove is reduced if the gap dimension exceeds the mean free path of gas (see Section 5.3.4). Finally, placing grooves too close to the wafer edge to improve temperature uniformity can lead to gas leakage into the process volume. Grooves also reduce the area of physical contact between chuck and wafer, which reduces the potential for tribologically generated particles.

Well-designed ESCs are capable of holding 5 Torr or more of BSA with Ar leakage < 0.1 or 0.2 sccm, which is much less than the $\approx$ 50 sccm of Ar typically used during PVD. On the other hand, leakage of even ppm levels of contamination into the chamber is to be avoided, and this suggests that ultrahigh-purity Ar gas should be used for the chuck. Also, the leak rate around the chuck may rule out the use of gases such as He for backside gas heat transfer since a sustained leak rate of even 0.1 sccm of He may be sufficient to dump the cyropump on the process chamber.

The simple chuck in Fig. 5.35 is a monopolar design; more common for PVD is a bipolar design (Fig. 5.37) in which different portions of the wafer are clamped by oppositely charged electrodes. While the bipolar design is more complicated, one advantage is that for equal electrode areas with equal and opposite voltage drops to the wafer, no net charge need flow to the wafer. The wafer can then be held at any time in the process step or in moving parts such as a robotic handler.

In addition to the usual considerations of using an ESC in vacuum processing (cost, reliability, and particles), there are three materials-related issues that are especially relevant to PVD: (1) high vacuum compatibility, (2) high-temperature operation, and (3) dechucking.

High Vacuum and High Temperature Electrostatic chucks were first introduced into microelectronics in the early 1970s for holding wafers flat during photolithography [5.38] and have become common on advanced plasma etching systems for wafer cooling. Their application to PVD has been more difficult because the materials of construction must be compatible with both UHV base pressure (e.g., low outgassing rate) and high temperature ($\approx$ 500°C). This rules out both polymeric coatings as well as a number of high-k ceramics that are compatible with high vacuum but are too leaky at high temperature. The most common ESC dielectric candidate for PVD is alumina (Al_2O_3) — either single crystal (i.e., sapphire) or as a plasma-sprayed or anodized film — with other ceramics of interest including aluminosilicates, AlN, BN, and diamond.

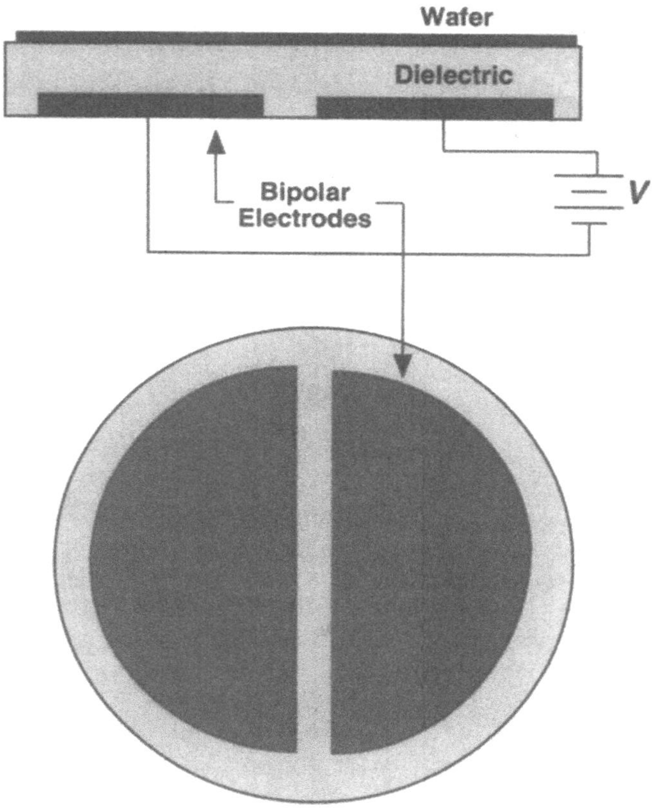

FIG. 5.37 Bipolar ESC design.

De-chucking De-chucking refers to the controlled removal of the attractive force. When the applied voltage is turned off, a wafer can still stick to the chuck due to residual forces — e.g., from permanent bulk polarization of the dielectric (many ceramics of interest for an ESC are highly polarizable) or from charge trapped at the dielectric surface or back of a wafer with an insulating oxide or nitride film. If the time required for this charge to leak off is large ($\gg$ 1 sec), wafer throughput will be reduced. In this regard, one approach that has been used successfully for ion implant applications is the six-electrode "hexapolar" design shown in Fig. 5.38. Application of a square wave of opposite polarity to each of the three pairs of dielectric sectors creates three bipolar chucks. By choosing the voltage

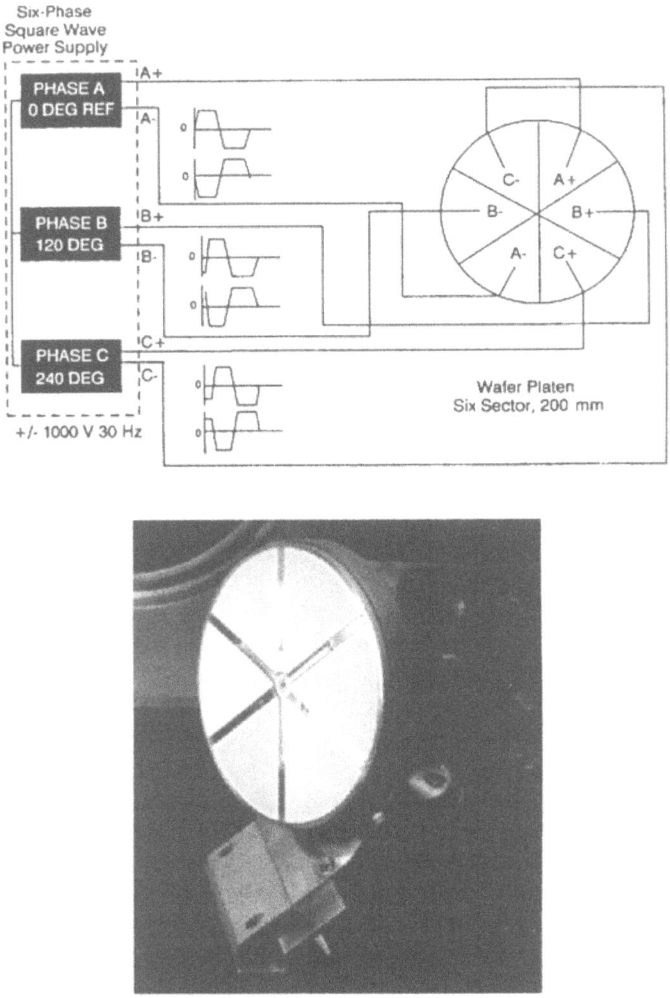

FIG. 5.38 Six-electrode ESC that combines the high holding force of a DC e-chuck with the rapid release time of an AC e-chuck [5.37].

to each bipolar chuck to be 120° out of phase, when the voltage goes through zero for any one bipolar chuck, the other two are at full holding force. This allows a large holding force to be produced as with a DC chuck, but because the applied voltage is AC, no significant DC polarization occurs. This design has permitted dechucking in < 80 msec.

While having a wafer stuck to the chuck after deposition is clearly undesirable, a related concern is the possibility of not having a wafer present on the e-chuck during deposition. That is, if the PVD tool fails to sense that a wafer is not on the e-chuck and begins metal deposition, the surface insulation of the chuck will be shorted out and the chuck will have to be removed and cleaned or replaced — a costly proposition. For example, in the chuck shown in Fig. 5.38, the transient current flowing in the wafer when the chuck is first energized is used to sense whether the wafer is on or off the chuck, since the magnitude of this current depends strongly on the capacitance between wafer and chuck.

5.3.6 Particles and Other Foreign Matter

Given the critical influence of device yield on functional cost (e.g., the cost per bit of memory on a DRAM), improvements in overall defect density will be needed to allow cost-effective production of ever-smaller devices on ever-larger die (see Fig. 5.39). Because the minimum feature size of the typical IC has decreased so rapidly over time, particles of a size considered harmless in past device generations are now potential "killer" defects. For example, 0.25-μm device roadmaps call for contamination-free manufacturing at the level of below ≈ 0.016 defects/cm^2 = 160 defects/m^2 for both DRAMs and microprocessors [5.39], where a defect is any structural flaw, contamination, particle, etc. that causes a chip to fail electrically. This

Year of First DRAM Shipment	1997	1999	2001	2003	2006
Minimum Feature Size (μm)	0.25	0.18	0.15	0.13	0.10
DRAM Device Generation (bits)	256M	1G	1G	4G	16G
DRAM Cell Size (μm^2)	1.50	0.60	0.24	0.096	0.038
DRAM Chip Size (cm^2)	2.8	4.0	4.5	5.6	7.9
Microprocessor Transistor Count (10^6)	12	28	64	150	350
Microprocessor Chip Size (cm^2)	3.0	3.4	3.9	4.3	5.2
Particulate Size (minimum size nm)	125	90	75	65	50
Particles per Thin Film Process Step (m^{-2})	3.5	2	1.8	1.5	1
Electrical Defect Density (m^{-2})	160	140	13	120	100

FIG. 5.39 Allowable particulate contamination for each generation of DRAM and microprocessor ICs (after ref. 5.39).

level corresponds to < 5 such defects over a 200-mm wafer. Defects associated with fine submicron size particles ($\leq 0.12\ \mu m$) are a critical yield-reducing agent in ULSI devices, and increasingly these are introduced within the process tool itself [5.40].

Particles added onto the wafer surface as a result of either robotic transport (mechanical adders) or processing (process adders) within a PVD cluster tool can have a significant effect on device yield, which in turn leads to a higher cost-of-ownership. A representative PVD particle specification for 0.25-μm device production is < 125 particles/m^2 of size $\geq 0.08\ \mu m$. For a 200-mm wafer without edge exclusion (314 cm^2), this allows only about 4 such particles to be added per wafer. While the particle density allowed for next-generation 0.18-μm technology is expected to be similar to that at 0.25 μm, the size limit on the added particles will be reduced from 0.08 μm to 0.06 μm. In this regard, there is concern whether metrology will be available to map and quantify such low particle levels on metal films, which is a challenge for conventional laser light scattering methods. We next briefly discuss common sources of particles and related foreign matter (FM) encountered in PVD cluster tool processing as well as practices that tool suppliers and users have used to reduce their levels.

Clean Room

When an open wafer cassette is moved between process tools within an ultrahigh-quality clean room, the level of exposure to particulate contamination is very low (e.g., a Class 1 clean room has less than about 35 particles/ft^3 of size 0.1 μm and greater). On the other hand, a non-negligible concentration of volatile hydrocarbons can exist in clean-room air, which can lead to surface adsorption during wafer transport or storage. The resulting hydrocarbon film may be quite thin — only a monolayer or so — but can give rise to reliability issues during subsequent PVD processing. For example, if the wafer undergoes a rapid heating step, the hydrocarbon film may not have time to thermally desorb, but instead may crack or react with exposed Si to form SiC. To prevent such a situation, suitable precleaning of the wafer surface is necessary before high-temperature processing. Alternatively, one can effectively reduce airborne contamination by transporting wafers in a closed, loadlocked "pod" having a controlled environment such as dry, inert gas or even a moderate vacuum. The front end of advanced PVD tools are configured with a mechanical interface whereby an operator or robotic vehicle can attach the pod, and the wafers can be moved out of the pod and into the PVD tool without ever being exposed to the clean-room ambient.

Loadlocks

Loadlocks must pump from atmospheric pressure (p_0 = 760 Torr) to moderate vacuum ($p \approx$ 1 mTorr) as rapidly as possible to maintain high wafer throughput. Rapid pump down can cool the gas by adiabatic expansion, with a change in temperature given by the expression [5.41]

$$\frac{T}{T_0} = \left(\frac{p}{p_0}\right)^{(\gamma-1)/\gamma} \tag{5.9}$$

where γ is the ratio of the specific heat of the gas at constant pressure to that at constant volume. Figure 5.40 shows the measured gas temperature vs time for a 40-sec pump down from 650 Torr to $\approx$ 30 mTorr. As seen, the reduction in temperature after a few seconds of pumping is sufficient to condense out water droplets that can then serve as nuclei for subsequent particle formation. The problem can be avoided by using a slow, or "soft," pump procedure and/or purging the loadlock with clean, dry N_2 to remove the water vapor before pump down. Soft pumping and venting also tends to reduce turbulence that can release particles trapped within mechanical fixtures.

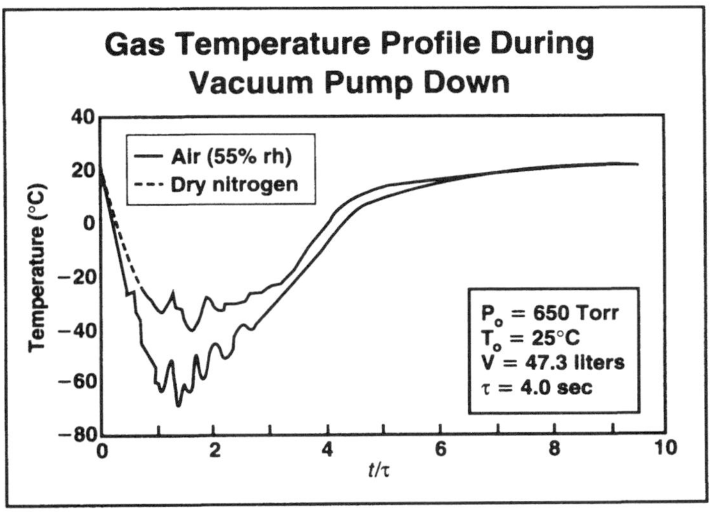

FIG. 5.40 Measured gas temperature versus reduced time for a t = 40 sec pumpdown of a 47 liter volume from 560 Torr to $\approx$ 30 mTorr (*Source:* B. Y.-H. Liu, *Semiconductor Int'l*, p. 75, March 1994).

Clamp Ring

Mechanical clamp rings get coated along with the wafer during PVD and are also subject to periodic heating (e.g., by the substrate heater table) and cooling when lifted to release the wafer. These rapid thermal cycles can cause particle shedding in poorly adhering materials. It is possible to dampen such fluctuations by using a higher-mass clamp (e.g., stainless steel) in which temperature swings will be minor. The surface finish of the wafer edge, if sufficiently rough, can abrade during clamping and generate Si dust. To the extent that this can be controlled by the user, wafers with smooth edge finish are preferred. These specific issues can be addressed by using an electrostatic chuck; however, ESCs can also generate particles by the microscopic-scale physical contact between the rough wafer backside and the ceramic dielectric. Also, the presence of backside particles can impede this physical contact and reduce heat transfer efficiency.

Shields

Shields are generally regarded as passive elements that protect the chamber walls from deposition; however, they can be an active source of particles. A shield in general will consist of several separate pieces (a shield set) held together with threaded fasteners or simply dropped into place and held in position by gravity. By design, threaded fasteners require surfaces that rub against each other and can therefore generate particles consisting of the fastener and/or the lubricant (e.g., TiS_2 or MoS_2). This is an ongoing process, as shields are periodically disassembled for cleaning or to access the chamber for maintenance. Therefore, shield sets with a minimum number of parts are desired. In general, all screws generate particulates no matter what dry lubrication is used, and threaded hardware of any kind should be minimized in PVD systems. Since it is not uncommon for the majority of the material sputtered from the target to wind up depositing onto the shields, these deposits can quickly build up to the point where particle generation and flaking are of concern — a particular problem with TiN deposits, which typically have high film stress and poor adhesion. Flaking is also enhanced by thermal stress of the shields, which are heated and cooled as the sputtering plasma goes on and off during PVD. Therefore, some designs incorporate internal lamp heating of the shields to maintain constant shield temperature during the off cycles of the sputtering plasma, as well as for more effective shield degassing during chamber bakeout. Proprietary coatings and surface texturing have also been used on shields to improve the adhesion of sputtered material and increase the time between shield cleaning to as much as 500 μm for difficult films such as TiN.

Tooling is generally reused, and this requires removal of the deposited material and cleaning of the parts back to their initial clean-room compatible condition. Many semiconductor fabs contract outside vendors for this task that could be done by sandblasting or wet chemical means such as dipping in an acid bath, although the latter creates environmental concerns with hazardous waste disposal. In a typical case, a stainless steel shield set might be sandblasted, ultrasonically cleaned, and then repackaged under clean-room conditions. The cost of the cleaning procedure plus the lost process time needed to change the tooling and recondition the process chamber for deposition is significant and might add as much as 5–10% to the overall cost of ownership of the PVD tool. In general, it is desirable that a shield set and/or collimator last about the same number of process hours as the target on the magnetron so that both can be changed at the same time. This time varies from material to material, but might typically be 3000 to 6000 wafers worth of deposition.

Gas Delivery System

It has become a standard practice for process gas lines to use electropolished tubing with orbital butt-welded joints, which prevents rough surfaces and internal crevices that might trap contaminants. Also, right-angle butt-welded elbows are used to prevent sharp bends in the gas line that can concentrate stress and thereby generate particles. The particle levels in the process gas as supplied are generally several orders of magnitude cleaner than the best clean rooms, or the gas can be filtered at the point of use to remove all but the finest particles ($<$ 100 Å). As a result of practices such as these on passive components, the majority of particles produced in the gas delivery system are from active components such as valves, mass flow controllers, and pressure regulators.

In general, it is desired to close-couple the gas flow system (e.g., flow controller, shut-off valve, particle filter) as close as possible to the sputtering chamber. This reduces the amount of tubing held at high vacuum as well as shortens the response time. Needless to say, each modular process chamber typically has its own dedicated gas flow system; however, these systems may share a common gas bottle or tank. Each process chamber is also outfitted with a capacitance manometer (usually 0.1 or 1.0 Torr full range) as well as an ion gauge. The gas control system on the process chamber can then be feedback-controlled from the capacitance manometer. Depending on the tool manufacturer, the gas operating system will specify either an absolute pressure (e.g., 1.0 mTorr) or a fixed flow (e.g., 10 sccm) and then use feedback control to maintain that value during processing. In addition, if a gas mixture is used (e.g., Ar + N_2), control can be based on maintaining either a fixed total pressure and relative gas concentrations (e.g., 1 mTorr, 80% Ar, 20% N_2) or fixed flows of each species (e.g., 8 sccm of Ar, 2 sccm of N_2). Total pressure control is somewhat complicated by system-related changes in the net pressure, which may be caused by the initiation of a plasma, the breakdown of a gas in that plasma, heating and degassing of chamber and/or fixtures, or pump loading.

PVD Target and Source

There are number of particle generation mechanisms associated with target quality and related PVD source performance. For example, microbursts of gases that were trapped within the microvoids of a low-density target can be released as the target erodes, and the electrical arcing of the source caused by these high-pressure gas bursts can lead to particle generation. This is being addressed by target suppliers through improved manufacturing and by equipment suppliers through PVD sources with electronic arc suppression. These and target/source-related issues of particle generation are discussed in more detail in Chapter 11. Particle contamination during DC magnetron sputtering has been relatively unexplored compared to work on particle formation during reactive plasma etching or plasma-assisted CVD. Recent work using laser light scattering [5.42] suggests that the mechanisms of particle generation, transport, and trapping in PVD are different from those of plasma etching and CVD and that this is probably caused by the inherent spatial nonuniformity of the magnetically enhanced plasma of a DC magnetron.

5.3.7 Robotic Handling

Robotic wafer motion in a PVD cluster tool is a special challenge associated with simultaneous requirements of pressure, temperature, and contamination [5.43]. In particular, wafer handling within the tool must often be carried out under high or even ultrahigh vacuum ($\approx 10^{-8}$ to 10^{-9} Torr). Unlike motion under atmospheric pressure, this means that simple vacuum suction cannot be used to securely hold the wafer in place during rapid changes in position. Instead, mechanical or electrostatic clamping is required. Another concern is the robotic arm itself, which can be a source of unwanted particles and contamination. Unlubricated moving surfaces in contact (e.g., the bearing surfaces in the arm) can generate fine particles that reduce device yield. Unfortunately, wet oil-based lubricants tend to outgas and create molecular contamination, while dry lubricants can create as many or more particles as the bare contacting surfaces themselves. Also, an arm is often required to hand off or pick up a wafer in a process chamber at elevated temperature (> 400°C), and the radiant heating can cause grease-related outgassing.

With regard to wafer holding, it is useful to consider how rapidly a wafer clamped to a horizontal wafer platen can be accelerated or decelerated before it begins to slip since this has a strong influence on how rapidly the wafer can be moved from one stationary position in the PVD tool to another. Elementary mechanics shows this acceleration is $a = \mu F_c/M$, where μ is the coefficient of static friction between the wafer and platen (e.g., $\mu \approx 0.3$ between Si and Al_2O_3), M is the wafer mass ($\approx$ 50 gm for a 200-mm Si wafer), and F_c is the total vertical clamping force: clamping pressure × clamping area. For a wafer mechanically clamped at its edge, the maximum tangential acceleration is typically ≈ 2–3 g (1 g = the acceleration of gravity at the earth's surface = 980 cm/sec^2). The full-face holding of an electrostatic chuck (ESC) leads to a much greater clamping area nearly as large as the wafer itself, leading to perhaps a tenfold increase in maximum tangential acceleration. Therefore, while ESCs are generally thought of as a clampless way of holding wafers stationary during processing, they are compatible with rapid handling of wafers between process steps.

The typical application for a vacuum handler in a radial cluster tool is to transfer wafers between different process modules that are themselves vacuum-isolated from the transfer module by a slit valve. In view of the angular rotation and linear translation needed to effect this transfer, the simplest handler requires three rotating points: shoulder, elbow, and wrist. A representative handler of the "frog-leg" design is shown in Fig. 5.41, where a dual robotic arm has been incorporated to handle two wafers simultaneously for improved throughput. Wafer transfer from one module to another involves at least five separate motions (e.g., linear motion into and out of module 1; rotation to module 2 position; motion into and out of module 2). Since advanced PVD process sequences require multiple modules (3–5 or more), reliability of robotic arms is of great concern. It has been estimated that a mean-time-between-failure (MTBF) of > 10^6 cycles is required to avoid impacting overall PVD cluster tool performance.

Rotary motion is a particular challenge for a vacuum robot since this requires coupling of the arm, which is under high vacuum, to the motor, which is out of the chamber at ambient pressure (760 Torr). Vacuum-tight

FIG. 5.41 Representative vacuum robotic handler of the "frog-leg" design (courtesy of Brooks Automation, Lowell, MA).

sealing of the rotating-shaft connecting arm and motor is often accomplished with a Ferrofluidic™ seal, in which a concentrated magnetic field is used to retain a ferrofluid (ferrite particles suspended in a low vapor pressure fluid) in an annular gap between the shaft and the magnetic components surrounding it (see Fig. 5.42). Direct-coupled rotary feedthroughs with Ferrofluidic seals allow rotary operation in vacuum at high speed and high torque. Another approach to rotary motion is to indirectly link the motor and arm by means of magnetic coupling. For example, a permanent internal magnet fixed to the shaft can be used to track the motion of a rotating external magnet in air (see Fig. 5.43). The simplicity of the linkage is offset to some extent by limited torque transmission, backlash, and the difficulty of coupling at high rotational speed.

A variety of robot designs have been implemented for handling wafers in the high-vacuum ambient of a PVD cluster tool (10^{-8}–10^{-9} Torr); however, they all share a common concern with wearing surfaces (e.g.,

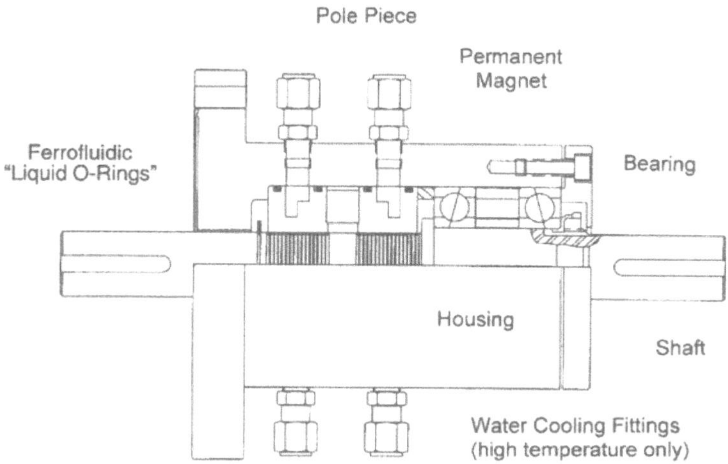

FIG. 5.42 Schematic of a Ferrofluidic™ seal used to make a vacuum seal to a rotating shaft (courtesy of Ferrofluidics Corp., Nashua, NH)

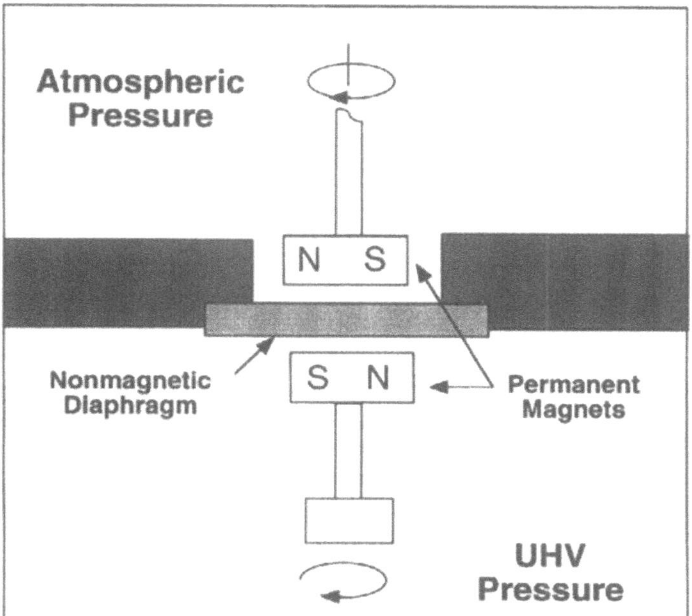

FIG. 5.43 Schematic of a magnetic approach used to couple rotary motion into a vacuum ambient (after Fig. 6 in ref. 5.43).

stainless steel or ceramic ball bearings) that shed particles. As noted earlier, contamination and outgassing generated by surface lubricants complicates the issue. Familiar dry film lubricants with a platelet-type microstructure (e.g., sulfides such as MoS_2 and WS_2) have ultralow vapor pressure even at moderate temperatures (typically $< 5 \times 10^{-12}$ Torr at 20°C and $< 5 \times 10^{-9}$ Torr at 100°C) but shed particles at levels comparable to plain dry bearings. While high-vacuum lubricants designed for the stringent particle and contamination requirements of cluster tool processing are a relatively new development, there is growing interest in Teflon-type dry lubrication. One such formulation of note is poly-tetra fluoroethylene (PTFE), which has been incorporated into ball bearing assemblies with a reduction of several orders of magnitude in particle generation rate.

5.4 300-mm PVD

> The terms "8 inch" and "200 mm" are often used to describe the same wafer diameter as if there were 25 mm in an inch instead of 25.4 mm. Actually, all wafer diameters since 6 inch have been metric. Therefore, although a 4-inch wafer is in fact 4 inches in diameter, referring to an 8-inch wafer overstates the actual diameter by about 1.5% (8 inch $\approx$ 203 mm).

When PVD was introduced into microelectronic production in the late 1970s, wafer diameter was predominantly 3 inches. By 1997, however, the total area of Si used to make ICs ($\approx 4 \times 10^9$ in^2 per year) was more or less equally divided between 6-inch (150-mm) wafers and 8-inch (200-mm) ones. The IC industry has agreed that the next step will be to 300 mm, and this transition will present a significant technical and economic challenge for PVD — and most other processing as well.

The motivation for chip makers in going from 200-mm to 300-mm Si is cost reduction — 2.5 times more die can be obtained per wafer. This gain is due to a 2.3 times larger wafer area and a larger edge-to-area ratio that allows large rectangular die to be more effectively packed on the wafer. Overall, chip makers hope to lower the cost per cm^2 of processed Si by 15–40%. We will not provide an in-depth treatment of the hardware implications of processing tools for 300-mm wafers (refs. 5.44–5.48 provide useful background information on growth, handling, and processing of these wafers); however, several comments can be made with regard to the specific use of 300-mm wafers for PVD.

1. *Wafer Cost.* When 300-mm wafers of test grade were introduced around 1993, they cost ≈ \$1500. Prime device-quality material will be more costly, and unless production-volume usage greatly reduces 300-mm wafer price, this will be an issue for PVD. In particular, the use of test wafers for equipment development or process qualification will be more limited, leading to more use of hardware modeling and in-situ metrology to qualify hardware performance.

2. *Wafer Dimensions.* Proposed dimensions for a 300-mm Si wafer are diameter = 300 mm (± 0.2 mm) and thickness = 775 μm (± 25 μm). It is unlikely that nonuniformity of film properties such as thickness, sheet resistance, and step coverage will be relaxed from their current 200-mm levels. Retaining such levels of uniformity ($3\sigma < 5\%$) over an area 2.3 times as great will be a major challenge to DC magnetron design. This could lead to the use of rectangular sources with relative substrate motion similar to what is done when using PVD to coat extremely large-area glass panels for flat panel display or architectural applications. Also, gas injection and pumping ports for reactive PVD will need to be designed to produce uniform films such as TiN over these larger areas. Nevertheless, one expects PVD processes to scale more easily to 300 mm than do chemistry-dominated processes such as CVD and reactive ion etching (RIE).

Since the thickness planned for 300-mm wafers is virtually the same as that currently used for 200-mm ones, the area-to-thickness ratio ($\pi r^2/t$) will increase by a factor of 2. Such wafers will be very fragile to handle and particularly susceptible to thermal or mechanical stress. For example, from Eq. (5.4), the central deflection of an edge-clamped wafer for a given pressure of backside gas and wafer thickness is proportional to r^4, leading to 16 times more bow in 300-mm wafers. This strongly argues for the full-face holding and temperature uniformity provided by an ESC. The backside holding of an ESC also avoids the frontside edge exclusion associated with a typical mechanical clamp ring (≈ 6 mm), which for a 300-mm wafer would exclude ≈ 8% of the area.

3. *Cluster Tool.* Because many PVD processes are enabled by such steps as degas and preclean, it is clear that these processes (and their hardware) must also be scaled up on a 300-mm PVD cluster tool. While scaling up an entire 200-mm tool by $(300/200)^{1/2} \approx 1.2$ in all directions is possible but probably not required, the footprint of a PVD cluster tool for 300 mm is still expected to exceed that of 200 mm. Also, when the IC industry

switched to 200 mm, PVD equipment capable of 8-inch processing was purchased and used initially for 6-inch product wafers. The cost premium of a 300-mm tool (by some estimates 70% or more) will probably not give rise to a similar retrofit equipment market. More likely, 300-mm tools will initially be used for production-proven 200-mm processes (e.g., 0.25 μm) before adding the extra risk of next-generation device technology (e.g., 0.18 μm). This also allows for the problem of "missing" 300-mm technologies that are needed to establish a complete IC fab line but may not be available early on in the conversion from 200 mm to 300 mm. Using a 300-mm PVD tool for 200-mm production relaxes this problem but adds considerable cost per wafer due to the increased 300-mm tool cost.

5.5 PVD Process Mapping

In Chapters 9 and 11 a number of materials and process issues associated with PVD contacts, barriers, and interconnects used in MLM will be discussed. Regardless of their specific metallurgy, all these materials are deposited using vacuum-integrated multistep recipes that can be broken down for analysis into separable "process blocks" (e.g., degas, preclean, PVD deposition, anneal), separated in time from one another by a combination of robotic wafer handling, vacuum valving, and pumping steps. Mapping a PVD process sequence in this graphical way allows one to assess the impact of non-value-added time (e.g., time associated with wafer handling or valve sequencing) and identify ways to improve wafer throughput. To illustrate process mapping, we consider a representative PVD Al process for simultaneous via fill and interconnect planarization of an advanced 0.25-μm geometry device (Fig. 5.44). A two-step process (TSP) consisting of sequential cold Al and hot Al deposition in a single PVD module is used for this purpose. The detailed steps and timing of this process are dependent on the specific cluster tool used. Therefore, the information shown on the process map in Fig. 5.44 is intended for illustration only and should not be taken literally as a process recipe.

Figure 5.44a shows the basic sequence in which an incoming wafer is degassed, precleaned by a sputter etch, and then sequentially deposited with a collimated PVD Ti wetting layer, via-filled using a cold-hot PVD Al deposition in the same chamber, and then coated with a TiN ARC layer (see Chapter 7). Fig. 5.44b maps the wafer's progress through the steps of degas and preclean etch preceding PVD Ti deposition. As can be seen, the time spent on actual processing (degas = 50 sec; preclean = 30 sec) is comparable to the total overhead time associated with wafer motion,

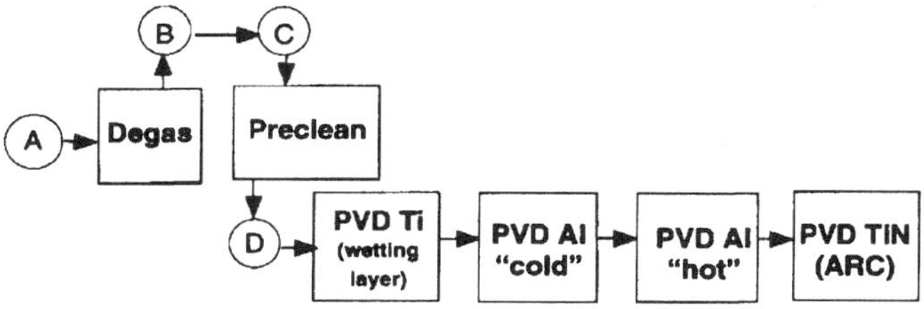

(a)

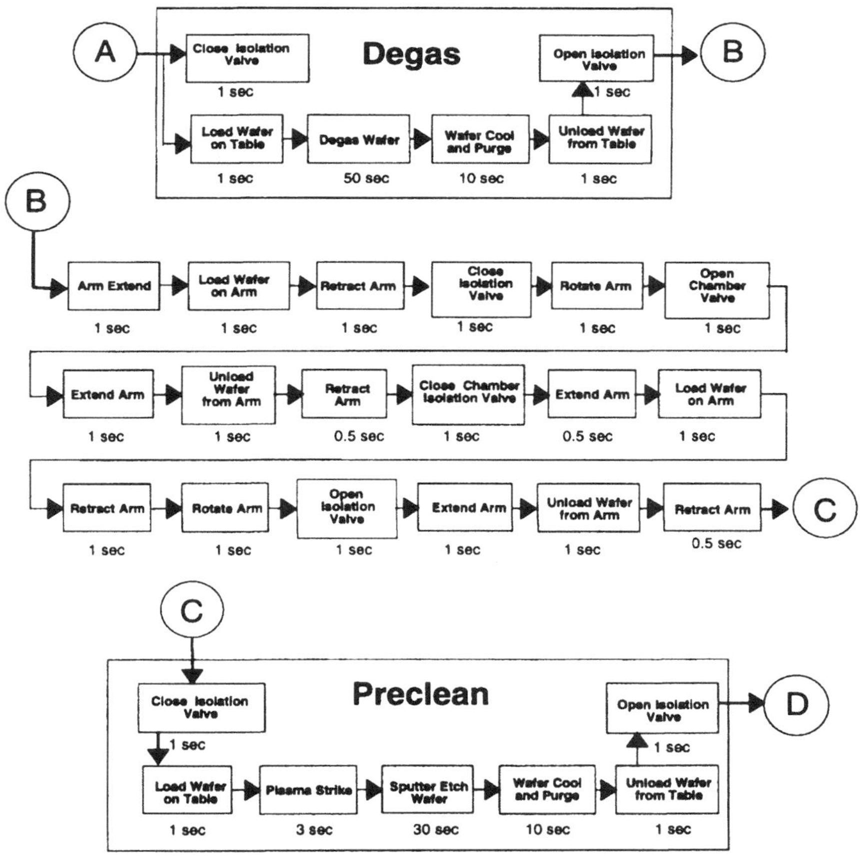

(b)

FIG. 5.44 Map for a representative PVD Al planarization integrated process sequence, with details shown from degas through preclean.

vacuum valving and pumping, and wafer thermal management (e.g., a 10-sec cooldown after preclean etching). While not mapped out in detail in Fig. 5.44, the remainder of the planarized cold-hot Al process (PVD Ti + PVD Al cold + PVD Al hot + PVD TiN) would consume about 270 sec of process time but only about 80 sec of handling time. In this part of the process sequence, the overhead time is much less than the time associated with actual PVD deposition — primarily because the two-step Al process time is relatively long (≈ 150–200 sec). An analysis of this kind might lead to the use of two modules to carry out parallel PVD cold-hot Al, which appears to be the rate-limiting process step on overall tool throughput.

5.6 Cost-of-Ownership (CoO)

It has been predicted that the single largest increase in IC manufacturing costs through the year 2000 will come from capital equipment (Fig. 5.45), and this in turn has focused considerable attention on cost-of-ownership modeling and reduction [5.49–5.51]. When PVD was introduced into microelectronic manufacturing in the 1970s, equipment selection was weighted heavily (≈ 60%) by process performance — e.g., the improved

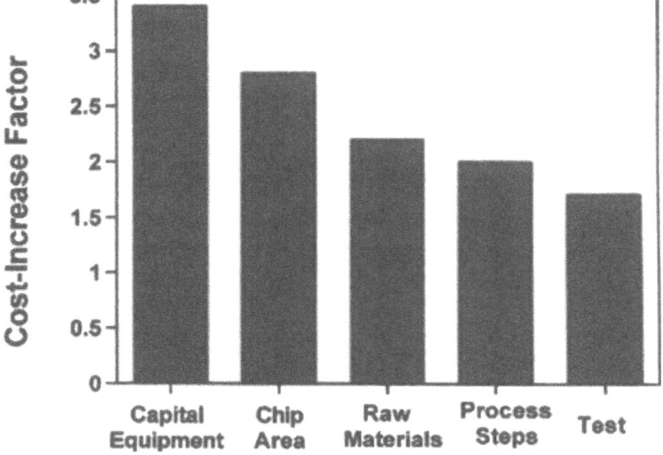

FIG. 5.45 The single largest increase in IC manufacturing cost from 1991 through 2000 is estimated to come from capital equipment (*Source:* W. Rhines, Texas Instruments).

step coverage of PVD over e-beam evaporation or the fact that one PVD system had vacuum loadlocks for improved Al film quality. However, by the 1990s, tool selection was based on more or less equal considerations of cost-of-ownership, equipment price, support, and performance (see Fig. 5.46). For commodity products like DRAMs that are produced with a low profit margin (e.g., 20%), even a slight increase in return on investment of capital assets (+1%) can have a significant impact on gross margins (+5%). Therefore, increasing tool productivity has become a key element in the strategic business plans for both equipment suppliers and users.

The strategic importance of tool productivity can be quantified by considering the cost per function of a chip, which has historically been reduced over time by 25–30% per year (Fig. 5.47). For example, in 1975 the cost per bit of memory in a 4K DRAM was ≈ 0.2 cents, while in 1995 the cost per bit of a 64M DRAM was $\approx 2 \times 10^{-5}$ cents. This exponential decrease reflects the IC industry's strategy of shrinking device dimensions on

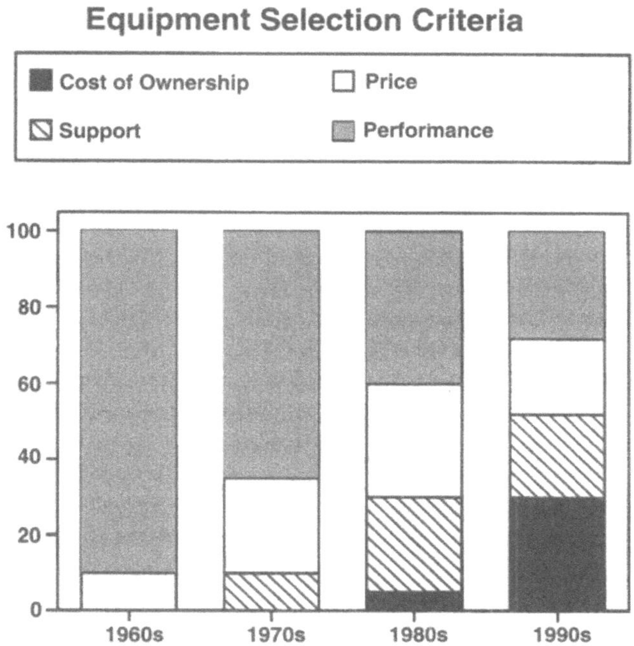

FIG. 5.46 Relative importance of cost-of-ownership in equipment purchase decisions has increased significantly over the last 20 years (after Fig. 2 in ref. 5.52) (*Source*: W. Rhines, Texas Instruments).

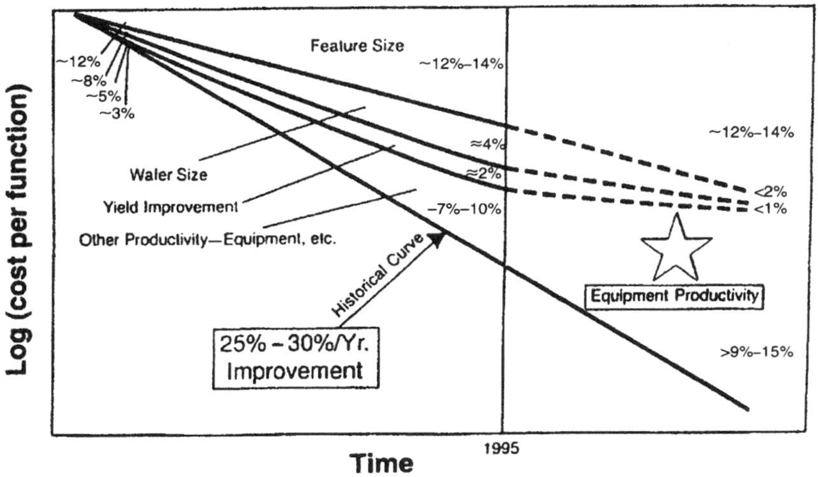

FIG. 5.47 Through about 1995, a combination of feature size reduction, wafer size increase, yield improvement, and "other" productivity enhancements enabled a 25–30% annual reduction in cost per function. To continue on this historical curve well into the future, equipment productivity must improve to compensate for diminishing returns in other areas (*Source:* B. Owens, SEMATECH).

average every 3 years and increasing wafer size every 7 years or so. As a result, the cost of a chip has increased much more slowly than the total computing power or memory it provides, and cost per function has decreased dramatically. Figure 5.47 also breaks out the four factors that have accounted for the 25–30% annual improvement in cost per function: feature size reduction, wafer size increase, yield improvement, and "other" enhancements [5.43]. SEMATECH has analyzed the further gains expected from going to ULSI devices and 300-mm wafers, leading to the extrapolated curve in Fig. 5.46. Since the first three factors are predicted to flatten out, the IC industry can only keep on its historical cost productivity curve through dramatic improvements (9–15% per year) in the "other" category — a category dominated by process tool productivity.

Cost-of-ownership is a quantifiable measure of the actual investment represented by a process tool. Being a high-level parameter, CoO is dependent on a large number of subfactors related to equipment design and operation. Hence, using spreadsheet analysis, the sensitivity of CoO to changes in tool configuration or use can be assessed. Not surprisingly, major contributions to CoO generally turn out to be tool selling price, throughput of functional die and tool reliability (e.g., mean-(productive) time-between-failures = $MTBF_p$), availability (e.g., uptime), and maintainability (e.g., mean-time-to-repair = MTTR). Given the importance of the latter three attributes and the potential confusion over how to quantify them, an industry standard — SEMI Standard E10 — for "Definition and Measurement of Equipment Reliability, Availability, and Maintainability" was created in 1996 [5.54]. For example, $MTBF_p$ is defined as productive time divided by the number of failures that occur during this time. Time when the tool is not productive (e.g., tool is down for scheduled maintenance or is being used to run qualification wafers) would not be included in the calculation of $MTBF_p$.

As a simple illustration of PVD cost-of-ownership, consider that the selling price of a high-end PVD cluster tool in 1997 was on the order of \$5M. Assuming this tool were processing 200-mm product wafers 16 hrs/day (2 production shifts) for 5 days/wk at a throughput of $\approx$ 45 wafers per hour (e.g., for an Al slab interconnect application), then $\approx$ 190,000 wafers per year are produced, which over a depreciated tool lifetime of 5 years translates into $\approx 1 \times 10^6$ wafers. Dividing the initial tool cost by the total wafers processed gives \$5 per wafer, providing an extremely crude estimate of the cost-of-ownership for a PVD tool — and one that is absolutely a lower bound.

This simple calculation underestimates the real cost of owning and operating a PVD tool in several respects.

(1) *Occupancy Costs.* Even if the tool is not in use it occupies space, and one needs to consider the fixed costs associated with the area taken up by the frontend of the tool in the clean room (face print) as well as the footprint of the tool in the adjoining, equipment maintenance bay.
(2) *Consumables.* When the tool is running, one needs to include the cost of consumables such as sputter targets and process gases. For an integrated process such as slab Al, where one could be depositing a Ti/TiN/Al/TiN stack, a mix of targets and gases need to be factored into the equation.
(3) *Maintenance and Repair.* The cost of spare parts such as shields or replacement pumps will also add to the operating cost over the life of the tool.
(4) *Labor.* There will also be labor costs (salary and fringe benefits) associated with running and maintaining the tool, which can be significant. For example, two 8-hour shifts/day $\times$ 200 work days/year $\times$ 5 years = 16,000 hours of labor, assuming one shift operator per tool.

(5) *Utilization.* Tool utilization will be less than 100% because of scheduled downtime for preventive maintenance and target and shield changes; unscheduled downtime to correct operator errors, repair broken equipment, etc.; running test wafers for process requalification after the tool has been brought back to operational status; and carrying out processes such as "Ti-pasting" in a TiN deposition chamber (see Section 11.8) that can reduce Ti target life by 20–30%. This means that instead of 16 hrs/day, the tool may really be utilized to run product wafers 12.8 hrs/day (80% of the time).
(6) *Yield.* Finally, since no process has 100% yield, some of these product wafers will contain die that fail electrically when probed due to one or more defects introduced by the PVD tool (e.g., an open interconnect line caused by a large particle that landed on the wafer during PVD and masked the subsequent deposition).

Sophisticated modeling software, such as the widely used SEMATECH Cost of Ownership Model, take these and other economic considerations into account. The cost per wafer calculated using a sophisticated CoO

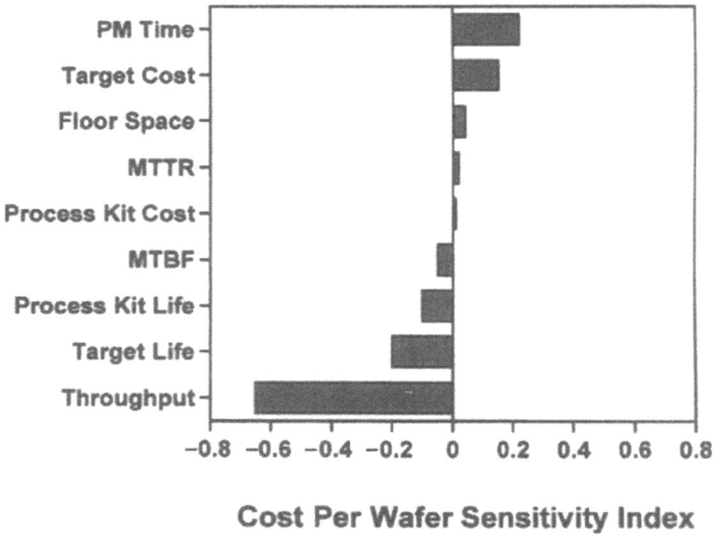

FIG. 5.48 Cost-of-ownership sensitivity analysis of an advanced PVD cluster tool showing the importance of throughput and preventive maintenance (PM) time. (*Source:* P. Singer in *Semiconductor Int'l.*, p. 113, July 1995).

model could be much higher (perhaps 50–100% more) than the $5 per wafer calculated by simply dividing the selling price of the tool by the total wafers run through it. Equally important to CoO is the sensitivity of the value to changes in input parameters, since this identifies where engineering or process improvements will have the greatest return on investment. Figure 5.48 shows a CoO sensitivity analysis reported for an advanced PVD cluster tool; of the variables considered, wafer throughput and preventive maintenance time had the greatest impact on cost per wafer, or CPW [5.55]. For example, a 10% increase in throughput might lower overall CPW by 6% (sensitivity index of −0.6), while a 10% increase in preventive maintenance time might raise CPW by 2% (sensitivity index of +0.2) This type of analysis is also useful in evaluating interrelated economic impacts of a given engineering improvement. For example, using a larger diameter target might allow one to achieve an erosion profile leading to a greater target life for a given PVD film uniformity; however, this reduction in CPW might be largely offset by increased target cost.

References

5.1. P. Clarke, "Sputtering Apparatus: Low-Pressure Operation," U.S. Patent No. 3,616,450 (Oct. 26, 1971).

5.2. R. W. Wilson and L. E. Terry, "Application of high-rate **E** × **B** or magnetron sputtering in the metallization of semiconductor devices," *J. Vac. Sci. & Tech.* **13**(1): 157–164 (1976).

5.3. Anonymous, "Sputtering deposition trends," *Semicond. Int.*, 43–55 (1979).

5.4. P. E. Guise and R. Blanchard, *Semiconductor and Integrated Circuit Fabrication Techniques*, p. 127, Reston Publishing, Reston VA, 1979.

5.5. T. G. O'Neill, "Evaporation systems remain competitive," *Semicond. Int.* **3**(8): 85–100 (1980).

5.6. P. S. Burggraaf, "Magnetron sputtering systems," *Semicond. Int.*, 37–52 (Oct. 1982).

5.7. P. Burggraaf, "Advances in metallization technology," *Semicond. Int.*, 73–79 (Nov. 1985).

5.8. G. Birkmaier, A. Tampon, and H. Grunes, "Ultrahigh vacuum in production applications," *Semicond. Int.*, 108–109 (Apr. 1991).

5.9. I. Hashim, I. J. Raajimakers, S.-E. Park, and K.-B. Kim, "Vacuum requirements for next wafer size physical vapor deposition system," *J. Vac. Sci. & Tech.*, **A15**(3): 1305–1311 (1997).

5.10. A. McGeown, "Sputter deposition via single-wafer, multichamber systems," *Microelectronic Manufacturing and Testing*, 11–13 (Aug. 1989).

5.11. S. Dushman, *Scientific Foundations of Vacuum Technique*, 2nd ed., J. M. Lafferty, Ed., Wiley, New York, 1962.

5.12. A. Roth, *Vacuum Technology*, 3rd ed., Elsevier North-Holland, New York, 1990.

5.13. J. F. O'Hanlon, *A User's Guide to Vacuum Technology*, 2nd ed., Wiley-Interscience, New York, 1989.

5.14. J. F. O'Hanlon, "Ultrahigh vacuum in the semiconductor industry," *J. Vac. Sci. & Tech.* **A12**(4): 921–927 (1994).

5.15. E. H. A. Granneman, "Film interface control in integrated processing systems," *J. Vac. Sci. & Tech.*, **B12**(4): 2741–2748 (1994).

5.16. P. Singer, "Vacuum pump technology leaps ahead," *Semicond. Int.*, 52–54 (Sept. 1993).

5.17. R. Heyder, L. Watson, R. Jackson, G. Krueger, and A. Conte, "Nonevaporable gettering technology for in-situ vacuum processes," *Solid State Tech.*, 71–73 (Aug. 1996).

5.18. R. Giannantonio, M. Succi, and C. Solcia, "Combination of a cryopump and a non-evaporable getter pump in applications," *J. Vac. Sci. & Tech.* **A15**(1): 187–191 (1997).

5.19. N. Lifshitz, W. Y. C. Lai, and G. Smolinsky, "Water related degradation of contacts in the multilevel MOS IC with spin-on glasses as interlevel dielectric," *IEEE Electron Device Lett.* **10**: 562–563 (1989).

5.20. T. Tokunaga and N. Owada, "Effects of multichamber processing on reliability of submicron vias," in *Proc. Conf. on Multichamber and In-Situ Processing of Electronic Materials*, SPIE vol. 1188, pp. 61–68 (1989).

5.21. Y. Tanaka, H. Suzuki, F. Yanagawa, B. Cohen, H. Hanawa, T. Taniguchi, M. Togashi, and K. Watanabe, "Damage free, low via resistance sputtering cleaning technology for ULSI devices," in *Technical Proc. 183rd Electrochem. Soc. Symp.* vol. 93-1, abstract no. 314, p. 481 (1993).

5.22. R. A. M. Wolters and W. C. J. Heesters, "Experimental study of metal-metal contact properties using spin on glass," in *Proc. VLSI Multilevel Interconnection Conf.*, pp. 447–449 (1990).

5.23. C. T. Gabriel and J. P. McVittie, "How plasma etching damages thin gate oxides," *Solid State Tech.*, 81–87 (June 1992).

5.24. J. P. McVittie, "Plasma charging damage: An overview," in *Proc. 1st Int. Symp. on Plasma-Induced Damage*, 7–10, San Jose, CA, May 1996.

5.25. W. Class and R. Hieronymi, "The measurement and sources of substrate heat flux encountered with magnetron sputtering," *Solid State Tech.*, 55–61 (Dec. 1982).

5.26. A. N. Pargellis, "Evaporating and sputtering: Substrate heating dependence on deposition rate," *J. Vac. Sci. & Tech.* **A7**(1): 27–30 (1989).

5.27. L. T. Lamont Jr., "Thermal history of substrates during sputtering and sputter etching," *Solid State Tech.*, 107–112 (Sept. 1979).

5.28. W. Eckstein and J. P. Biersack, "Reflection of heavy ions," *Zeitschrift fur Physik* **B63**(4): 471–478 (1986).

5.29. D. R. Wright, D. C. Hartman, U. C. Sridharan, M. Kent, T. Jasinski, and S. Kang, "Low temperature etch chuck: Modeling and experimental results of heat transfer and wafer temperature," *J. Vac. Sci. & Tech.* **A10**(4): 1065–1070 (1992).

5.30. D. C. Evans, "A generalized mathematical model for wafer cooling with gas," *Nuclear Instruments and Methods in Physics Research* **B21**: 385–390 (1987).

5.31. M. E. Mack, "Wafer cooling and wafer charging in ion implantation," in *Handbook of Ion Implantation Technology*, pp. 599-646 J. F. Ziegler, Ed., Elsevier, North-Holland, Amsterdam, 1992.

5.32. A. Clarke, "Contamination control and thermal management of aluminum sputtering," *Semicond. Int.*, 189–196 (June 1997).

5.33. G. J. van Kolk, M. J. Verkerk, and W. A. M. C. Brankaert, "Effects of contamination on aluminum films, Part I: Room temperature deposition," *Semicond. Int.*, 224–227 (May 1988); M. J. Verkerk, G. J. van der Kolk, and W. A. M. C. Brankaert, "Effects of contamination on aluminum films, Part II: Elevated temperature deposition," *Semicond. Int.*, 106–111 (June 1988).

5.34. P. S. McLeod and L. D. Hartsough, "High rate sputtering of aluminum for metallization of integrated circuits," *J. Vac. Sci. & Tech.* **A14**(1): 263–265 (1977).

5.35. D. R. Wright, L. Chen, P. Federlin, and K. Forbes, "Manufacturing issues of electrostatic chucks," *J. Vac. Sci. & Tech.* **B13**(4): 1910–1916 (1995).

5.36. J.-F. Daviet, L. Peccoud, and F. Mondon, "Electrostatic clamping applied to semiconductor plasma processing," *J. Electrochem. Soc.* **140**(11): 3245–3261 (1993).

5.37. B. Frutiger, R. Eddy, D. Brown and M. Mack, "Production proven electrostatic platen for medium current implantation," *Proc. 11th Int. Conf. on Ion Implant Techn. — IIT'96*, abstract T20 (Austin, TX, June 16–21, 1996).

5.38. G. A. Wardly, "Electrostatic wafer chuck for electron beam microfabrication," *Rev. Sci. Instr.* **44**(10): 1506–1509 (1973).

5.39. *The National Technology Roadmap for Semiconductors*, p. 126, Semiconductor Industry Association, San Jose, CA 1994.

5.40. J. A. Cunningham, "The remarkable trend in defect densities and chip yields," *Semicond. Int.*, 86–90 (June 1992).

5.41. B. Y.-H. Liu, "How particles form during vacuum pump down," *Semicond. Int.*, 75–80 (Mar. 1994).

5.42. G. S. Selwyn, C. A. Weiss, F. Sequeda, and C. Huang, "Particle contamination formation in magnetron sputtering processes," *J. Vac. Soc. & Tech.* **A15**(4): 2023–2028 (1997).

5.43. E. Korczynski, "Design challenges in vacuum robotics," *Solid State Tech.*, 62–70 (Oct. 1996).

5.44. *Proc. First Forum on 300 mm Equipment Design*, SEMATECH, Santa Clara, CA, May 12, 1995.

5.45. K.-M. Kim, "Growing improved silicon crystals for VLSI/ULSI applications," *Solid State Tech.*, 70–80 (Nov. 1996).

5.46. W. Fosnight, R. Martin, and A. Bonora, "300 mm — A new frontier," *Solid State Tech.*, 77–81 (Feb. 1996).

5.47. M. A. Drew, M. G. Hanssmann, and D. Camporese, "Automation and control for 300-mm process tools," *Solid State Tech.*, 51–64 (Jan. 1997).

5.48. C. Van Leeuwen, "Implications of 300 mm for fab design and automation," *Semicond. Int.*, 91–96 (Apr. 1996).

5.49. J. Secrest and P. Burggraaf, "The reasoning behind 'cost of ownership'," *Semicond. Int.*, 56–60 (May 1993).

5.50. R. L. LaFrance and S. B. Westrate, "Cost of ownership: The supplier's view," *Solid State Tech.*, 33–37 (July 1993).

5.51. P. Singer, "1996: A new focus on equipment effectiveness," *Semicond. Int.*, 70–74 (Jan. 1996).

5.52. W. Rhines, Texas Instruments (Fig. 1 in ref. 5.49).

5.53. J. Owens, SEMATECH (Fig. 1 in ref. 5.51).

5.54. V. H. Dhudshia, "SEMI E10 — Equipment reliability, availability and maintainability," *Semicond. Int.*, 167–174 (June 1997).

5.55. P. Singer, "The driving forces in cluster tool development," *Semicond. Int.*, 113–118 (July 1995).

Chapter 6 Directional Deposition

The sputter emission process was described in Chapter 2 in terms of a cosine-like angular distribution for the sputtered atoms. Although variations in this distribution are routinely observed, the general trend is that sputtered atoms are ejected from the target surface in a broad range of angles. Practical sputter deposition tools, described in Chapter 5, are usually characterized by short target-to-sample (throw) distances of a few centimeters and relatively large target diameters or lateral dimensions, often 50% larger than the (wafer) sample to be deposited on.

These two features — the broad angular emission distribution and the short throw distance — plus any in-flight gas scattering that might deflect the sputtered atoms even more, result in a depositing flux to the sample surface that is nearly isotropic. This differs significantly from the case of low-pressure evaporation where all the atoms arrive at normal incidence. The isotropic sputtered flux results in good coverage and film continuity over bumps and steps on the film surface (Fig. 6.1); this was originally one of the great advantages of sputter deposition for microelectronic applications. The term *step coverage* was originally defined to describe this ability to continuously cover vertical steps on the surface. The sputtered films also contributed to the planarization of the surface. This effect could be enhanced by the imposition of a sample bias (usually RF) during deposition, which would cause resputtering of the growing film. This resulted due to the enhanced sputter yield at moderate angles in a smoother, more planar surface that could be more easily patterned lithographically.

The deposition of interconnect features on semiconductor wafers relies on this smooth, nearly planar metal layer deposited by sputtering. Photoresist layers are then deposited on the metal layer, patterned by optical exposure, and developed onto a contact-surface mask layer that will protect and delineate the underlying metal circuit pattern. The unwanted metal between the photoresist-protected lines is then etched away using a technique known as reactive ion etching (RIE), in which chemically active ions, such as Fl^- or Cl^-, bombard the surface and form volatile metal fluorides or chlorides, which are then pumped away in the vacuum system.

Once the unwanted metal is completely etched away, the resist layer can be chemically stripped. This leaves behind the underlying, protected metal circuit lines on the surface. Subsequent layers of oxide (deposited from a liquid, from a chemical vapor process, or even by sputtering) are deposited onto the metal lines providing the base for the next metallization level.

There are two, reasonably fundamental problems with this RIE metallization technology. The first is due to the intrinsic nonplanarity of the

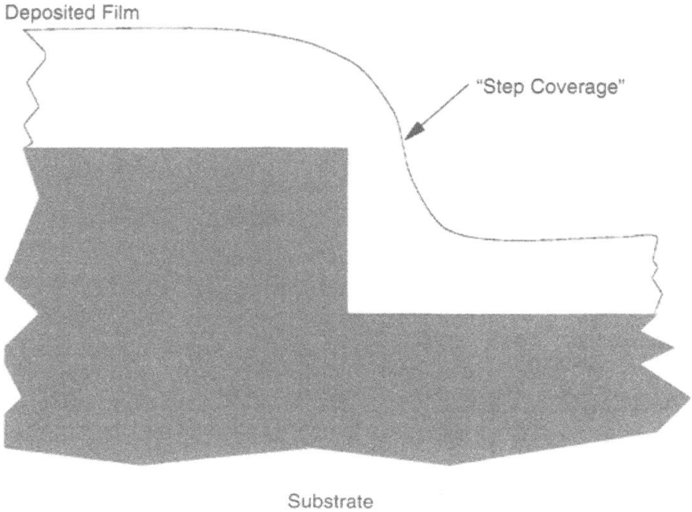

FIG. 6.1 Schematic of sputter deposition over a step.

deposition. Even though the metal layers are reasonably smooth, etching them into lines and pads results in a topographically varied surface (i.e., lines and spaces between the lines). The subsequent dielectric deposition is moderately smooth but can result in various undulations and bumps as the oxide covers over the metal lines. This surface undulation makes subsequent photoresist patterning more difficult due to depth-of-focus problems, and the undulations become more severe as the number of layers increases. This limits RIE metallization interconnect schemes to typically 3 to 4 layers.

The second, more practical problem is that whereas RIE of Al is characterized by a reasonably high vapor pressure for the product molecules at a typical processing temperature of 200°C, the same cannot be said for RIE of Cu. The vapor pressure for the Cu-chlorides is about 2 orders of magnitude lower than for the Al products. For the case of etching AlCu, this can leave behind an enriched Cu layer unless additional ion bombardment is used during the RIE. This may not be desirable due to redeposition and/or beveling problems intrinsic to resputtering. This problem with Cu also precludes the eventual transition over to a Cu-based interconnect metallization, which is desirable for its lower resistance and subsequently lower RC propagation delays.

6.1 Damascene Processing

In the early 1990s an alternate lithography process for ULSI applications was developed at IBM — *damascene processing* [6.1]. This process is shown in several steps in Fig. 6.2. The first step is the deposition of a planar dielectric film, typically either silicon dioxide (or a related glassy structure) or a polymer such as polyimide. Next, features are lithographically patterned into the dielectric by means of photoresist deposition, exposure, and processing and then reactive ion etching (RIE) through the resist mask. The features produced are typically holes, known as vias, or trenches that function as interconnect conduction lines. Once the trenches or vias are etched, the mask is removed.

The next step (Fig. 6.2c) is the deposition of metal into the feature such that it is completely filled to the top of the original dielectric surface. It may be appropriate to use several levels of metal in this feature, perhaps with thin layers for adhesion, interdiffusion resistance, or seeding. The final step (Fig 6.2d) is to remove the overdeposit of metal from the "field"

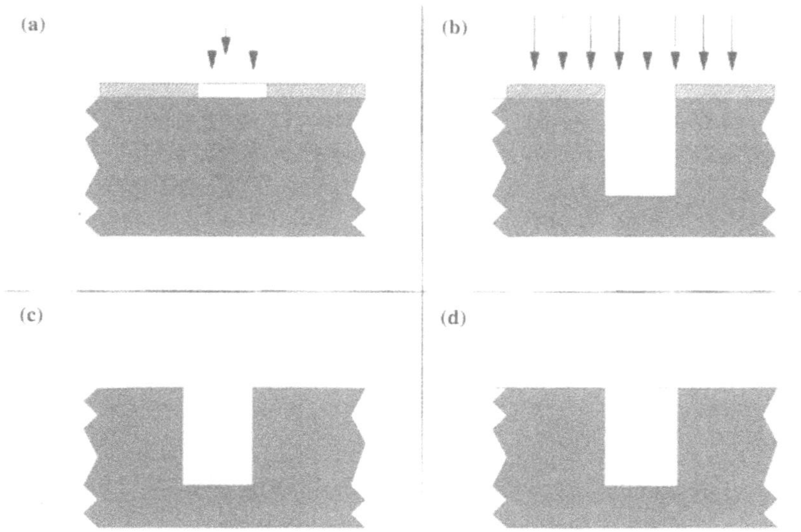

FIG. 6.2 Damascene process: (a) oxide deposition and resist exposure, (b) RIE through mask to form via, (c) metal deposition, (d) CMP removal of overdeposit.

or top areas (the "overburden") by means of a chemical-mechanical polish (CMP) process. This last process is mostly a physical abrasion process but may also have a weak chemical etching component in some metal systems. The result of CMP is that the surface is completely planarized and that the metal-filled feature is embedded in the dielectric.

Damascene wiring gets its name from a method of making jewelry in ancient Damascus, whereby pieces of glass or precious stones were pressed into holes in a metal and were then polished down to give an inlaid look. In the ULSI version, it is the glass that is patterned and filled with precious metals.

The damascene process can be repeated any number of times (12 layers is the current record) for multilevel interconnect structures because the technique is intrinsically self-planarizing and the deposited metal features are well encapsulated within the stable dielectric. In addition, it is possible to etch two-layer features into the oxide with a sequential patterning and etch process. This typically takes the form of a trench or pad with vias extending down from the pad to a lower conductor level (Fig. 6.3). This is often called *dual damascene*. Several process steps as well as at least one or two interfaces can be eliminated by the use of a dual damascene process, resulting in significant savings as well as an increase in reliability.

While this process has great promise for changing the way semiconductors are made, the metal deposition step (Fig 6.2c) is incompatible with

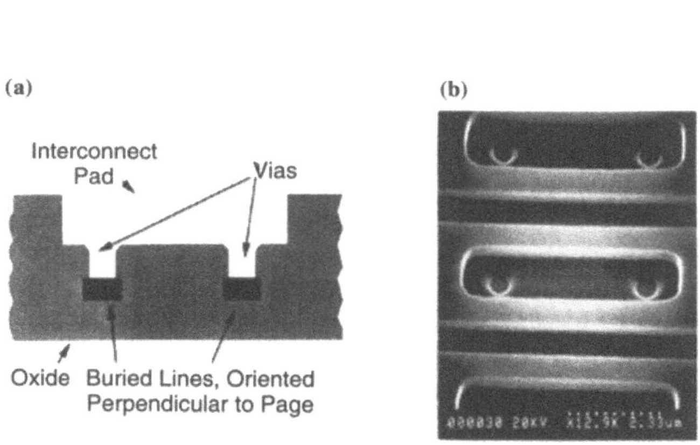

FIG. 6.3 Schematic and SEM of a dual damascene structure consisting of a rectangular pad with two circular vias at each end that connect to a lower conductor.

conventional sputtering for AR > 0.5. Due to the wide range of arrival angles of the depositing atoms discussed earlier, the deposition starts to coat the upper sidewalls and corner of the feature, which shadows the lower area from deposition (Fig 6.4). This has been described by some authors as "bread-loafing." Continued deposition results in a closed-off feature and a buried void, often called a *keyhole void.*

This overall problem has been addressed in many ways, using both PVD technologies and non-PVD technologies such as chemical vapor deposition (CVD) and electroplating. The PVD approaches can be grouped into conventional and directional approaches (Fig. 6.5). (The non-PVD technologies will not be discussed.)

The conventional PVD approaches make use of either the rearrangement of deposited atoms on the surface or the removal of atoms from the surface during deposition by ion bombardment. In the first case, either elevated sample temperature or very high pressure on the deposited film (and elevated temperature) are used to provide sufficient surface or bulk diffusion

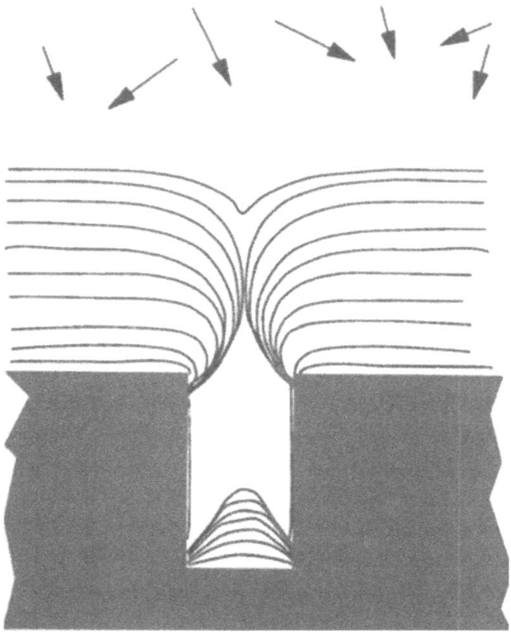

FIG. 6.4 Conventional magnetron sputter-deposited films onto a high aspect ratio via showing the formation of a keyhole void.

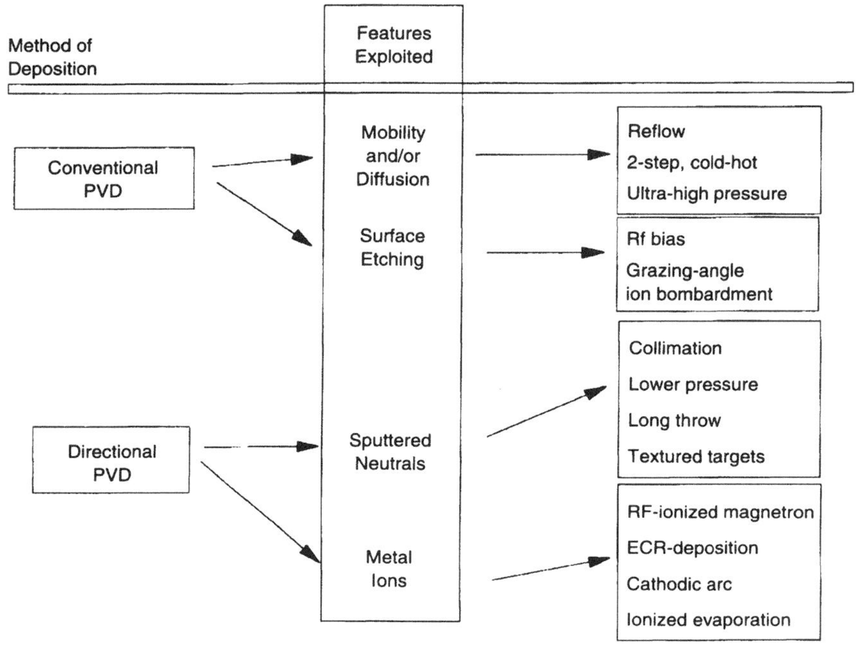

FIG. 6.5 Chart of PVD approaches to deposition into high aspect ratio features (courtesy of K. Lai, Varian Associates, Palo Alto, CA).

such that film atoms can move into high aspect ratio features. These topics will be discussed in detail in Chapter 7.

It is also possible to bombard and sputter the film surface during deposition. The simplest approach to this is bias sputtering, in which an RF bias is imposed on the sample during deposition and the surface is sputtered by normal-incidence inert gas ions from the plasma. For low aspect ratio features, this can tend to preferentially resputter the film's atoms from the sharp, high-angle features. However, for aspect ratios above about 0.5, the effect of normal-incidence bombardment is to sputter the features closed, leaving entrapped voids.

A slightly different approach uses grazing-angle ion beam bombardment during deposition to selectively etch the deposited films from the top areas of the sample, while the shadowed areas within the deeper features are protected [6.2]. This technique overcomes the problem of sputtering shut features but requires either multiple ion sources or sample rotation to provide uniform erosion. This is shown schematically in Fig. 6.6. However, this

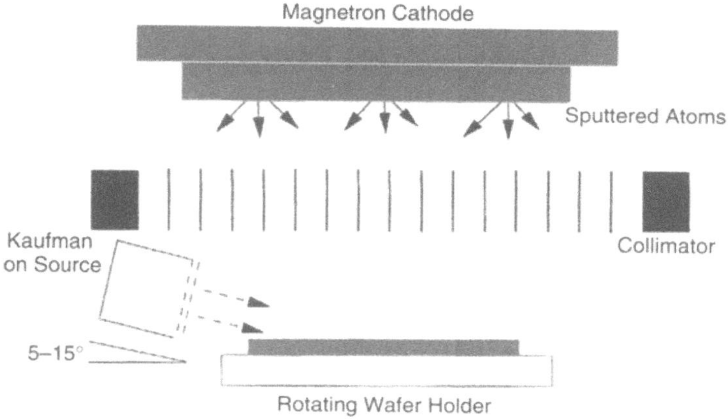

FIG. 6.6 Deposition system using grazing-angle ion bombardment to reduce the deposition onto planar areas of the sample [6.2].

technique has only been used at pressures lower than those at which most magnetrons operate (0.1 mTorr) and has been combined with a collimator (discussed below) to help separate the magnetron plasma from the grazing-angle ion beam.

The directional approaches (lower half, Fig. 6.5) use either directional neutrals or ions. Neutral sputtered atoms are not easily controlled, so the most general path toward a controlled, directional flux of neutrals is one of subtraction: Atoms with the wrong trajectories are removed, leaving the correctly directed atoms to form the deposit. The deposition of directional neutrals uses either extended cathode-to-sample throw distances or a directional filter, or collimator, between the target and the sample. It is also possible to alter the sputtering process somewhat to affect the net direction of the sputtered atoms. Two examples of this are the machined target surface and the oriented-crystal target. The ionized deposition techniques will be discussed in Chapter 10.

6.2 Long-Throw Deposition Techniques

Sputter deposition at pressures below 1 mTorr (0.13 Pa) results in a virtually collision-free trajectory for the sputtered atoms from the target to the sample. If the throw distance is increased at these low pressures, atoms that have trajectories at low angles (i.e., close to parallel to the sample surface) will be deposited on the chamber sidewalls and only atoms that have close to normal-incidence trajectories will make it to the sample. Therefore, increasing the target-to-sample distance results in a geometrical filtering of the sideways-moving atoms and results in a deposition that is more vertical. This feature can be used to project a higher fraction of the sputtered atoms deposited on the sample into deep features, and it reduces the build-out characteristic of conventional sputtering (Fig. 6.4).

Long-throw techniques were first attempted for an earlier patterning scheme known as *lift-off*, in which directional deposition is necessary to allow removal of a slightly overhanging resist mask on the sample surface [6.3]. It was necessary in that work to use a hollow cathode electron source to augment the magnetron discharge to allow low enough pressures. It was found that operating pressures in the low 10^{-4}-Torr range were necessary to reduce gas scattering sufficiently to have a nearly directional deposition process.

Current-day long-throw sputter deposition tools use throw distances of 25 to 30 cm, coupled with cathode diameters on the order of 30 cm [6.5]. This results in relatively poor directional filtering and as such is only appropriate for features with low aspect ratios (in the range of 1.0). It is usually necessary to alter the erosion and emission profile of the target when using these long throw distances. Generally, the edge regions of the target must be more highly etched than the center regions because significant numbers of atoms are lost from the edge region onto the chamber walls. This is necessary to make the net deposited flux constant across the wafer. The planar uniformity may not necessarily correlate with the directional uniformity due to the system geometry and as such may not be a good measure of the effectiveness of the directional deposition.

The deposition rates in long-throw sputtering suffer significantly compared to shorter, more conventional distances. When the throw distance is increased from a typical 30–50 mm to a long throw distance of 250–300 mm, the planar deposition rate can fall 5–10 times. This rate reduction is coupled with increased deposition on the sidewalls and tooling within a chamber.

There is a fundamental geometrical problem with long-throw sputter deposition, shown schematically in Fig. 6.7. The directionality of the arriving flux is not uniform across the wafer surface [6.6]. In the center region of the wafer, the flux is symmetric and the angular divergence is limited by the solid angle that encloses the edges of the target source. However, near the edge regions of the sample, the arriving flux becomes asymmetric, again due to the geometrical solid angle subtended by the target. As a

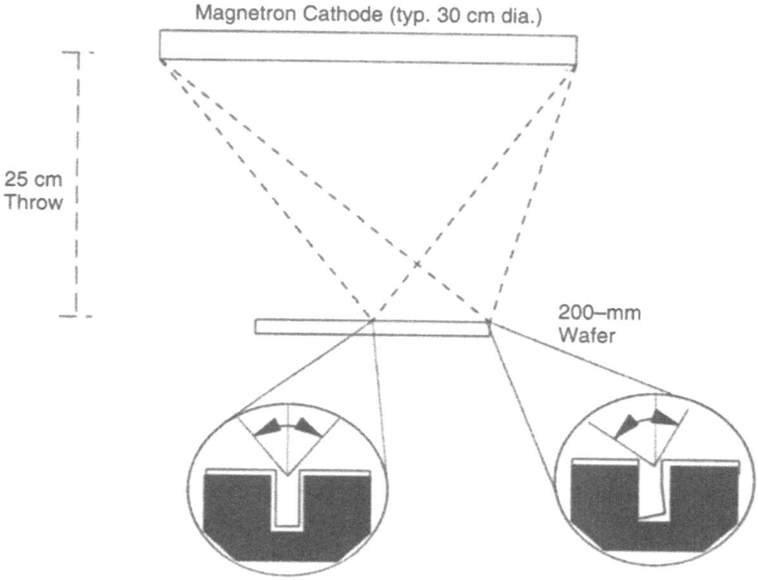

FIG. 6.7 Schematic of the cross-wafer changes in directionality for the deposited flux using long-throw deposition.

result, more atoms arrive incident on the outermost sidewalls and relatively few arrive incident on the inner-radius sidewalls. SEM photos of this effect are shown in Fig. 6.8.

This deposition asymmetry is intrinsic to the long-throw deposition geometry and is difficult to overcome. Typically, the inner side of the trench receives 1/3 or less of the film deposited on the outer side of the trench, resulting in both increased deposition time necessary for complete coverage as well as a build-out of the deposition on the outer side, which will shadow later deposition [6.6, 6.7].

To overcome this problem, it is tempting to increase the target diameter but this results in more isotropic deposition (i.e., wider angle deposition) in the center regions. It should be noted that increasing the target diameter is geometrically equivalent to reducing the throw distance. A second possibility is to increase the erosion rate at the cathode edge to increase the inward-moving flux. This results in a less uniform planar deposition (i.e., on the top surface), which is exaggerated toward the wafer edge. Although this is often undesirable from a process-control point of view (i.e., the edge

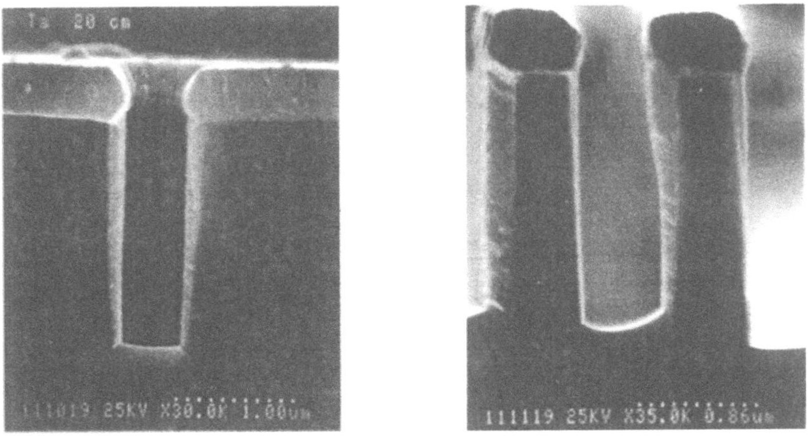

FIG. 6.8 SEM photos of trenches near the outermost region of a 200-mm wafer: (a) the deposition cross section for a trench that runs tangent to the edge of the wafer, (b) the deposition cross section for a trench that is parallel to the radial direction on the wafer [6.4].

of the wafer gets a thicker planar deposit than the center), it may result in a roughly uniform deposition *within* a feature. It should be noted that in a damascene process, the films deposited on the top, planar areas are completely removed with the CMP process. Nevertheless, CMP works best with uniform, planar films, and it is preferable not to have large center–edge asymmetries.

A final possibility for overcoming the deposition asymmetry problem is to increase the throw distance even more. As the wafer is moved farther and farther away from the target, the depositing flux that arrives at the wafer becomes more nearly normal incidence. As a practical matter, though, the increased throw distance results in a much-reduced deposition rate, increased gas scattering, and potentially a less directional deposition. The mean free path for gas atoms is approximately given for most species by:

$$\Lambda = \frac{1}{n\sigma} \approx 5\,\frac{\text{cm}}{P}$$

where n is the gas density, σ is the cross section for a momentum-transfer collision [6.8], and P is the pressure in mTorr. Although sputtered atoms are more energetic than background gas atoms and have a 50% smaller cross section, significant increases in throw distance require reductions in the operating pressure of the system. Conventional magnetrons can be operated well in the 0.5-mTorr range but require some degree of enhancement (e.g., hollow cathode) to reach the 0.1-mTorr region or below that would be required for throw distances much beyond 25 cm.

Another technical concern related to long-throw sputtering is in the realm of manufacturing tool operation. Because the long-throw chamber can be typically 2 times the height of a conventional chamber, the volume of the chamber is increased as is the interior surface area. This results in slower pump-down times following a chamber vent and adds a minor expense to the cost of operating these tools. Conversely, though, since the chamber is much larger than conventional sputter chambers, there is more room for additional flanges on the chamber, allowing increased pumping capability and/or diagnostic access.

Given these rather practical, yet fundamental, problems with long-throw sputter deposition, the process is used somewhat sparingly. It will be appropriate for aspect ratios of < 1.0, but will not work well with higher ratios. In addition, this technique scales poorly to 300-mm wafer diameters and beyond. For a 300-mm wafer system, the cathode diameter would need to be on the order of 45 cm, requiring a 45-cm throw distance to provide directionality equal to the 200-mm wafer case. This puts significant additional stress on the need to attain a low working pressure, which would need to be now in the high 10^{-5}-Torr range. This pressure range requires significant modification of the magnetron to use auxiliary electron sources, and these modifications are mostly incompatible with manufacturing applications.

6.3 Collimated Sputter Deposition

A collimator, simply a directional filter, may consist of a simple aperture for a particle beam application to an array of aligned tubes for a plasma application. For sputter deposition applications, collimators are typically close-packed arrays of tubes or cylinders configured in the space between a sputter target and a sample [6.9]. The tubes are oriented such that they are perpendicular to the plane of the target and sample (Fig. 6.9).

In general, the collimator is positioned a few centimeters from the cathode surface (> 3 cm) so that it does not interact directly with the magnetron plasma. In addition, the energy deposited on the collimator from the plasma can be many tens to a hundred or more watts, and increasing the cathode-to-collimator distance reduces the heating. On the lower side of the collimator, it is necessary to locate the sample at least

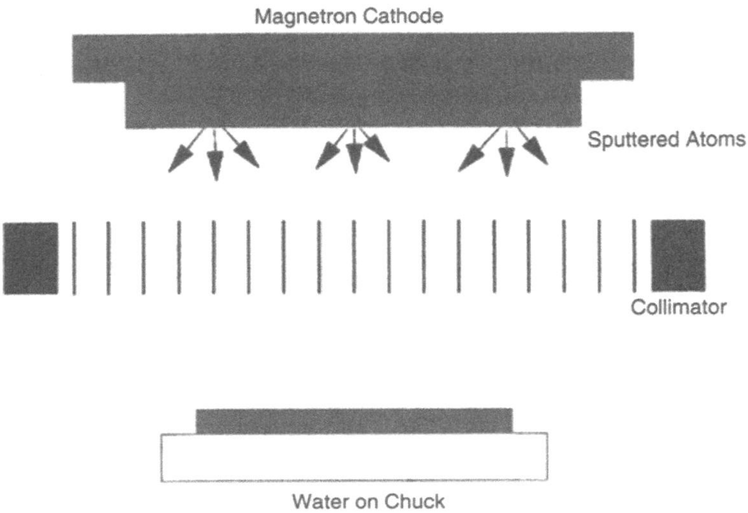

FIG. 6.9 General configuration used for collimated sputter deposition.

one hole diameter away for collimator aspect ratios of < 2, and perhaps two hole diameters away for higher aspect ratios. This reduces shadowing of the sample by the walls of the collimator. (This shadowing effect has been modeled and measured experimentally by the Alberta group [6.10].)

A collimator cell has an aspect ratio defined as the length of the tube divided by its diameter. For practical sputtering systems, this aspect ratio ranges from about 1/2 to 4. The aspect ratio of the collimator limits the flux of sputtered atoms ejected from the target by simply absorbing the atoms that impinge on the collimator walls. This selectively filters atoms that are not moving along the axis of the collimator cell. The amount of filtering is aspect-ratio-dependent; the transmitted solid angles for the deposited flux as a function of aspect ratio are given in Fig. 6.10. As is obvious from the figure, increasing the aspect ratio narrows the divergence of the transmitted flux, but at the expense of the net deposition rate. Geometrically, this can be viewed as a small cone drawn within the emission "sphere" of sputtered atoms (Fig. 6.10). The higher the aspect ratio the smaller the cone, but also the smaller the volume enclosed by the cone. This relative volume would correlate directly with deposition rate.

The net deposition rate at the sample is strongly reduced due to this filtering. Figure 6.11 shows the effect of both collimator aspect ratio as well

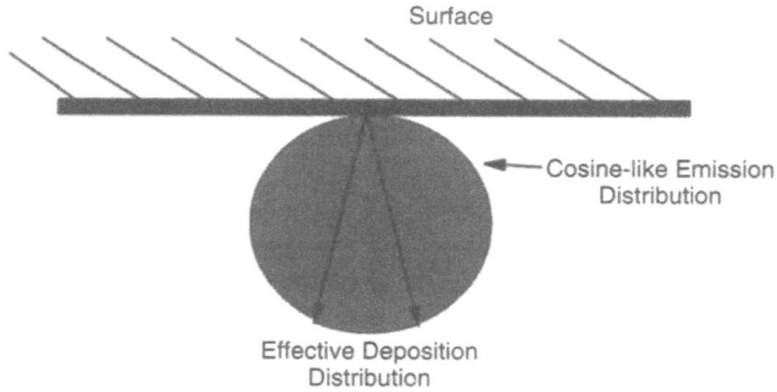

For a 2–cm high collimator located 2 cm from cathode:

Aspect ratio	Emission width (degrees)
1:1	28 i.e., +/– 14)
2:1	14
3:1	11
4:1	7

FIG. 6.10 Geometrical representation of the filtering effect of a collimator. The area within the arrows is the range of angles that are transmitted through the collimator. The solid angles (actually, in three dimensions this is a solid cone) transmitted as a function of aspect ratio are given in the table.

as chamber pressure on the net, planar deposition rate below a collimator. Generally, for each increase in the collimator aspect ratio of 1.0, the deposition rate is reduced about 3 times. In addition, the effect of increasing pressure is such that gas scattering at higher pressures reduces the rate even more, as atoms are scattered within the collimator itself and land on the collimator cell walls.

Collimated sputtering was first applied to wafer patterning in the mid-1980s, again pointed toward lift-off processing [6.9]. As with the early long-throw work, it was necessary to augment the discharge with electrons from a hollow cathode to allow low-pressure operation. In the late 1980s, magnetrons became available that operated well at 1 mTorr and below, and it was no longer necessary to use the hollow cathode enhancement. Early work also showed the capabilities for filling moderate aspect ratio features (Fig. 6.12).

Collimated sputter deposition has been used on a wide scale for the deposition of thin diffusion barriers or "liners" within vias or trenches. First shown by Joshi and Brodsky [6.11] (Fig. 6.13), this has been widely

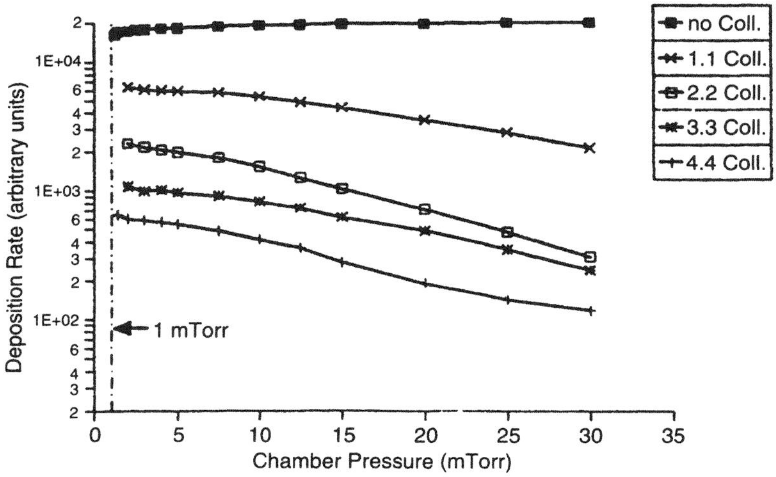

FIG. 6.11 Deposition rate through a collimator as a function of collimator aspect ratio and system pressure.

described in various references [6.12–6.16]. The most common applications are for the deposition of Ti layers at the bottom of vias that are used to decrease the contact resistance of the subsequent metal used to fill the via (typically W). Ti and primarily TiN are also valuable as diffusion barriers, which are then used within a via to provide a barrier for the interaction of Si and Al, and also to provide a nucleation layer for W-CVD. The TiN also functions to protect the SiO_2 walls from attack by the WF_6 gas used for W-CVD [6.17] (Fig. 6.14). These materials will be discussed in much greater detail in Chapter 10.

The step coverage for interconnect metallization has been functionally redefined in the recent past to mean the relative thickness at a specific point compared to the thickness of the film on the top areas. The bottom coverage in a contact hole or via can be significantly enhanced by the use of collimated sputter deposition. Figure 6.15 shows data measuring the bottom step coverage as a function of the aspect ratio of the contact hole for 1:1 and 1.5:1 AR collimated sputter deposition of Ti [6.18]. The step coverage takes a significant drop at contact hole aspect ratios roughly equal to the collimator aspect ratio. However, even at via ARs much greater than the collimator AR, bottom step coverage is still significantly increased. This may, in part, be due to some slight forward scattering of the deposited atoms down the hole. Additional data for step coverage and

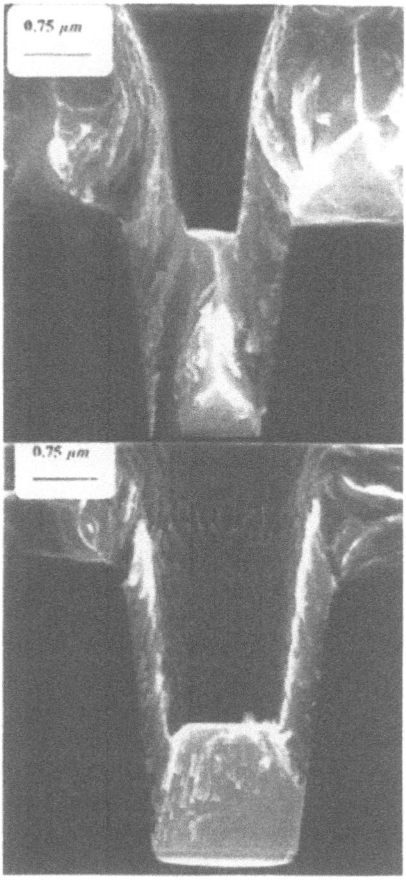

FIG. 6.12 Fully filled (top) and partially filled (lower) via features using collimated deposition of Cu. The aspect ratio of the collimator was 4.0, and the AR of the feature is about 2.7 [6.9].

deposition rate as a function of collimator aspect ratio are shown in Fig. 6.16 [6.11]. The step coverages shown in this figure are primarily for bottom coverage. As the aspect ratio of the collimator is increased and the deposition becomes more directional, sidewall coverage will drop off rapidly. Therefore, high aspect ratio collimators are best used for depositing the bottom-of-the-via contact layer with low resistance (Fig. 6.17) [6.11] and perhaps are less valuable for conformal liner deposition.

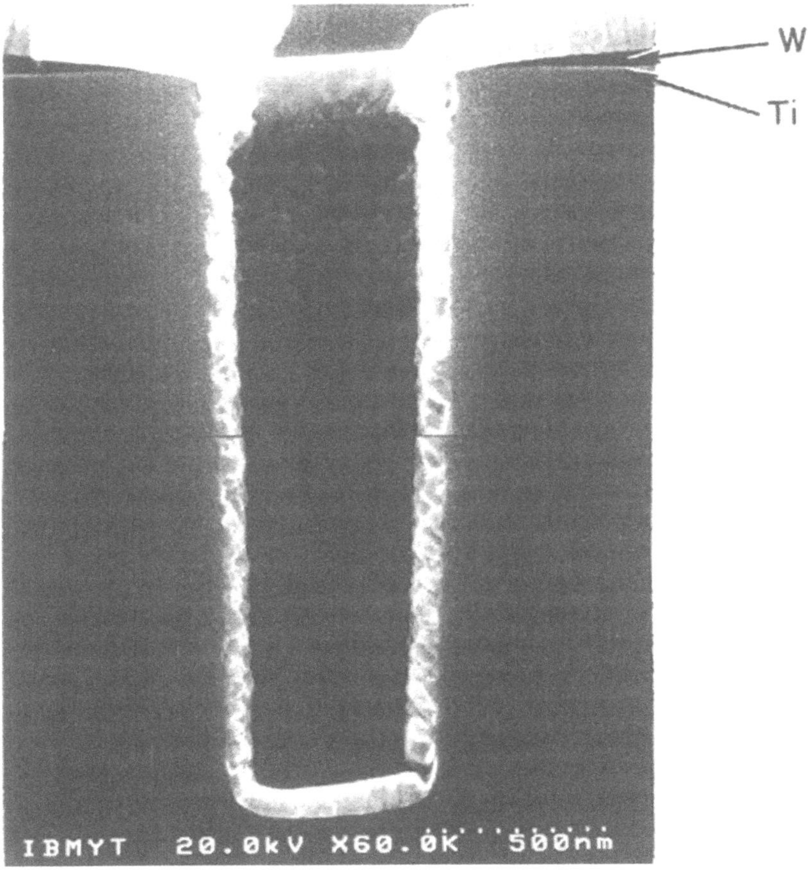

FIG. 6.13 SEM cross section of TiN liner in deep via [6.11].

Still, the use of collimated sputtering for the deposition of liners or diffusion barriers has several advantages over alternative techniques. Compared to CVD deposition, sputtered Ti and TiN are reasonably pure and stable and have low resistance. Collimated sputtering is also compatible with the general design of the PVD manufacturing tool sets, and several tool manufacturers offer collimation in their tools. The ability to use the existing PVD tool base is one of the intrinsic advantages of collimated sputtering, and it allows the introduction of collimation as a simple tool option rather than a completely new tool system, as would be necessary for CVD, for example.

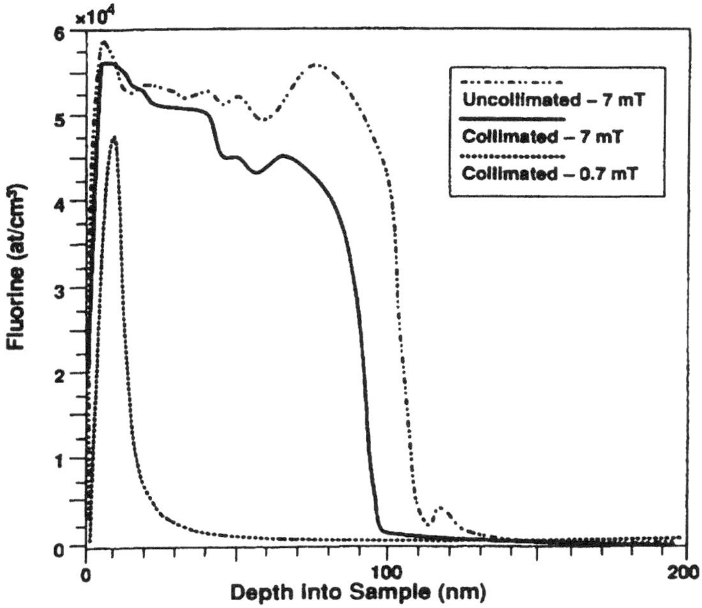

FIG. 6.14 SIMS results measuring the penetration of F (from WF_6 gas, 6 min at 450°C) in TiN films made from various processes [6.17].

Collimated sputtering has several drawbacks that have limited its application to semiconductor processing. These include a slightly overhanging profile; low rates; uniformity concerns; columnar microstructure; stress; high cost; and tool issues such as flaking, collimator-induced uniformity changes, and target utilization; and construction and maintainence issues.

Profile The collimator serves to reduce the angular divergence of the depositing flux, but it does not make the depositing flux entirely perpendicular. As a result, the overhang formation found with conventional sputtering is reduced significantly but not eliminated (Fig. 6.18). As seen in the figure, the sidewall profiles are slowly undercutting, and they are thinnest at the bottom corners of the deposited film. The step coverage in this case, which is now defined as the local thickness relative to the top, flat plane (field) thickness, can be as low as a few percent. This requires, then, depositions of perhaps 1000 Å on the top areas to reach a film of perhaps 50 Å (5 nm) in the bottom corner. In addition, often a crack or seam is observed in the bottom corner between the films deposited on the sidewall and on the bottom. This crack is a weak point for diffusion resistance.

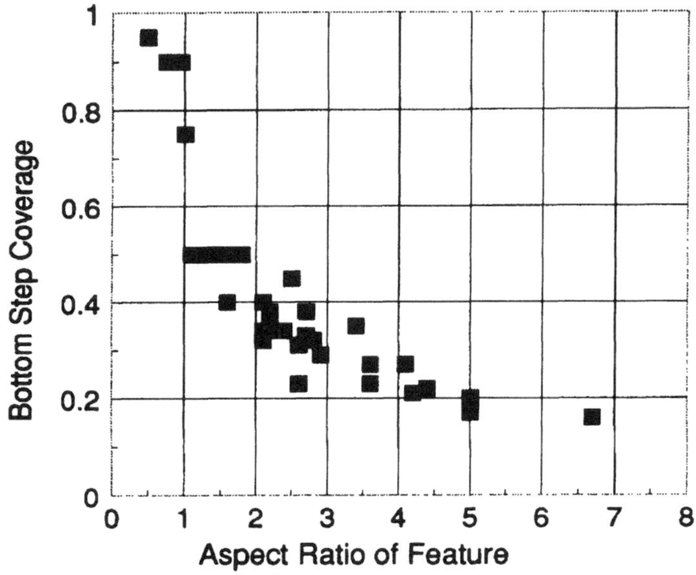

FIG. 6.15 Step coverage on the center-bottom of vias as a function of via aspect ratio for conventional and collimated deposition of Ti [6.18].

Low Rates The collimator is a filter and as such has less than 100% transmission. In fact, the transmission of a collimator drops by roughly a factor of 3 for each unit increase in the collimator aspect ratio (see Fig. 6.11). This means that a collimator of aspect ratio 1.0 has only about 30% of the open system deposition rate, and a collimator of aspect ratio 2.0 has a transmission of about 10%. The shape of the collimator is also fairly unimportant to the rate fall-off (Fig. 6.19 [6.21]), and the rate reduction is simply due to the subtractive filtering of the wide-angle sputtered atoms. This low rate, as well as a low effective efficiency in the use of the sputtered atoms, suggests that collimated sputtering can be significantly more expensive than conventional deposition. The low rates may also be a concern for materials that are very sensitive to background gas contamination. For example, the grain size of deposited AlCu has been empirically correlated with chamber base pressure: low pressures correlate with larger grain sizes. Reducing the deposition rate by 10 times is equivalent to an increase in the effective chamber base pressure of the same magnitude in terms of the relative arrival rates of metal and background gas atoms.

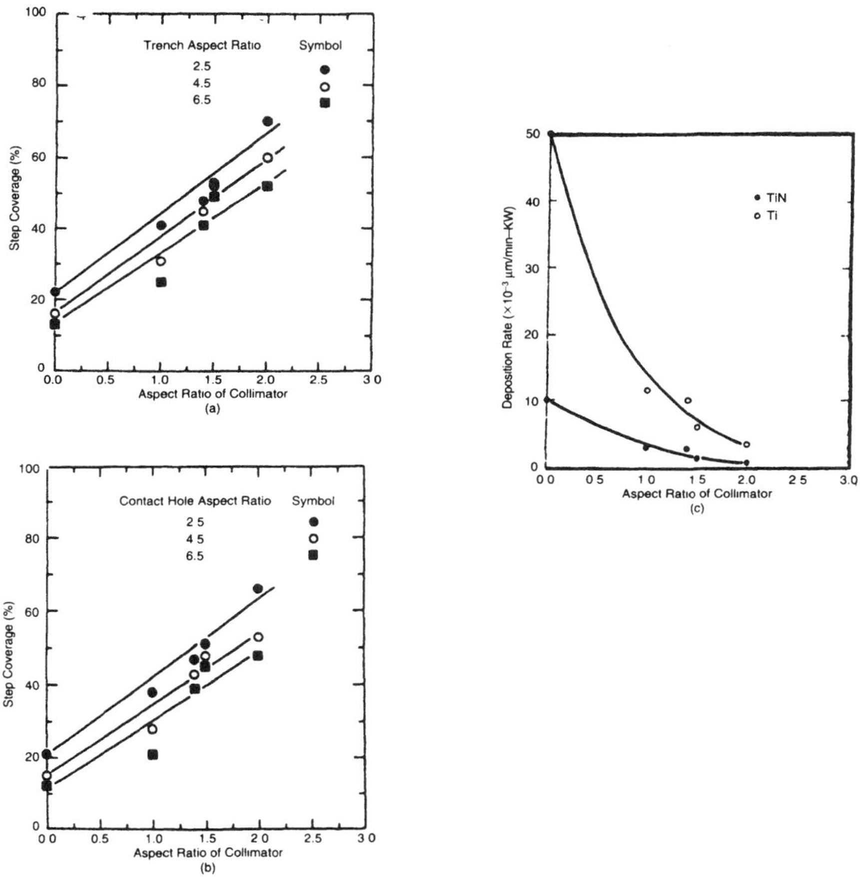

FIG. 6.16 Experimental data for (a) step coverage of lines, (b) step coverage of vias, and (c) deposition rate per unit power, all as a function of the aspect ratio of the collimator used [6.11].

Uniformity Concerns The collimator functions as an array of pinhole cameras for the sputtered atoms. At low pressure, each collimator cell can be considered to "image" a small region of the cathode onto the sample surface. The collimator itself blocks deposition onto that same area from any other part of the cathode surface. For best deposition uniformity, then, the uniformity of the cathode emission should be as flat as possible across the width of the cathode. This is different from the noncollimated case, where typically the edge regions of the cathode are more highly eroded to compensate for edge losses to the chamber walls.

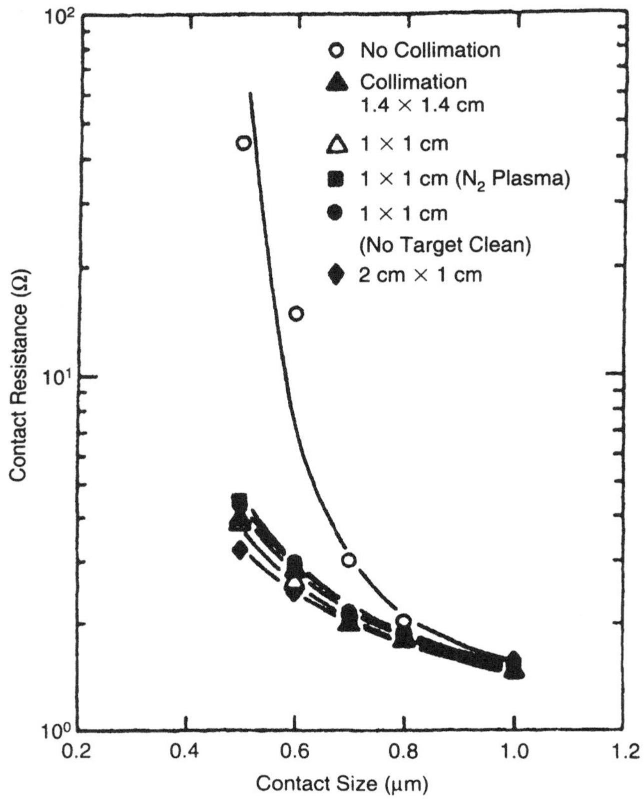

FIG. 6.17 Contact resistance of Ti structures with and without collimation [6.11].

To shape the uniformity profile, generally the magnet set behind the cathode surface is redesigned. This is somewhat quantitative (Chapters 4 & 5) and also somewhat of an art. It also depends on the aspect ratio of the collimator used. Very high aspect ratio collimators will have more of a direct imaging effect, whereas very low aspect ratio collimators (< 0.5) will be closer to the noncollimated case. To compensate for these geometrical difficulties, many equipment manufacturers supply a range of magnet designs for their basic magnetron source.

Columnar Microstructure Because of the directionality of the deposition, the sidewall deposits with collimated sputtering in some material sys-

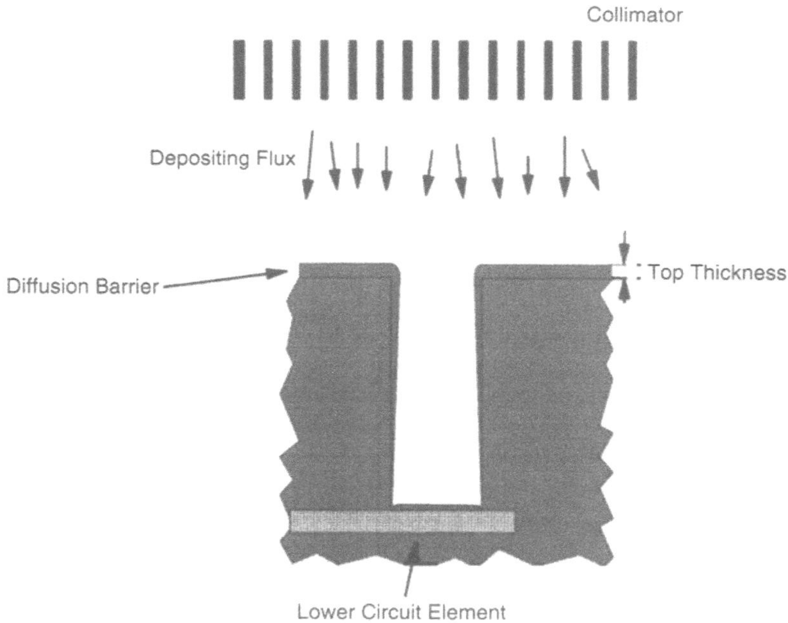

FIG. 6.18 Schematic of liner film deposited with collimated sputtering, showing undercutting profile and corner crack.

tems are not smooth, but nodular or columnar. This is particularly an issue for TiN, which shows columnar grains at a slight upward angle toward the opening of the via. The rough film cross section is less effective at forming a complete diffusion barrier because of the large number of grain boundaries (Fig. 6.20). There are several practical ways to overcome this problem. Postdeposition annealing in nitrogen or even oxygen is one option. Other scenarios use a two-step deposition process in which the gas chemistry is altered between steps, resulting in a discontinuous grain growth. This means that the grain boundaries from the first part of the deposition do not line up with those from the second, resulting in a better barrier [6.19].

Stress One aspect of collimated sputtering intrinsic to the deposition process is that it is likely to be more energetic than conventional sputter deposition. The operating pressures are low by design to reduce possible gas-phase collisions so that the full kinetic energy of the sputtered atoms

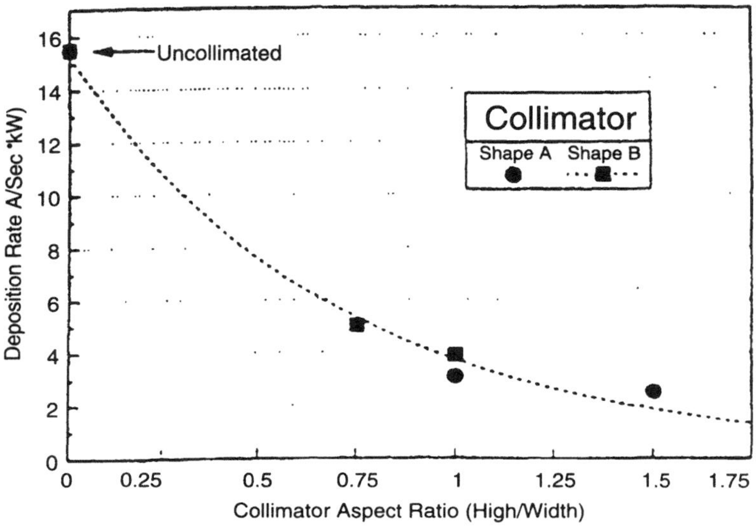

FIG. 6.19 Deposition rate dependence as a function of collimator aspect ratio for round and square collimator holes [6.21].

is preserved. In addition, since the sputtered atoms are spatially filtered to arrive at the surface at normal incidence, their kinetic energy is deposited in a small region around the impact sight. Therefore, increasing the aspect ratio of the collimator may result in increased film stress in the compressive direction; this has been observed experimentally (Fig. 6.21 [6.22]).

Cost For manufacturing applications, the net cost per layer as well as the long-term reliability of the system are very important. Many of these issues are discussed in Chapter 5 on system design. Collimated sputtering results in higher cost per layer for many reasons, although the most significant is the reduced deposition rate.

Tool Issues In any application of collimated sputtering, an eventual problem will be the result of thick deposits on the collimator itself. Since the transmission of the collimator can be as low as a few percent, the rest of the sputtered atoms remain on the collimator. As a rough measure of this problem; consider a typical sputtering cathode, which has a working thickness of about 1 cm; i.e., the high-purity target is at least 1 cm thick. Typical collimator diameters are on the order of 1–2 cm. So it can easily

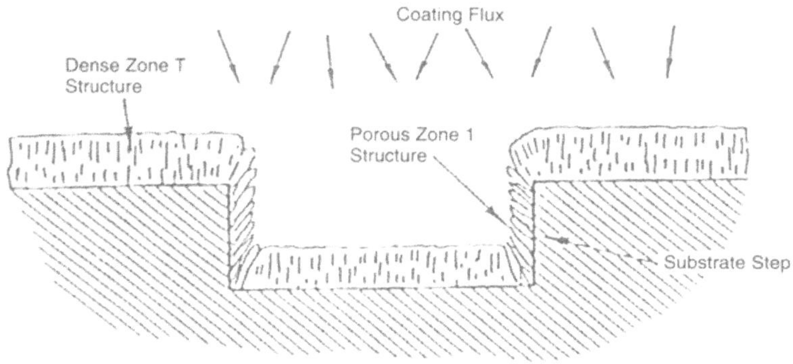

(a)

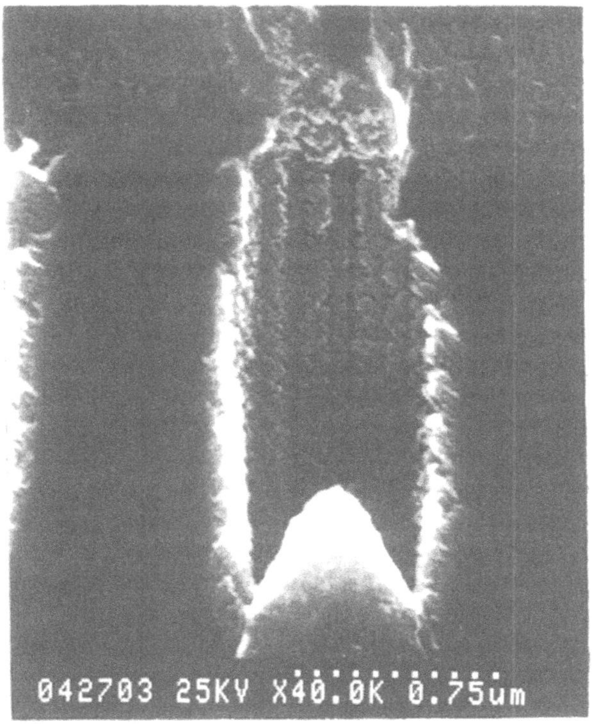

(b)

FIG. 6.20 (a) Sketch of microstructure of sidewall of a via deposited with collimated sputtering showing columnar microstructure, (b) SEM of sidewall.

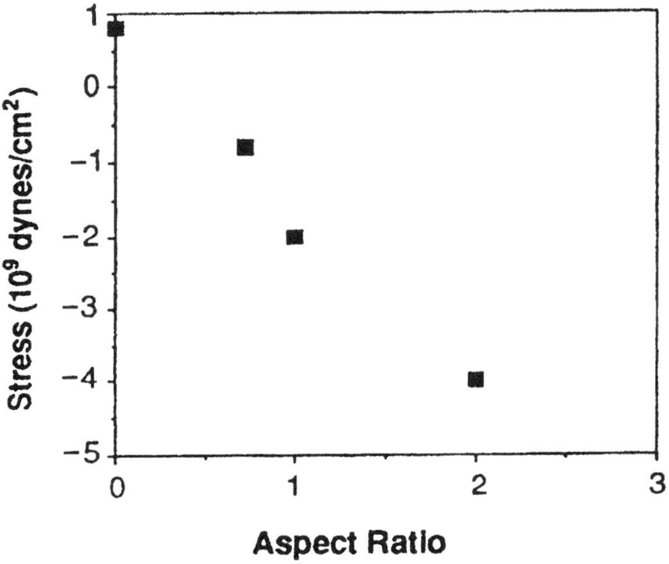

FIG. 6.21 The stress in deposited Ti films as a function of collimator aspect ratio [6.22].

be expected that if the majority of the atoms from the cathode land on the collimator, thicknesses of many millimeters can be deposited during the lifetime of the cathode, and these deposits will induce problems.

As the deposits on the collimator build up, the effective aspect ratio of the cell increases, which not only changes the angular profile of the transmitted atoms but also leads to even faster build up on the collimator walls and a net reduction in the deposition rate (Fig. 6.22). (This effect is described in detail in ref. 6.23). An example of the change in the angular profile of the transmitted atoms is shown in Fig. 6.23. The time before a collimator must be removed due to concerns about flaking and/or clogging is typically 25 to 50% of the target lifetime.

The eventual flaking of the deposited film results in contamination of the wafer. Flaking can occur either when the films simply become too thick or due to thermal cycling that might occur during a deposition, wafer/chamber heating, or chamber venting.

Another concern with collimator clogging is that it does not occur uniformly across the collimator. As a collimator clogs, its effective diameter decreases and its aspect ratio increases. This results in even lower trans-

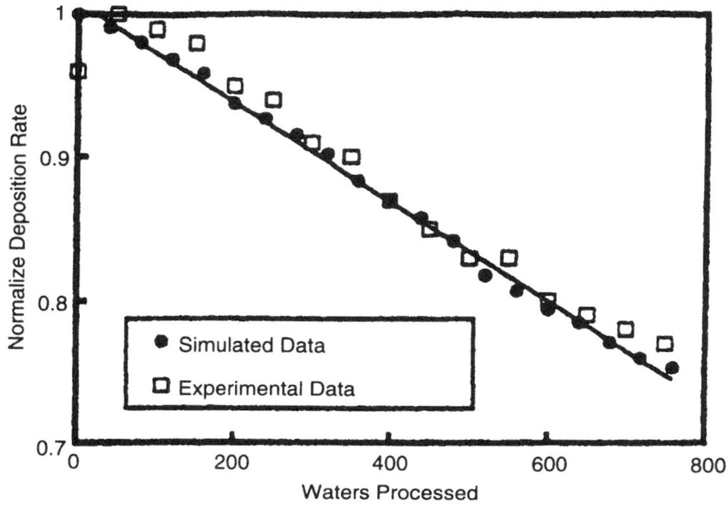

FIG. 6.22 Calculated and experimental deposition rates for a 1.5:1 collimator as a function of the number of wafers processed [6.23].

mission and faster clogging. During this whole process the collimator cells in the center of the wafer tend to clog first, resulting in a net change in the uniformity of deposition across the wafer (Fig. 6.24).

Collimator Construction The design and construction of a collimator may have an effect on its operation and/or lifetime. The earliest collimators were constructed by clamping together arrays of short tubes. This moved rapidly to a machined approach, in which close-packed arrays of round holes were machined into solid plates of Al or Cu. This approach was designed for water cooling, although it was only possible to extract the heat at the perimeter of the collimator plate. Under high power (> 10 kW), the center of a 30-cm-diameter Al collimator plate could be 60–100°C hotter than the edge. With no water cooling, collimator temperatures could easily reach 400°C due to the combined effects of deposition on the collimator and energetic electrons and photons from the plasma. Later, plate-type collimators were milled with hexagonal holes to minimize the amount of geometrical blocking by the walls of the collimator holes. These were still cooled, though, and had similar thermal performance to the round-hole collimators.

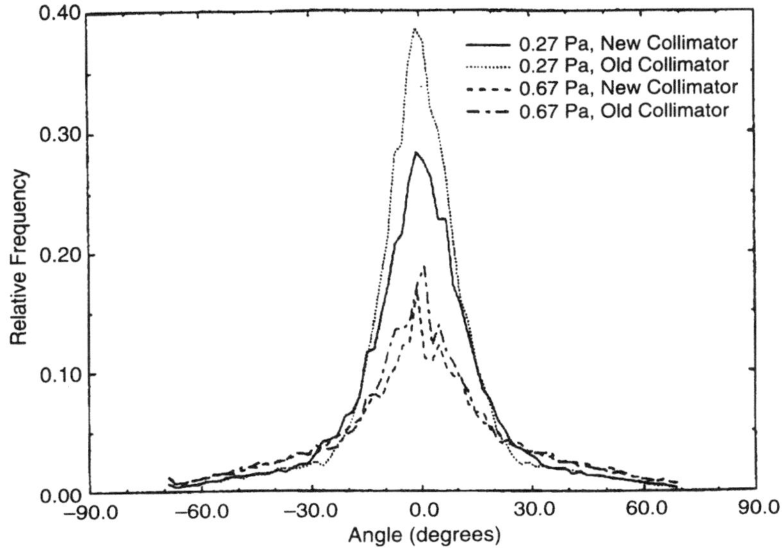

FIG. 6.23 Angular flux distributions of depositing atoms at the wafer surface as a function of collimator filling and also of operating pressure. Note: 0.27 Pa = 2 mTorr, 0.67 Pa = 5 mTorr. It would be unusual to operate a collimated sputtering system at pressures much above 2 mTorr [6.13, 6.14].

In parallel to the plate-collimator approach, work at Varian Associates centered on the assembly of collimators from thin sheet metal [6.20]. The metal strips (approx. 0.2- to 0.5-mm thick) were spot-welded into a hexagonal array. Since the heat transfer across these arrays of thin stainless steel strips is very low, there was no attempt to cool this type of collimator, and during high-power operation the temperature of the collimator could approach 500°C or more. This high temperature, when used during the sputtering of a low melting point material such as Al, could actually reduce the net deposition on the collimator due to reevaporation. However, this effect with Ti and/or TiN, which are the most widely used cathodes for collimated sputtering, is low.

Currently (1998), the sheet metal approach has been adopted by most tool manufacturers, and there is very little attempt to draw heat from the collimator. Because most collimator applications use Ti, often the collimator is constructed from Ti sheets. This matches the thermal expansion coefficients of the collimator and the depositing film, which results in low thermal stress on the films deposited on the collimator during the various

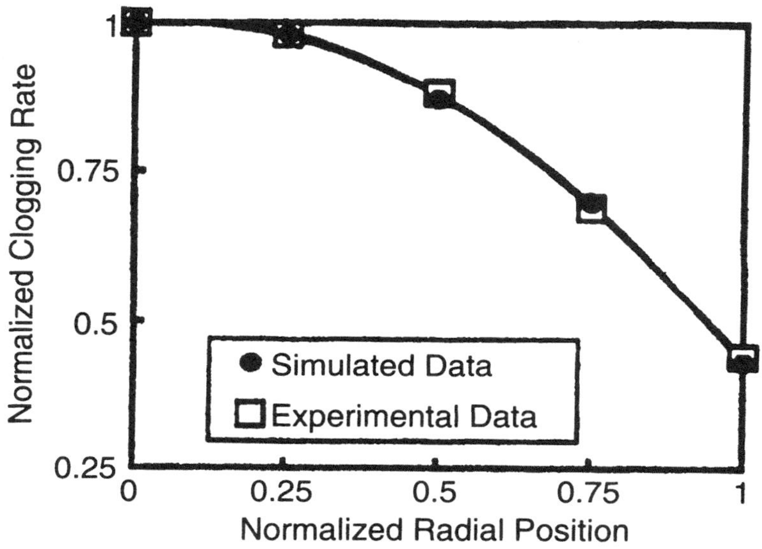

FIG. 6.24 Normalized clogging rate of a collimator as a function of the radial position of the collimator [6.23].

temperature cycling that can occur in manufacturing. The reduction in thermal stress is critical to better adhesion of the films deposited on the collimator and to less flaking.

Collimator Cleaning Historically, there have been two approaches to the problem of what to do with a heavily deposited collimator. In the early development days, collimator materials were chosen such that the deposited films could be chemically etched and the collimator reused. This might be as simple as using an Al machined collimator for the deposition of Cu, which could then be easily removed from the collimator with nitric acid. The cost of a machined collimator for 200-mm wafer applications was about $2000–$5000, so cleaning was worthwhile. Manufacturing applications of collimators have put more stringent requirements on tooling cost as well as the environmental problems associated with cleaning. Currently, the widespread use of sheet metal collimators has allowed a more disposable approach; there is no attempt to clean the collimator after usage and it is simply replaced. Sheet metal collimators can be fabricated much less expensively ($500–$1000/collimator), which makes the expense of cleaning too high.

References

6.1. C. W. Kaanta, S. Bombardier, W. Cote, W. Hill, G. Korszykowski, H. Landis, D. Poindexter, C. Pollard, G. Ross, J. Ryan, J. Wolff, and J. Gonin, "Dual damascene: A ULSI wiring technology," in *Proc. IEEE VMIC*, Santa Clara, CA 1991, p. 144–152 (unpublished).

6.2. S. M. Rossnagel and R. Sward, "Collimated magnetron sputter deposition with grazing angle ion bombardment," *J. Vac. Sci. & Tech.* **A13**(1): 156 (1995).

6.3. J. J. Cuomo and S. M. Rossnagel, "Hollow cathode enhanced magnetron sputtering," *J. Vac. Sci. & Tech.* **A4**: 393–396 (1986).

6.4. S. M. Rossnagel, C. A. Nichols, S. Hamaguchi, D. Ruzic, and R. Turkot, "Thin, high atomic weight refractory film deposition for diffusion barrier, adhesion layer and seed layer applications," *J. Vac. Sci. & Tech.* **B14**: 1819 (1996).

6.5. J. N. Broughton, C. J. Backhouse, M. J. Brett, S. K. Dew, and G. Este, "Long throw sputter deposition of Ti at low pressure," in *Proc. VLSI Multilevel Integration Conf.*, p. 201. (1995).

6.6. I. Wagner, "Sputter deposition of Ti and TiN films with variable target-to-substrate distance," in *Proc. VLSI Multilevel Integration Conf.*, p. 226 (1995).

6.7. A. A. Mayo, S. Hamaguchi, J. H. Joo, and S. M. Rossnagel, "Across-wafer nonuniformity of long throw sputter deposition," *J. Vac. Sci. & Tech.* **B15** (1997).

6.8. R. S. Robinson, "Energetic binary collisions in rare gas plasmas," *J. Vac. Sci. & Tech.* **16**: 179–185 (1979).

6.9. S. M. Rossnagel, D. Mikalsen, H. Kinoshita, and J. J. Cuomo, "Collimated magnetron sputter deposition," *J. Vac. Sci. & Tech.* **A9**: 261–265 (1991).

6.10. R. N. Tait, S. K Dew, W. Tsai, D. Hodul, T. Smy, and M. J. Brett, "Simulation of uniformity and lifetime effects in collimated sputtering," *J. Vac. Sci. & Tech.* **B14**: 679 (1996).

6.11. R. V. Joshi and S. Brodsky, "Collimated sputtering of TiN/Ti liners into sub-half-micrometer high aspect ratio contacts/lines," *Appl. Phys. Lett.* **61**: 2613–2615 (1992), and R. V. Joshi and S. Brodsky, in *Proc. VMIC*, Santa Clara, CA, 1992 (unpublished) p. 253.

6.12. T. Janacek, D. Liu, S. K. Dew, M. J. Brett, and T. J. Smy, "The effects of collimation on intrinsic stress in sputter-deposited metallic thin films," *Thin Solid Films* **253**: 372 (1994).

6.13. D. Liu, S. K. Dew, M. J. Brett, T. Janacek, T. Smy, and W. Tsai, "Experimental study and computer simulation of collimated sputtering of Ti thin films over topographical features," *J. Appl. Phys.* **74**: 1339 (1993).

6.14. D. Liu, S. K. Dew, M. J. Brett, T. Janacek, T. Smy, and W. Tsai, "Properties of Ti and Al thin films deposited by collimated sputtering," *Thin Solid Films* **236**: 267 (1993).

6.15. S. Meikle, S. Kim, and T. Doan, "Semiconductor process considerations for collimated source sputtering of Ti films," *Proc. VMIC*, Santa Clara, CA, 1992, pp. 289–291 (unpublished).

6.16. T. Hara, T. Nomura, and S. C. Chen, "Properties of titanium layers deposited by collimation sputtering," *Jpn. J. Appl. Phys.* **31**: L1746–L1749 (1992).

6.17. J. G. Ryan, S. Brodsky, T. Katata, M. Honda, N. Shoda, and H. Aochi, "Collimated sputtering of Ti and TiN films," *MRS Bulletin*, 42–45 (November 1995).

6.18. Varian Associates, Palo Alto, CA.

6.19. "Bipolar 212TiN," from Sputtered Films, Inc, 320 Nopal St., Santa Barbara, CA 93103.

6.20. E. Demeray (formerly of Varian Assoc.), 1989.

6.21. S. Roehl, L. Camilletti, W. Cote, D. Cote, E. Eckstein, K. H. Froehner, P. I. Lee, D. Restaino, G. Roeska, V. Vynorius, S. Wolff, and B. Vollmer, "High density damascene wiring and borderless contacts for 64M DRAM," in *Proc. VMIC*, Santa Clara, CA, 1992, pp. 22–28 (unpublished).

6.22. C. C. Fang, R. V. Joshi, V. Prasad and C. Ouyang, "Modeling of intrinsic stresses of titanium thin films deposited by collimated sputtering," *Advanced Metallization and Interconnect*

Systems for ULSI Applications in 1995, R. C. Ellwanger and S.-Q. Wang eds., Materials Research Society, Pittsburgh PA, 1996, p. 423.

6.23. D. S. Bang, J. P. McVittie, M. M. Islamraja, K. C. Saraswat, Z. Krivokapic, S. Ramaswami, and R. Cheung, "Dynamic modeling of collimator clogging in physical vapor deposition systems," *Proc. VMIC*, Santa Clara, CA, 1994, p. 554 (unpublished).

Chapter 7 Planarized PVD: Use of Elevated Temperature and/or High Pressure

The diffusion rate of sputter-deposited atoms in a thin metal film, either along the film surface or through the bulk, is highly sensitive to temperature. Therefore, temperature provides the PVD user with a "process knob" that can be used to control metal atom mobility, which in turn allows one to engineer the profile of a PVD film for a given application. For example, the use of elevated temperature ($\approx$ 350–550°C) either during or after PVD deposition has successfully been exploited for improved step coverage and even complete filling of PVD Al alloy and Cu films in high aspect ratio structures. Most of the work to date has focused on Al metallurgy with the intention of replacing CVD W plugs with more conducting PVD Al plugs in multilevel metallization schemes, allowing an all-aluminum solution with vertical Al contact/via plugs and horizontal Al interconnect wires. In addition, since PVD is a blanket and not a selective process, Al is deposited on both the field regions and in the via holes. This opens the possibility of using a "hot Al" process to simultaneously fill the plugs and planarize the free Al surface (Fig. 7.1), with the benefit of fewer overall process steps. As shown in Fig. 7.1a, since the thicknesses of the field and the via hole are usually comparable, the volume of the incident slug of material that enters the hole is not great enough to both fill the hole and form a planarized surface. There is a "missing mass" that must be provided. Figure 7.1b shows a more realistic shape of the missing volume. Hot Al processing induces mass to migrate into the hole from the much larger volume of the field and also redistribute within the hole (Fig. 7.1c). The net result is a simultaneously plugged hole and planarized surface (Fig. 7.1d).

This chapter discusses underlying physics and practical aspects of hot PVD processing with an emphasis on Al. The trend in hot PVD, as with MLM processing in general, is toward lower process temperatures — e.g., to reduce thermal stress voiding in metal lines and to ensure compatibility with future low-*k* dielectrics that are likely to be highly temperature-sensitive. Therefore, we also discuss several methods that have been developed to lower the temperature needed to reliably coat or fill a given aspect ratio structure, including multistep processes such as the two-step or "cold-hot" Al process, the use of unconventional alloys such as Al-Si-Ge, and the application of ultrahigh-pressure annealing after PVD deposition to force the as-deposited metal to fill the feature by enhanced plastic flow — the Forcefill™ process.

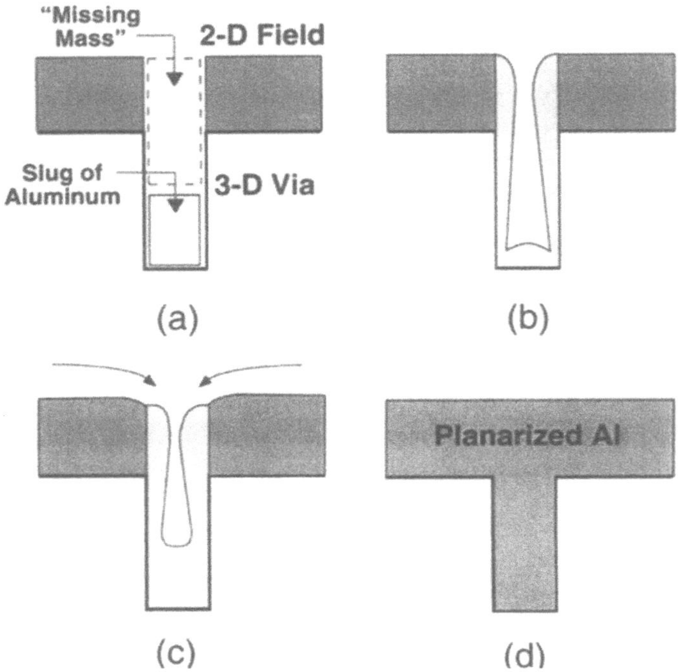

FIG. 7.1 Hot PVD Al process allows simultaneous filling of a three-dimensional plug and planarization of the two-dimensional interconnect surface over the hole.

7.1 Physics of Hot PVD

Figure 7.2 shows the general case of hot deposition of a metal film over a surface with nonplanar topography, although similar considerations apply when the PVD film is annealed after deposition. Although gas-phase transport accounts for the transfer of sputtered atoms from the face of the target to the surface of the growing film, it is solid-phase transport that accounts for the operation of hot PVD. In particular, the fundamental idea behind hot PVD is to increase the self-diffusion rate of the metal (e.g., the diffusion rate of Al atoms in an Al film) to rapidly build up the film in areas of low coverage at the surface and/or fill up voids within the bulk. In the first case, transport of material occurs by the mechanism of surface diffusion; in the second case, the mechanism is by bulk diffusion of vacancies with grain boundary diffusion also possible in a real polycrystalline film.

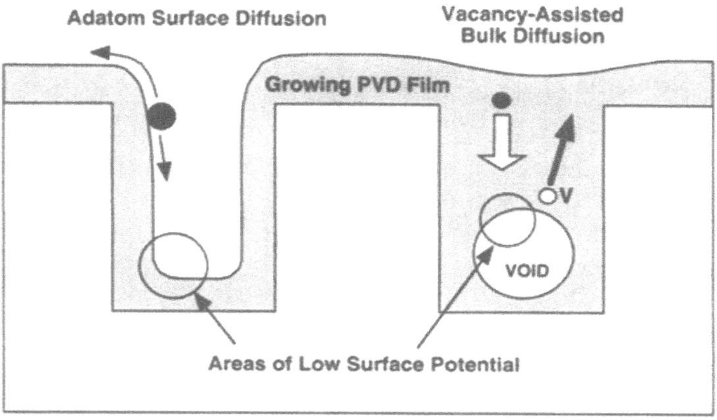

FIG. 7.2 During hot PVD, surface diffusion redistributes depositing adatoms along the surface, while bulk diffusion transports activated atoms into the bulk.

In either case, the driving force is the reduction of the surface potential of the film, which is determined by the surface curvature of the film (or subsurface curvature in the case of a buried void). The surface potential is negative for concave surfaces and positive for convex ones with the overall effect that matter is transported from the upper part of the film to the lower part of the film growing nearer to the substrate. This behavior is analogous to that of a fluid in that regions of lowest curvature (concave) will be filled first. Assuming no film stress and an isotropic material, the surface potential in two dimensions can then be written as $\mu = \mu_0 + \gamma R/\Omega$, where μ_0 is the chemical potential of a flat surface, γ is the surface tension, Ω is the atomic volume, and R is the radius of curvature of the surface [7.1]. It is the gradient of this curvature that will cause the preferential diffusion of adatoms into areas where the surface is concave, such as the inside of a contact or via hole.

For the materials and temperatures typically encountered with reflow processing (Al alloys and Cu at 350–550°C), the metal vapor pressure is so low that transport of matter by the mechanism of gas evaporation-condensation need not be considered. If the process temperature is much greater than one-half of the melting point of the metal in degrees Kelvin (i.e., $\gg$ 470°C for Al), then plastic flow may contribute to the bulk movement of the film — an effect that occurs more readily in films that are under compressive stress. On the other hand, process temperatures are rarely so close to the melting point that viscous flow is significant. In this case, it is the more conventional mechanisms of surface and bulk diffusion that dominate film transport during high-temperature deposition and/or postdeposition annealing. Due to the relatively low activation energy for diffusion at a free metal surface, surface diffusion rates are usually much larger than bulk diffusion rates. For example, representative activation energies for surface and bulk diffusion in a clean Al film at moderate temperatures (450–550°C) are about E_a = 0.40 eV and 1.45 eV, respectively. Also, diffusion theory shows that the $1/e$ decay time for a sinusoidal surface feature of frequency ω goes as $1/\omega^3$ for bulk diffusion but $1/\omega^4$ for surface self-diffusion [7.2]. Hence, one expects "high-frequency" submicron scale surface features to be smoothed out far more quickly by transport of material along the surface than through the bulk.

The surface diffusivity D, in cm^2/sec, depends exponentially on temperature through an expression of the form

$$D = D_0 \exp\left(\frac{-E_s}{kT}\right) \tag{7.1}$$

where E_s is the activation energy for surface diffusion ($E_s \approx 0.4$ eV for Al), and the prefactor D_0 is a strong function of material and deposition conditions (the bulk diffusivity of Al has the same form as Eq. (7.1) but with $D_0 \approx 1.7$ cm^2/sec and $E_a \approx 1.45$ eV). In modeling of hot Al PVD processing, it has been useful to further refine Eq. (7.1) by defining a diffusion length along the surface given by $L = (D\tau)^{1/2}$, where τ is the mean lifetime for a mobile surface adatom [7.1]. This lifetime is limited by such factors as the deposition rate — which determines the time it takes to deposit a monolayer over the atom and effectively bury it — and by the presence of reactive gas species that can chemically bind with the atom and fix its position on the surface. Using Eq. (7.1), we can then write the diffusion distance of the mobile adatom as

$$L = (D_0\tau)^{1/2} \exp\left(\frac{-E_s}{kT}\right) = (D_0\tau)^{1/2} \exp\left(\frac{-5T_m}{2T}\right) \tag{7.2}$$

where we have used the fact that the activation energy of a metal below its melting point can be empirically approximated as $E_s = 5kT_m$ [7.3]. Eq. (7.2) shows the quantitative influence on reflow of higher temperature (larger ratio of T/T_m), longer time at high temperature, and conditions that reduce the effective value of D_0 (such as surface oxidation). In addition, the relatively low eutectic point of alloys such as Al-Cu (577°C) or Al-Si (548°C) can lead to more flow at a given temperature than pure Al, which melts at 660°C.

Figure 7.3 shows a simulation of elevated-temperature Al deposition at four different surface diffusion lengths from $L = 0.06\ \mu m$ to $0.6\ \mu m$. Perfect wetting of the film was assumed, so that a continuous film without voids developed. In spite of this simplification, Fig. 7.3 graphically shows how even a factor-of-10 change in surface diffusion length can greatly impact the profile. The diffusion lengths indicated in Fig. 7.3 were not calculated *ab initio* but were determined as the value of L giving the best fit of the modeled profile with the experimentally measured profile.

To make this more quantitative, consider the hot Al profile ($T = 523$ K) in Fig. 7.3b for which a value of $L = 0.18\ \mu m$ gave the best agreement with experiment. Since $T_m/T = (933\ K/523\ K) = 1.78$, we can use Eq. (7.2) to calculate that $D_0\tau \approx 240\ \mu m^2 = 2.4 \times 10^{-6}\ cm^2$. An educated guess for τ can then be made by considering τ to be the time to immobilize a surface Al adatom by burial with incident Al or reaction with a reactive residual gas. For example, if the Al deposition rate were 1 μm/min (167 Å/sec) and burial occurred after 5 Å were deposited following adatom arrival, then $\tau \approx 5$ msec. Alternatively, assuming that residual gas contamination pressure (e.g., water vapor) is $\approx 10^{-8}$ Torr, then the time to form a monolayer of contamination is 0.02 sec. If a coverage of 10% on the surface is sufficient to immobilize the Al atom with reasonable probability, then $\tau \approx 10\% \times 0.02$ sec = 2 msec. Based on the agreement of these

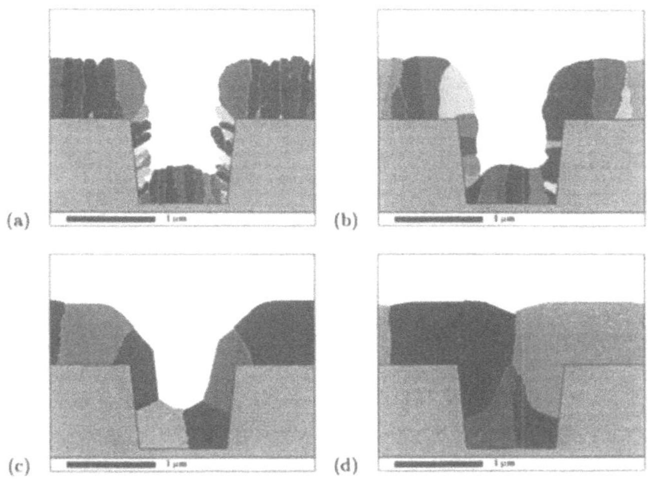

FIG. 7.3 SIMBAD™ simulation of $T = 250°C$ (523 K) Al deposition at four different diffusion lengths: (a) 0.06 μm, (b) 0.18 μm, (c) 0.6 μm, and (d) 1.2 μm.

admittedly hand-waving arguments, a value of $\tau \approx 3$ msec seems reasonable. Since we calculated that $D_0\tau \approx 2.4 \times 10^{-6}$ cm^2 for the profile in Fig. 7.3b, taking $\tau = 3$ msec then leads to an order-of-magnitude estimate for D_0 of $\approx 8 \times 10^{-4}$ cm^2/sec.

7.2 Elevated-Temperature PVD Al

7.2.1 Reflow Al

In general, one expects hot PVD processes to be most effective with materials that have high surface and/or bulk diffusivity at typical PVD process temperature, which turns out to be the case for Al. M. Inoue *et al.* of Fujitsu reported in 1988 that sputtering Al onto very hot patterned substrates (> 500°C) could be used to produce smooth, planarized metal films [7.4], and in 1991 a refinement involving "aluminum planarization by post heating" was presented by C. S. Park *et al.* of Samsung. The latter method, now commonly called *reflow*, involved the deposition of Al at room temperature followed by a vacuum-integrated, high-temperature annealing step at ≈ 500°C [7.5]. Submicron contacts with aspect ratio > 1:1 could be completely filled with Al alloys by this process. Figure 7.4 shows typical process conditions for a single reflow (room-temperature PVD Al + high-temperature anneal) and a double reflow in which the single reflow process is applied twice in succession without a vacuum break. In the double reflow case, the thickness of the initial Al deposition is typically ≈ 30–50% of the desired final film thickness. Both viscous flow of Al and bulk diffusion are relatively small at 500°C, and material transport is primarily by surface diffusion.

Reflow requires an extremely clean system because both surface contamination on the wafer and gas-phase impurities such as oxygen or water vapor that oxidize the surface will significantly reduce the diffusion of Al adatoms (see Eq. (7.2)), particularly at high temperature where the reactivity of Al is greater. Surface diffusion is suppressed even if aluminum oxide only forms in islands. Therefore, providing a proper wafer degas before reflow and maintaining an ultralow partial pressure of oxidants during reflow are essential to success. For example, exposure of Al to oxidizing gases above that needed to form a monolayer of Al_2O_3 (≈ 5–10 Langmuirs) can suppress the reflow process. For an oxidizing gas partial pressure of 10^{-8} Torr, this would take about 10–20 minutes. In this regard, the time between the initial Al deposition and the reflow must be kept short enough to prevent oxidation of the Al surface that could poison the

Single Step Reflow Process

Degas
Preclean Sputter Etch (450°C)
Collimated PVD Ti (300Å, 100°C)
PVD Al-Cu (8000Å, 50°C)
Reflow Anneal (525°C, 1.5 mTorr)

Double Step Reflow Process

Degas
Preclean Sputter Etch (25°C)
Collimated PVD Ti (300Å, 25°C)
PVD Al-Cu (1670Å, 25°C)
Reflow (580°C, 1.5 mTorr)
PVD Al-Cu (3330Å, 25°C)
Reflow (580°C, 1.5 mTorr)

FIG. 7.4 Typical process conditions for single and double Al reflow.

process. Therefore, prompt transfer of the wafer between the PVD Al module and annealing module is required. Also, since the self-diffusion constant of Al is exponentially dependent on temperature ($D \propto \exp(-E_a/kT)$), the uniformity of the reflow over the wafer surface will be strongly affected by the uniformity of the wafer temperature.

It has also been found that the reflow process is enhanced by deposition in vacuum of a thin underlayer of Ti (e.g., ≈ 300 Å) immediately prior to the Al deposition, although an underlayer of TiN deposited without vacuum break has also been found effective for this purpose [7.4, 7.6, 7.7]. The underlayer serves as both a wetting layer and adhesion layer for the Al and is sometimes deposited by collimated PVD to improve its conformality in high aspect ratio structures. A Ti wetting layer is particularly important to use if a TiN barrier is present. Since TiN is often air-exposed to improve its barrier properties by oxygen-stuffing of grain boundaries, the oxygenated TiN surface is readily reduced by the Al, and the resulting oxygen contamination can then poison the reflow. The Ti layer prevents this problem by gettering oxygen at the surface of the TiN and forming lower oxides of Ti that are not easily reduced by Al. At the completion of the high-temperature Al reflow, the Ti wetting layer has typically been converted into a refractory Ti-aluminide ($TiAl_3$) that serves to prevent stress and/or electromigration-induced voiding of the Al lines. On the other hand, $TiAl_3$ has a high resistivity ($\approx$ 35 $\mu\Omega$-cm) and takes up volume in the plug that could otherwise have been occupied by the much lower resistivity Al or Al alloy ($\approx$ 3 $\mu\Omega$-cm).

> Assuming a gas has a sticking coefficient of unity on a given surface and adsorbs uniformly, a monolayer of gas will cover the surface after about 2 sec at 10^{-6} Torr. The Langmuir is a unit of gas exposure defined such that 1 L = 10^{-6} Torr-sec, and therefore corresponds to an exposure of about 0.5 monolayer. Because the units of a Langmuir are pressure $\times$ time, 1 L corresponds to 1 sec exposure at 10^{-6} Torr, 100 sec at 10^{-8} Torr, etc.

While reflow Al processing has been implemented for 0.5-μm devices, there are difficulties in making it work in production at much smaller geometries. Since reflow relies on the mobility of Al over an underlying wetting layer, conformal coverage of this thin layer is desired, with particular attention to the sidewalls. Sidewall coverage can be facilitated by using sloped or even champagne-glass-shaped hole profiles; however, the high packing density of sub-0.5-μm devices requires straight-walled contacts and vias that are much more difficult to coat and fill.

Another concern about reflow Al is the relatively high wafer temperature, which is contrary to the trend toward lower process temperature ($<$ 400°C) in advanced device fabrication. Clearly, heating the wafer above the melting point of Al (660°C) is to be avoided because of potential interactions with other materials; however, even temperatures of 500–550°C can produce films with large grains and grain boundary grooving with resulting high surface roughness and poor reflectivity. Such films are difficult to optically align for submicron lithographic patterning. In addition, because the reflow process is extremely sensitive to oxidation, wafer degas is required, which in turn requires subjecting the wafer to temperatures 50°C hotter than the reflow process itself (i.e., $>$ 550°C). The effect of repeated high-temperature cycling on lower levels of metal raises concerns about stress voiding and could limit the use of high-temperature reflow to the lowermost levels — i.e., the contact hole and lowest-level via — of a multilevel metal interconnect. Consequently, the application of a hot Al PVD process to the upper via levels of advanced devices will probably require wafer temperature < 400°C. In addition, reflow temperature must also be compatible with the increased thermal sensitivity of polymeric insulators that are being considered as a low-*k* interlevel dielectric replacement for CVD silicon oxides [7.8].

7.2.2 Two-Step Process (TSP) Al PVD

Reflow Al involves the high-temperature annealing of an Al film deposited cold, i.e., at room temperature. It is natural to ask if the process could be improved if the sputter deposition of Al were done hot. This turns out not to be the case. In the absence of perfect wetting of the Al to the substrate, the increased mobility of the Al adatoms at high temperature quickly leads to agglomeration of the Al into islands. As the Al continues to deposit and nucleate on this discontinuous seed layer, the high mobility of the Al atoms leads to the nuclei growing larger and more widely spaced before they coalesce to form a continuous film. Large nuclei forming at the top edge of the via can shadow the walls, thereby preventing deposition deeper into the structure and exacerbating the situation. The net result is incomplete filling or complete filling but with a buried void (see Fig. 7.5).

What is required is a fine-grained, continuous seed layer of Al upon which the hot Al can flow. This concept is the basis of the two-step process (TSP) for PVD Al illustrated in Fig. 7.6. TSP consists of cold deposition of a fine-grained Al seed layer upon which hot Al deposition and flow proceeds; therefore TSP is also referred to as the cold-hot Al process. Technically, the two-step process actually consists of three steps since, as with reflow, a wetting layer is deposited immediately prior to the cold Al step to reduce the surface tension of the cold Al and prevent voiding at the sidewalls during the hot Al step. In addition, the cold and hot steps are conveniently done in the same PVD module by using backside gas to "switch on" the hot Al PVD step.

In this case, the TSP Al process can be summarized as follows. After deposition of a PVD Ti wetting layer, the wafer is handed off without vacuum break into a separate process chamber where both cold and hot PVD Al take place. The heater table in the PVD Al module is maintained at elevated temperature (e.g., 475°C) — with the gas turned off. The cold Al seed layer is then deposited under conditions of high power (e.g., 10 kW), which can be done sufficiently fast to limit the rise in wafer temperature. At this point the backside gas is switched on, and the wafer quickly increases in temperature due to the gas-assisted heat transfer between wafer and heater table. At the same time, the sputtering power to the target is also

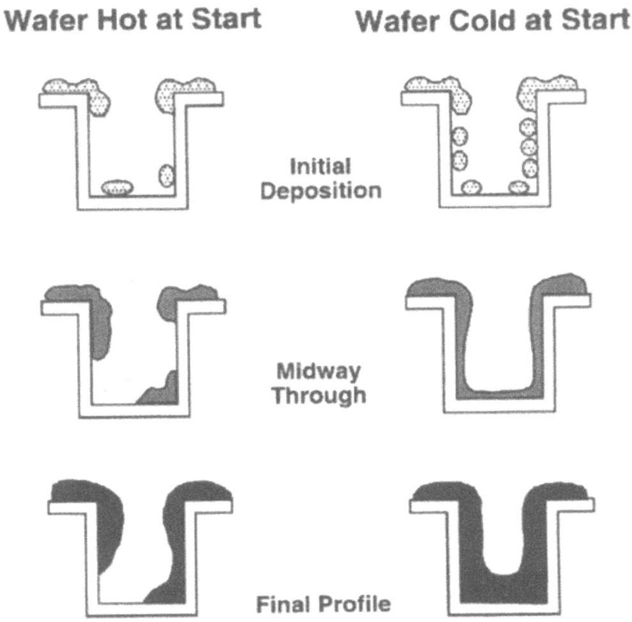

FIG. 7.5 Deposition of Al onto a hot substrate can lead to islanding of the Al, which in turn gives rise to a discontinuous film that prevents void-free filling. Reprinted from the March 1990 edition of Solid State Technology (copyright 1990 by PennWell).

decreased (e.g., 1 kW) and the remainder of the Al thickness is slowly deposited onto the hot wafer.

This slow deposition allows sufficient time for thermal diffusion to planarize the surface and fill the structure. Since the as-deposited cold Al layer has a large degree of structural disorder, it provides a ready source of vacancies and point defects that act to enhance the bulk transport of material from the surface. Ideally, one wants to prevent the top of the feature from closing during deposition since this "chokes off" surface diffusion of material down into the feature and creates buried voids that can only be filled by the slower process of bulk diffusion. Slow deposition helps prevent this closure by allowing time for material that would otherwise grow laterally at the top of the feature to diffuse deeper into the structure. Unfortunately, formation of voids, albeit small ones, cannot always be prevented before complete filling of very high aspect ratio holes. In this case,

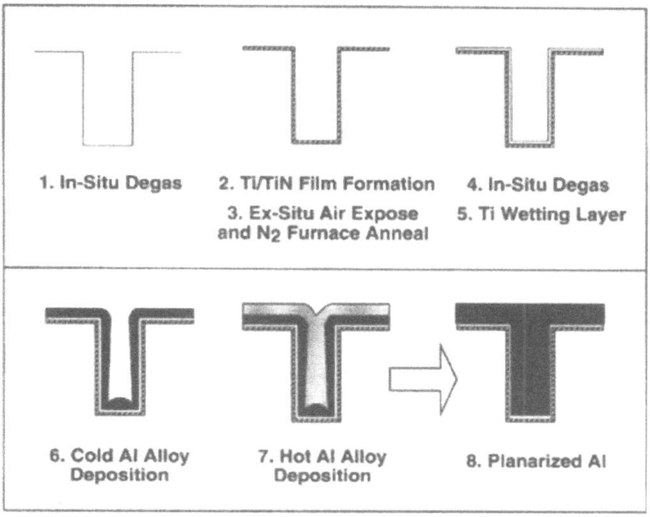

(a)

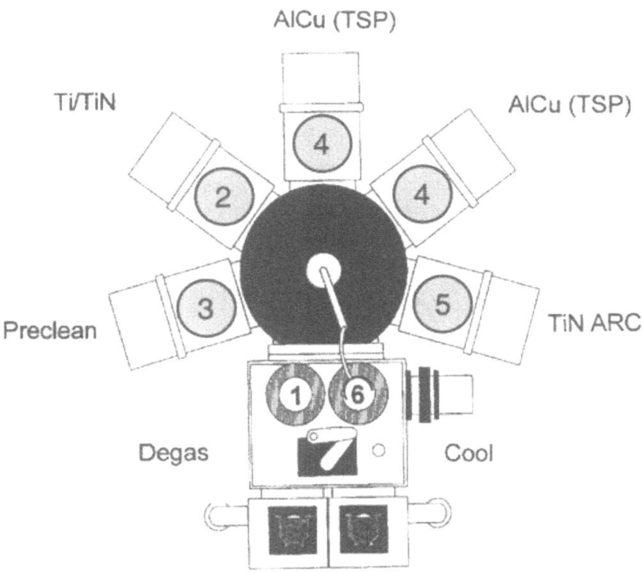

(b)

FIG. 7.6 (a) Two-step process (TSP) for Al, (b) illustration of how TSP might be implemented on a PVD cluster tool.

the slow deposition then allows time for the buried voids to fill prior to the desired Al field thickness being deposited. Because the total time of the cold-hot Al step can be relatively long, two modules are sometimes dedicated to this step to balance the throughput of the overall cluster tool (see Fig. 7.6b).

It should be noted that the concepts underlying two-step processing (and multistep PVD processing in general) can and have been implemented in many different ways, so there is really no such thing as *the* two-step process. With this understanding, a representative time-temperature profile for a TSP process is shown in Fig. 7.7. Note that the wafer chuck temperature was held constant through the process (heater set point ≈ 480°C), so that the changes in wafer temperature are the result of radiative coupling and, during the hot Al step, gas-assisted heat transfer to the chuck. Also, the steady-state temperature of the wafer is ≈ 50°C cooler than the chuck, so when discussing TSP it is important to indicate whether the temperature is that of the wafer or the heater table.

Modeling of the single-chamber TSP process using the SIMBAD™ code is shown in Fig. 7.8, where simulated filling of a 0.6-μm × 1-μm contact hole is compared with experiment. Initially, a 0.2-μm-thick layer of cold Al (20°C) was deposited for 10 sec at 11 kW, followed by 0.35 μm of hot

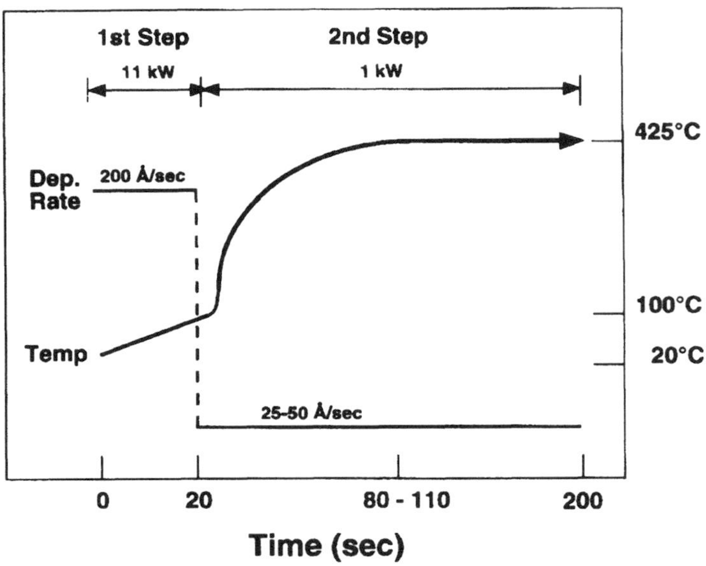

FIG. 7.7 Representative time–temperature profile for a two-step Al process.

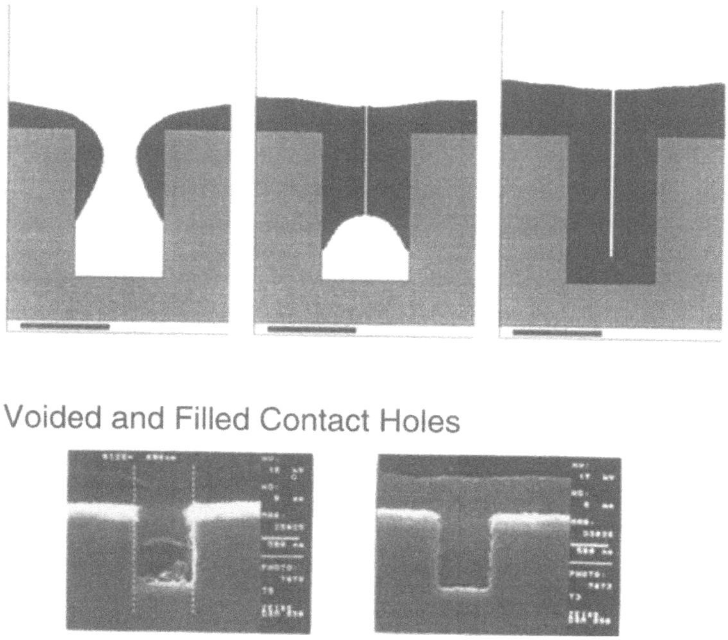

FIG. 7.8 SIMBAD™ simulation (above) and experimental SEMs (below) of a single-chamber TSP Al process used to fill a 0.6-μm × 1-μm contact hole. Initially, a 0.2-μm-thick layer of cold Al (20°C) was deposited for 10 sec at 11 kW, followed by 0.35 μm of hot Al (530°C) for 180 sec at 0.9 kW.

Al (530°C) for 180 sec at 0.9 kW. Figure 7.8 provides a snapshot of the TSP process after 60, 120, and 180 sec of hot Al deposition (left to right) progressing from voided to filled contact. The model is in excellent agreement with experiment and shows how the combined result of low-angle sputtered Al adatoms and surface diffusion leads to the formation of a large buried void 120 sec into the hot Al step, with the void being completely filled by vacancy-assisted bulk diffusion at the end of the 180-sec process.

7.2.3 Improvements to TSP Al PVD

The extension of TSP Al technology to higher aspect ratio structures and/or lower temperatures has concentrated primarily on vacuum quality, the microstructure and coverage of the Ti wetting layer, the conformality of the Al layer, and advanced Al alloys.

Vacuum Quality As discussed earlier, high partial pressures of oxidants during sputtering and reflow are to be avoided since this can suppress the TSP process by surface oxidation of the Al. However, it has also been found that ultralow partial pressures of oxidants allow TSP temperature to be reduced. For example, data by Kikuta *et al.* [7.9] on filling of a 3:1 aspect ratio contact hole with pure Al (Fig. 7.9) show that the temperature needed for complete filling could be reduced from 450°C to 430°C by lowering the partial pressure of water vapor during the PVD Al step from 10^{-7} Torr to 10^{-9} Torr. Using even lower partial pressures (2×10^{-11} Torr of O_2) and UHV conditions, Mukai *et al.* showed that evaporated Al can be reflowed at temperatures as low as 250°C. Although diffusive transport of Al proceeds more slowly at low temperature, the use of lower temperature can actually facilitate the process both because film contamination caused by wafer and hardware outgassing (e.g., the substrate heater) is less likely and because the Al film is less reactive with gases such as O_2, H_2O, and N_2 at lower temperatures.

The Ti Wetting Layer Wetting describes the extent to which two dissimilar materials are attracted at their interface. Wetting is represented in

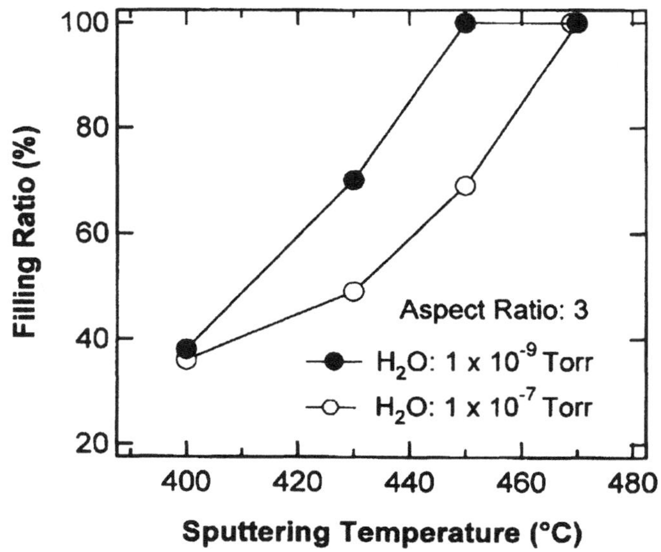

FIG. 7.9 Filling ratio of a 3:1 contact hole for hot Al PVD as a function of wafer temperature for different partial pressures of H_2O (taken from Kikuta *et al.*, ref. 7.9).

Fig. 7.10, which shows a vapor of sputtered Al atoms forming a droplet-shaped film on a substrate, where "substrate" is taken in the most general terms — e.g., bare Si, an interlevel dielectric such as SiO_2, a coating of Ti on the dielectric, etc. The contact, or wetting angle ϕ, between the film and substrate depends on the interfacial tensions between vapor and film (T_{vf}), film and substrate (T_{fs}), and substrate and vapor (T_{sv}). In equilibrium, these vector tensions are balanced as shown in Fig. 7.10 ($T_{sv} = T_{fs} + T_{vf} \cos \phi$) so that the wetting angle can be expressed as

$$\phi = \cos^{-1}\left[\frac{(T_{sv} - T_{fs})}{T_{vf}}\right] \tag{7.3}$$

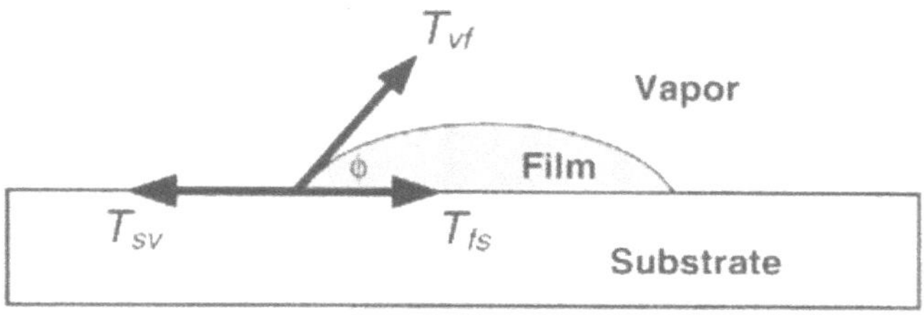

Surface and Interfacial Tensions (J/m²) T_{sv} T_{fs} T_{vf}	
1.6 0.1 1.5	$\phi = 10°$
1.1 0.2 1.5	$\phi = 45°$
1.0 2.0 1.5	$\phi = 135°$

FIG. 7.10 The relative values of the three interfacial tensions for a film–substrate combination determine the wetting angle ϕ of the film on the substrate. A substrate with a small wetting angle for Al is desired to facilitate the PVD Al cold-hot process.

The surface tension T_{vf} of Al is ≈ 1.5 J/m^2, while the other two tensions have values that depend on the specific substrate but are typically in the range of about 0.2–2 J/m^2. Figure 7.10 represents the wetting of an Al film for several values of T_{sv} and T_{fs}. To facilitate TSP Al processing, a small wetting angle is preferred (e.g., ϕ = 10°) since this results in the Al adatoms spreading out as a thin film instead of balling up (e.g., ϕ = 135°). The choice of a proper wetting layer is therefore a key to successful TSP processing — analogous to the addition of a surfactant to laundry water to improve its ability to wet the fine structures created by microscopic clothing fibers.

Ti turns out to be a good wetting layer primarily because the Ti-Al bond strength is relatively strong compared to either Al-Al or Ti-Ti. Use of Ti as a wetting layer has been further enhanced by controlling Ti directionality and deposition temperature. Namely, to ensure that this Ti film is continuous over steep structures, deposition by directional PVD methods such as collimation and long-throw sputtering have been used (see Chapter 6). In addition, it has been found that depositing the Ti with the wafer cold (e.g., at room temperature) forms a finer-grained nucleation/wetting layer for the Al that facilitates later reflow. Other materials such as TiN have also been explored as improved wetting layers for Al flow. Unlike Ti, TiN does not convert to high resistivity $TiAl_3$ during the hot cycle of the cold-hot TSP process. It has also been reported that TiN films deposited by ionized PVD methods can have a very smooth surface morphology that enables the two-step hot Al process to be done at lower temperature and/or in higher aspect ratio features.

Conformality of the Al Layer Conformality of the Al layer is equally important to avoid significant bread-loafing that could bridge the top of the structure with metal and form a large, buried void. The only way to fill a buried void is by bulk diffusion, which is in general much slower than surface diffusion. Therefore, the goal is to prevent a void from forming or, at least, to prevent its formation until late in the TSP process when it will be smaller and easier to remove. By using lower-pressure PVD (≈ 0.5 mTorr) to reduce gas-phase scattering and improve conformality, cold Al layers have been deposited for this purpose. This reduces the amount of Al migration needed for complete hole fill and metal layer planarization, and the low pressure helps maintain a clean environment. Another low-pressure PVD method — long-throw sputtering — has also been used to deposit Al for TSP applications. Although long-throw deposition generally leads to asymmetric sidewall coverage, this is compensated for to some extent by the smoothing effect of the reflow process. As long as a sufficiently thick, continuous Al seed layer is deposited (e.g., ≈ 500 Å in a 0.35-μm hole), successful reflow can be carried out. By employing a collimated Ti wetting layer deposited at room temperature and carrying out Al PVD at low pressure, it has been possible to use TSP for void-free filling of 4:1 aspect ratio 0.5-μm plugs at wafer temperature ≈ 380°C [7.10].

Advanced Al Alloys Although Al alloys based on the Al-Si-Cu system have dominated IC metallization, alternative alloys have been explored for improved performance. With respect to hot Al PVD, there is strong interest in alloys whose eutectic point is lower than that of either Al-Si (577°C) or Al-Cu (548°C) and that are therefore likely to flow more readily at lower temperature. Since Al-Si-Cu is well established in IC processing, Al-Ge-Cu alloys are of particular interest since Ge is isolectronic to Si, both being Group IV elements of the periodic table. Also, only a few weight percent of Ge is needed to significantly lower the reflow temperature [7.11–7.13]. For example, Kikawa *et al.* report that the addition of 0.5% of Ge to Al-0.5%Cu allowed the reflow temperature needed to fill a via to be reduced by ≈ 50°C, from 460°C to 410°C, reflecting a comparable decrease in the bulk melting point of the alloy [7.14]. With regard to electrical performance, Al-1%Ge-0.5%Cu alloys have exhibited electromigration reliability similar to that of conventional Al-1%Si-0.5%Cu alloys [7.12].

While the basic TSP process and its subsequent improvements have addressed many of the early concerns about hot Al PVD processing, susceptibility to geometry remains an issue. For example, a wide via fills more slowly than a small one, and the outermost features in an array of structures may fill faster than the innermost ones because they have more surrounding real estate to draw upon as a source of Al. This can lead to process variability whenever feature size and/or aspect ratio change over a given wafer or from wafer to wafer.

7.3 Elevated-Temperature PVD Cu

Although reflow, TSP, and Forcefill™ processing (to be discussed in Section 7.4) of Al alloys containing up to a few weight percent Cu have been well documented, much less has been reported on the hot PVD processing of pure Cu. This has begun to change based on the desire to replace or augment Al alloy interconnects in ULSI devices with lower-resistivity, more electromigration-resistant Cu wiring. Compared to Al, though, Cu has a much higher melting point (T_{mp} = 1356 K for Cu and 933 K for Al). Therefore, one might expect that hot PVD of Cu would be more difficult to implement than hot PVD of Al since the ratio of T/T_{mp} at typical reflow or TSP temperature (e.g., $T = 450°C = 723$ K) is so much smaller for Cu (≈ 0.5) than for Al (≈ 0.8). On the other hand, Cu exhibits a surface diffusion coefficient $D = D_0 \exp(-E_a/kT)$ with a relatively low activation energy $E_a = 0.83$ eV, allowing thermal flow to be exploited — although the flow is not as pronounced as with Al. Also, the choice of a suitable barrier and/or wetting layer for hot PVD Cu processing is still an open issue. Ti has worked well with Al and Al alloys but is not suitable for Cu. Other possibilities such as TiN, Ta, TaN, and PVD W or CVD W over TiN have not been sufficiently explored to make a clear recommendation, although work with Ta wetting layers has been quite promising [7.15]. For example, Cu has successfully been reflowed into high aspect ratio features (e.g., 0.1-μm × 0.5-μm trenches) at moderate times and temperatures (e.g., 5 min at 400°C). In this regard, it has been reported that an atomic hydrogen-enriched H/H_2 ambient during reflow annealing of Cu can enhance surface mobility and thereby lower the required process temperature [7.16]. The effect appears to be a combination of surface cleaning by active H (e.g., removal of O and C contamination) and reduction of the Cu-Cu binding energy of the top layer of Cu atoms to the bulk. As a result, void-free reflow of PVD Cu in 0.15-μm-wide × 0.5-μm-deep trenches could be achieved in 30 min at 320°C.

One of the most interesting aspects of hot Cu PVD relates to microstructural effects that have been found to have a strong influence on reflow process variability [7.17–7.19]. The evolution of the microstructure of Cu films is quite different than that of Al and may relate to fundamental differences in their metallurgy and oxidation kinetics. For example, the as-deposited grain size distribution of PVD Cu is usually not log-normal, but often contains a bimodal distribution of large and small grains. Under elevated-temperature annealing, the growth of small Cu grains occurs before that of the larger grains, and the microstructure and topography of the film evolves in a random way that may be very different from one via to the next. As a consequence, microstructural effects can influence the global uniformity of the process. For example, the deformation known as grain boundary grooving is a response to interfacial tension between two grains and serves to lower the surface free energy of the Cu below that of a free surface with the same curvature. As a result, surface diffusion slows in the vicinity of the groove and filling is held back. Since filling then depends on details of the local film microstructure, it can proceed at different rates in structures that are geometrically identical.

Figure 7.11 shows experimental data and simulations for PVD Cu on a 0.35-μm-wide, 2:1 aspect ratio trench with a Ta wetting/adhesion

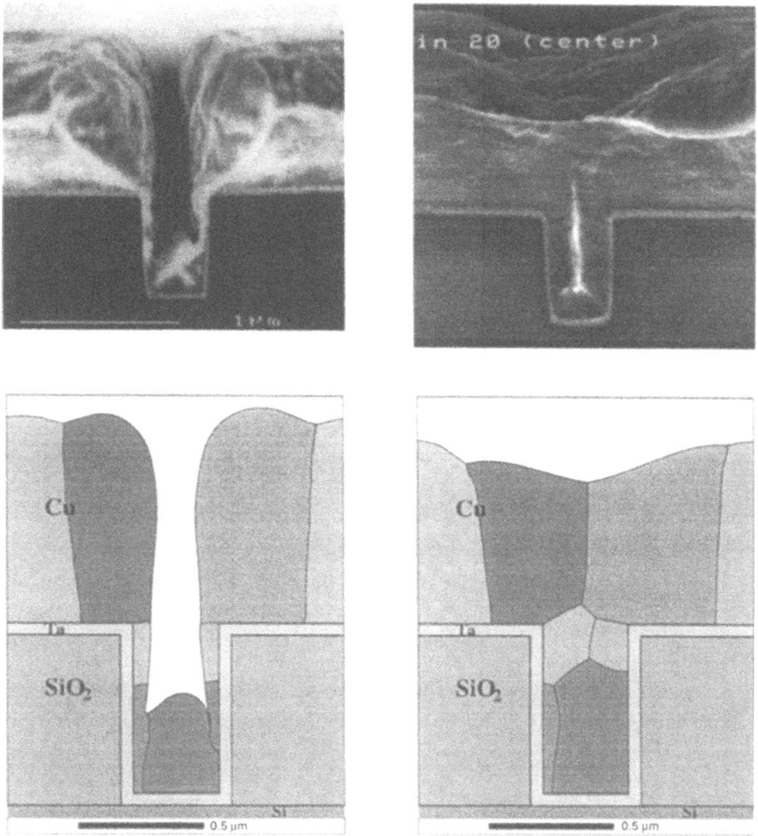

FIG. 7.11 Experimental SEM data (top) and GROFILMS™ simulation (bottom) of PVD Cu on a 0.35-μm wide, 2:1 aspect ratio trench with a Ta wetting/adhesion layer, before and after a reflow anneal (25 min at 450°C). The simulation took into account microstructural effects such as grain boundary grooving and faceting [7.17].

layer, both before and after a reflow anneal (25 min at 450°C). The GROFILMS™ simulation took into account microstructural effects such as grain boundary grooving and faceting, and is in good agreement with experiment. Figure 7.12 shows the same GROFILMS simulation of Cu reflow over a multiple-trench topography. When microstructural effects were omitted from the model, the time to completely fill each trench was the same. However, the process showed large variability when these

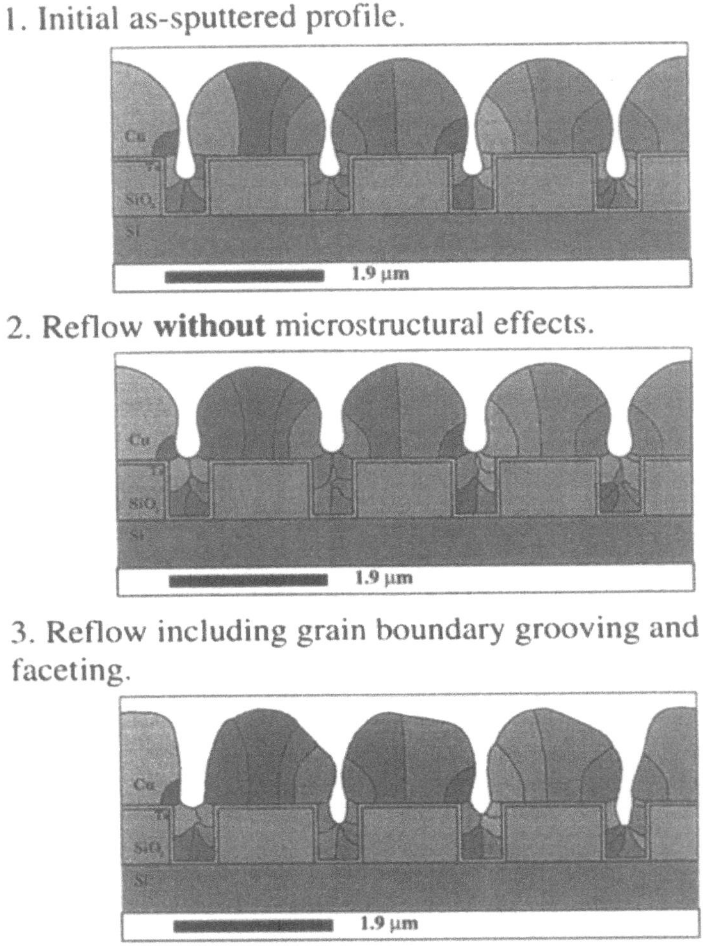

FIG. 7.12 GROFILMS™ simulation of PVD Cu reflow over a multiple trench topography: (1) as-sputtered, (2) reflowed without microstructural effects taken into account, and (3) reflowed with microstructural effects taken into account [7.17].

realistic effects were included, which is consistent with experimental data (Fig. 7.13). Other than engineering the PVD Cu grain structure or wetting layer for improved process uniformity, one should probably keep the reflow time sufficiently long and/or temperature high to accommodate the worst-case configuration.

FIG. 7.13 Cross-sectional TEM image of PVD Cu reflow in 0.5-μm, 1.5:1 aspect ratio trenches with a Ta liner, showing variability of the process due to local differences in Cu microstructure [7.17].

7.4 Application of High Pressure

The simultaneous application of ultrahigh pressure ($\approx$ 60 MPa or 600 atmospheres) and moderate heat ($\approx$ 400°C) after PVD have been exploited to enhance the filling of sub-0.5-μm geometries without the need for excessive temperature [7.20–7.22]. This novel process was introduced by Electrotech (now a part of Trikon Technologies, Inc.) and is called Forcefill™ or "Hi-Fill." The process is sometimes referred to as "high-pressure sputtering"; however, this is not accurate since the sputtering step is done at conventional mTorr-type pressures and is physically separated from the subsequent application of ultrahigh pressure.

Figure 7.14 shows a schematic diagram of the Forcefill process for an Al via-fill application. The wafer would typically be degassed, via cleaned, and have a Ti/TiN barrier deposited. An Al layer is then deposited at 400–450°C in a conventional sputtering system under high-deposition-rate conditions to generate rapid grain growth in the plane of the film. When the final Al thickness is greater than the hole diameter, it then forms a continuous bridge over the entrance of the via hole, resulting in a sealed gas cavity at the process pressure ($\approx$ 3 mTorr). The wafer is then transferred to another chamber (see Fig. 7.15) where high-pressure Ar is isostatically applied at moderate temperature to force the Al into the hole by an enhanced plastic flow process. Assuming that all of atoms from the sputtering ambient enclosed in the hole dissolve into the Al, this would represent only an insignificant impurity in the filled plug ($\approx$ 0.1 ppm Ar). The elevated temperature is employed to reduce the flow stress needed to achieve plastic deformation. It is also reported that that elevated temperature promotes a dynamic recrystallization, leading to strain-free grains with very large median grain size ($\approx$ 10 μm compared to 2 μm for TSP-processed

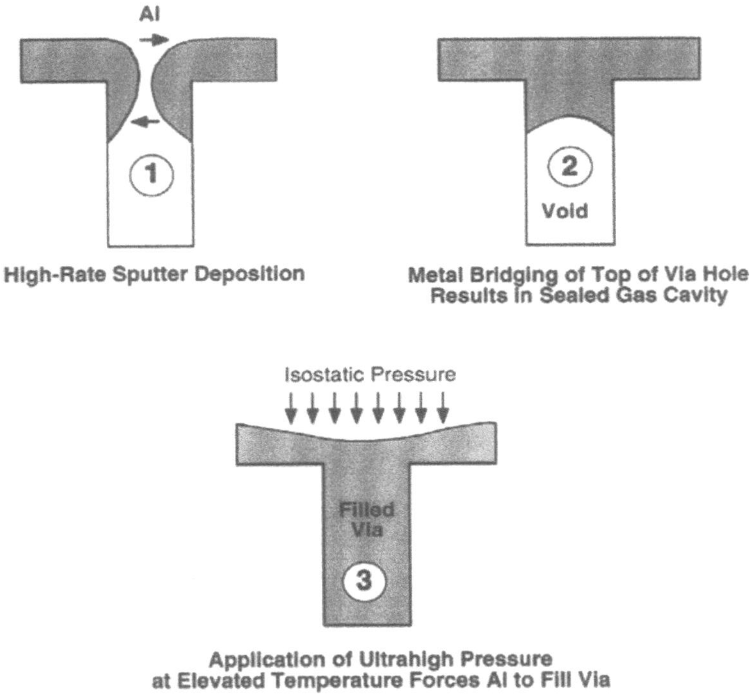

FIG. 7.14 Schematic of the Forcefill™ process applied for Al via filling.

Al), which could contribute to increased electromigration resistance [7.20]. The relative contributions of temperature and pressure to fill capability are illustrated in Fig. 7.16, which shows how one can fill at lower process temperature by applying higher pressure.

It has been reported that Forcefill Al provides global, complete filling of both straight-walled and reentrant 0.5-μm holes provided the metal is sufficiently thick on the field (e.g., 700 nm) to bridge the diameter of the hole and form an encapsulated void. That is, the fill capability of the method is less dependent on the amount of sputtered Al that reaches the bottom and sidewalls of the hole than on the amount deposited on the field. Since the former coverage depends on aspect ratio and the latter coverage does not, Forcefill is well suited for global filling even when features of different aspect ratio are present on the wafer surface. The method has been applied to 64Mb DRAMs and multilevel 0.35-μm logic devices.

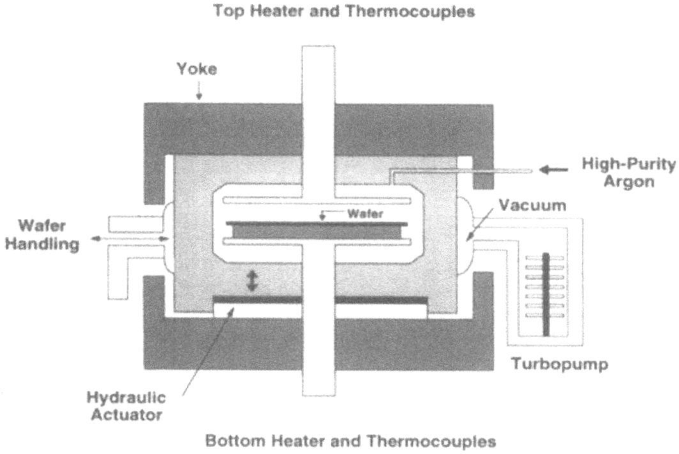

FIG. 7.15 Schematic of the high pressure chamber used as part of the Forcefill™ process. Two radiant heaters are used for heating the wafer (≈ 400°C) during the application of ultrahigh isostatic pressure (≈ 60 MPa). (Source: G. Dixet et al, *Semiconductor Int'l.*, p. 79, August 1995).

On the other hand, concerns have been raised about the compatibility of extremely high pressure with deformable, organic interlayer dielectrics. Also, while a reduced thermal budget is expected for Forcefill over reflow, it may not be significantly better than that of TSP since both utilize comparable temperature (≈ 400°C) and times. It has also been reported that deep grain boundaries can form in Al films when the Forcefill process is carried out over densely packed arrays of vias [7.23]. Such films have a cracked appearance where Al grains have separated from one another, which could lead to reliability or manufacturing difficulties if not controlled. The effect has been attributed to the fact that in such closely packed arrays, a single Al grain can serve as the source of material to fill several underlying vias. During high-pressure annealing, a large fraction of the grain's volume (> 25%) is rapidly consumed, which forces it to distort and physically separate from neighboring grains [7.23].

Forcefill shares elements in common with both reflow and TSP hot processing. For example, Forcefill and reflow both utilize a conformal PVD barrier/liner and a PVD Al step followed by transfer under high vacuum to a separate module where heat treatment is carried out. Of course, the ultrahigh pressure used results in a radically different module (Fig. 7.15). Both the Forcefill and the TSP processes have a common interest in buried voids, although Forcefill is designed to induce their formation while TSP

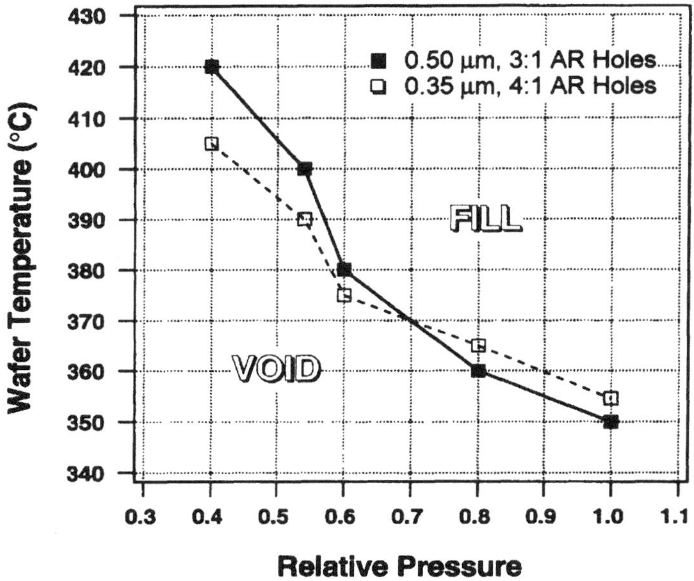

FIG. 7.16 The ability to completely fill a feature (void-free filling) using the Forcefill™ process depends on both pressure and temperature, as illustrated for a 0.5-μm, 3:1 aspect ratio hole (Source: G. Dixit et al, *Semiconductor Int'l.*, p. 79, August 1995).

seeks to prevent this. Also, both methods have been used for void-free filling of sub-0.5-μm features at 400°C and thus offer a lower-temperature process alternative to reflow.

7.5 Conclusions

As discussed above, elevated temperature and ultrahigh pressure have been used to promote material transport both during and after PVD. Improvements to these methods — e.g., higher vacuum base pressure, improved wetting/nucleation layers, more conformal cold layers for cold-hot processing — have allowed PVD to be applied for coating and filling of 0.25-μm ULSI structures. In spite of this success, it may not be possible to extend the same methods to deep submicron devices ($\leq$ 0.18 μm) due to reduced thermal budgets, much higher aspect ratios (e.g., simultaneous filling of a 5:1 via and 5:1 trench with the dual-damascene process is effectively like filling a 10:1 aspect ratio feature), and the introduction of interconnects and via plugs based on Cu which, compared to Al at a given temperature, has much smaller surface and bulk self-diffusion coefficients. Of the methods described above, the conventional reflow process is least likely to be deployed for 0.18-μm devices due to the relatively high temperatures and need to customize geometry (e.g., sloped or champagne-glass-shaped contacts). Likewise, the Forcefill method may not be compatible with temperature- and pressure-sensitive low-k polymers nor work as well with Cu as with Al. The most likely scenario for filling appears to be the use of slightly elevated temperature ($\leq$ 400°C) PVD in synergistic combination with ionized PVD or CVD (see Section 9.9). For example, a Ti or TiN wetting layer with improved conformality could be deposited using ionized PVD, the structure partially filled using CVD Al, and then filling and surface planarization completed using hot PVD AlCu in which the elevated temperature (or a subsequent anneal) serves to drive Cu from the AlCu into the underlying Al film for improved electromigration resistance.

References

7.1. S. K. Dew, T. Smy, and M. J. Brett, "Simulation of elevated temperature aluminum metallization using SIMBAD," *IEEE Trans. on Electron Devices* **39**(7): 1599–1606 (1992).

7.2. W. Mullins, "Flattening of a nearly planar solid surface due to capillarity," *J. of Appl. Phys.* **30**(1): 77–83 (1959).

7.3. G. Neumann and W. Hirschwald, "The mechanisms of surface self-diffusion," *Zeitschrift fur Physik Chemie. Neue Folge* **81**(1–4): 163–176 (1972).

7.4. M. Inoue, K. Hashizume, and H. Tsuchikawa, "The properties of aluminum thin films sputter deposited at elevated temperatures," *J. Vac. Sci. & Tech.* **A6**(3): 1636–1639 (1988).

7.5. C. S. Park, S. I. Lee, J. H. Park, J. H. Sohn, D. Chin, and J. G. Lee, "Al-PLAPH (*Al*uminum-*Pl*anarization by *P*ost-*H*eating) process for planarized double metal CMOS applications," in *Proc. VLSI Multilevel Interconnection Conf.*, pp. 326–328 (1991).

7.6. M. A. Biberger, V. Hoffman, C. H. Ting, B. Zhao, and L. Toa, "Process optimization of planarized Al via for sub-half micron ULSI interconnects," *Semicond. FABTECH* **2**: 241–247 (1995).

7.7. D. Pramanik and A. N. Saxena, "Aluminum metallization for ULSI," *Solid State Tech.* 73–79 (Mar. 1990).

7.8. P. Singer, "Low *k* Dielectrics: The search continues," *Semicond. Int.*, 88–96 (May 1996).

7.9. K. Kikuta, N. Itoh, and T. Kikkawa, "Low temperature aluminum reflow sputtering," in *Proc. Tech. Symp. of SEMICON-Korea*, pp. 281–287 (Nov. 9–10, 1993).

7.10. B. Zhao, M. A. Biberger, V. Hoffman, S.-Q. Wang, P. K. Vasudev, and T. E. Seidel, "A novel low temperature PVD planarized Al-Cu process for high aspect ratio sub-half micron interconnect," *Proc. Int. Electron Devices Meeting*, IEEE, New York) pp. 353–356 (1996).

7.11. K. Kikuta, T. Kikkawa, and H. Aoki, "0.25 μm contact hole filling by Al-Ge reflow sputtering," *Technical Digest of the Symp. on VLSI Tech.*, paper 5–2, pp. 35–36 (1991).

7.12. K. Kikuta and T. Kikkawa, "Electromigration Characteristics for Al-Ge-Cu," *J. Electrochem. Soc.* **143**(3): 1088–1092 (1996).

7.13. K. Kikuta, Y. Hayashi, T. Nakajima, K. Harashima, and T. Kikkawa, "Aluminum-germanium-copper multilevel damascene process using low-temperature reflow sputtering and chemical mechanical polishing," *IEEE Trans. on Electron Devices* **43**(5): 739–745 (1996).

7.14. K. Kikuta, "Al damascene technology for multilevel interconnections," *Technical Digest of the IEDMS Meeting*, Taiwan, December 16–20, 1996), abstract A4-3, pp. 161–167 (1996).

7.15. G. Bai, C. Chiang, J. Neal Cox, S. Fang, D. S. Gardner, A. Mack, T. Marieb, X-C. Mu, V. Ochoa, R. Villasol, and J. Yu, "Copper interconnection deposition techniques and integration," *Technical Digest of the Symp. on VLSI Tech.*, abstract 5-4, pp. 48–49 (1996).

7.16. T. Miyake, H. Petek, K. Takeda, and K. Hinode, "Atomic hydrogen enhanced reflow of copper," *Appl. Phys. Lett.* **70**(10): 1239–1240 (1997).

7.17. L. J. Friedrich, D. S. Gardner, S. K. Dew, M. J. Brett, and T. Smy, "Microstructural effects on the copper reflow process," in *Proc. of the VLSI Multilevel Interconnection Conf.*, pp. 213–218 (1996).

7.18. A. R. Sethuraman, J.-F. Wang, and L. M. Cook, "Copper vs aluminum: A planarization perspective," *Semicond. Int.*, 177–184 (June 1996).

7.19. R. A. Brain, D. S. Gardener, D. B. Fraser, and H. A. Atwater, "The effect of grain boundaries on surface diffusion mediated-planarization of polycrystalline Cu films," in *Proc. of the Materials Res. Soc. Symp.* **389**: 107–112 (1995).

7.20. G. Dixit, W. Y. Hsu, K. H. Hamamoto, M. K. Jain, L. M. Ting, R. H. Havemann, C. D. Dobson, A. I. Jeffreys, P. J. Holverson, P. Rich, D. C. Butler, and J. Hems, "Application of high pressure extruded aluminum to ULSI metallization," *Semicond. Int.*, 79–86 (Aug. 1995).

7.21. D. Butler, "Options for multilevel metallization," *Solid State Tech.*, s7–s10 (Mar. 1996).

7.22. P. Singer, "Filling contacts and vias: A progress report," *Semicond. Int.*, 89–94 (Feb. 1996).

7.23. M. Saran, "Grain-separation over vias filled by high-pressure aluminum extrusion," *Technical Proc. of the Electrochem. Soc. Symp.*, vol. 97-1, abstract no. 380, pp. 481–482 (1997).

Chapter 8 Ionized Magnetron Sputter Deposition: I-PVD

Physical sputtering from a magnetron cathode is dominated in most cases by the emission of single, neutral atoms. Cluster emission is rare and if a positive (sputtered) ion were formed at the cathode surface, it could not be emitted due to the strong, opposing electric field present in the sheath. Negative ions, which are formed during the sputtering of some compounds such as titenates, zirconates, and other oxides [8.1] are not a factor in most areas of inert gas sputtering of metal targets. The sputtered, neutral atoms have a wide variety of angular trajectories, and directional deposition is only possible by means of filtering, such as using collimators or long target-to-sample distances. This is due, of course, to the inability to controllably direct the trajectory of a neutral atom.

On the other hand, a metal ion, if it were formed away from the cathode sheath, it could be controlled by the presence of an electric field. The metal ion's trajectory would be "straightened out" in a suitably strong electric field located at the edge of the plasma, perhaps at the location of the sample. This concept is the basis of *ionized PVD*, also known as I-PVD, IMP (ionized metal plasma), or IMD (ionized magnetron deposition). The intrinsic advantage of I-PVD is that a large fraction of the sputtered metal atoms can be ionized and controllably directed to the sample surface at normal incidence. This compares to the relatively small fraction of sputtered atoms (typically a few percent) that pass through a directional filter such as a collimator.

In I-PVD, metal atoms are sputtered from a conventional magnetron source using an inert gas and a conventional magnetron power supply. A second plasma, nominally different from the magnetron source plasma, is produced in the region between the sputtering source and the sample using the same background gas. Some fraction of the sputtered metal atoms are ionized as they transit this second plasma. Finally, just above the sample surface, the metal ions are accelerated to the sample by the difference between the plasma potential (usually slightly positive) and the sample potential, which can be controlled externally to be either a zero or a negative voltage.

8.1 Experimental Systems

The deposition of metal ions to form films has been reported many times and is not a new phenomenon. This is particularly true of ECR (electron cyclotron resonance) plasmas, which can be quite dense and which if not tightly confined result in reasonable amounts of wall sputtering and subsequent metal ionization by the plasma. The practical application of this technology was demonstrated by Kidd *et al.* in the mid-1980s using a high-field ECR plasma in which the sputtering plasma doubled as the ionization plasma [8.2]. While this technique was not manufacturable, it led to work by Holber *et al.* at IBM Research [8.3] and by Barnes, Forster and Keller, at IBM East Fishkill [8.4]. The Holber group used evaporation from a metal boat of Cu or Al rather than sputtering into an ECR plasma chamber (Fig. 8.1) [8.2].

Sputtering in the axial geometry of Kidd's system results in the formation of high-energy secondary electrons that can directly impact the sample. In the Holber work, once the ECR plasma was started in an inert gas such as Ar, the metal evaporation source was turned on and the Ar gas supply turned off. This resulted in essentially a pure metal plasma, which was then used to deposit metal films onto negatively biased wafers just out of the field of view of the evaporation source. This work showed filling of various semiconductor features, although the overall tool design based on evaporation and ECR was not easily converted to a manufacturable system.

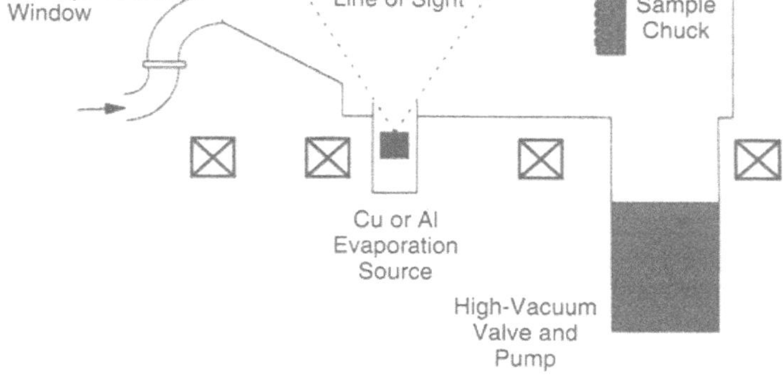

FIG. 8.1 Experiment of Holber *et al.* (8.3).

The Keller *et al.* group used a 4-cm metal sputtering source and arranged a second plasma antenna/electrode in the region between the cathode and the sample (Fig. 8.2) [8.4]. The plasma antenna was configured to be a multiple-turn metal coil, which was operated at 13.56 MHz and tuned to form an inductively coupled RF plasma. The deposition chamber was confined with a multipole magnetic bucket to enhance the degree of metal ionization as well as to increase the uniformity of the plasma and reduce edge losses. This work showed significant promise toward a manufacturing-scale system because each of the components was known to be scalable.

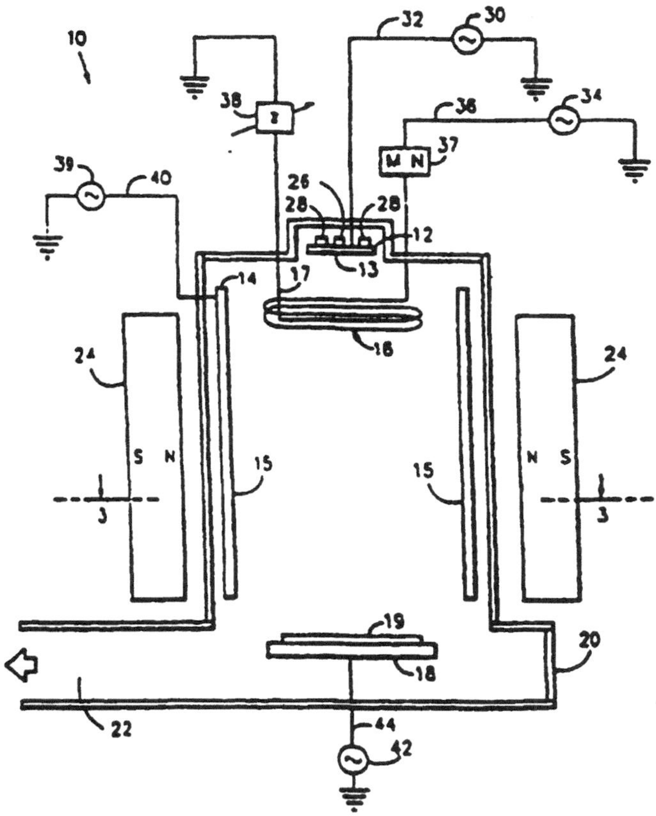

FIG. 8.2 I-PVD system from Barnes, Forster, and Keller patent [8.4].

Later work by Rossnagel and Hopwood examined the transition of this general scheme to manufacturing-scale proportions [8.5, 8.6]. This is shown schematically in Fig. 8.3. The magnetron was scaled first to 20 cm diameter and then 30 cm, the latter case using a rotating magnet design. A wide variety of RF coils were examined, with the eventual use of coils of 1–3 turns of diameter about 20% larger than the wafer, located about 1/4 of the cathode radius from both the cathode and the sample. The RF coils are operated at 13.56 MHz primarily due to the wide availability of power supplies at that frequency. Some commercial versions of this technology have moved to lower frequency — as low as 1.9 MHz — because this is believed to result in a slightly higher level of ionization [8.7].

A variety of RF coil designs have been explored that will have an effect on either the uniformity or the deposition processes. The RF antenna in an I-PVD system is immersed in the plasma, unlike inductively coupled RF coils used for etching applications. In that case, the coil is situated behind a quartz window outside the vacuum system. Since I-PVD tools are used in most cases for the deposition of metal films, any window used to separate the coil from the plasma would quickly become coated with metal and would no longer function as a window. One proposal is to shield the RF

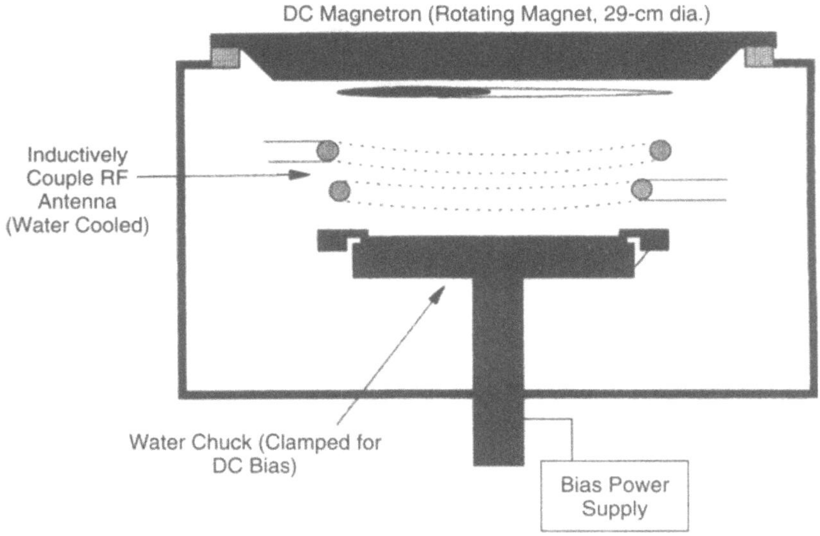

FIG. 8.3 Manufacturing-scale I-PVD tool design.

antenna/coil in the vacuum behind some sort of metal shielding. The RF fields could then penetrate into the plasma region through small holes, slits, or gaps in the shielding. Unfortunately, because of the high density of the plasmas involved ($10^{12}/cm^3$), the Debye length of the plasma is short enough (100 micron range) that the physical gaps would be quite difficult to build.

Since the RF coil is located within the plasma chamber, two significant processes occur: First, the coil develops a negative floating potential due to the higher intrinsic mobility of the plasma electrons compared to the ions. Second, the coil may receive substantial amounts of metal deposition from the magnetron source. The latter issue is of extreme importance to semiconductor applications in that it may lead to thick film buildup, subsequent flaking, and particle formation.

There are three approaches to this issue of film deposition on the RF coil. The first is to simply remove the coil from time to time for cleaning or replacement. This adds a regular maintenance task and the expense of downtime and base pressure recovery, which would be nominally similar to the expense of changing collimators.

A second approach is to tune the inductively coupled discharge to reduce the rate of film buildup on the coil. The RF coil antenna is coupled to the discharge both inductively as well as capacitively, the latter due to the physical presence of the metal coil antenna in the plasma. By altering the tuning of the coil, the ratio of the inductive coupling to the capacitive coupling can be altered. At high levels of inductive coupling, the negative bias on the coil is low and there is little sputtering of the coil surface. At high levels of capacitive coupling, the sputtering rate of the coil is high. By altering the ratio of the coupling, the level of sputtering can be adjusted to be just slightly lower than the deposition rate of metal from the magnetron. This results in a very slow buildup of metal on the coil surface and long lifetimes for the coil in the system.

The third approach overcomes the concern of deposition on the coil by always operating the coil in a net-erosion mode. By doing so, the net sputtering rate of the coil always exceeds the deposition rate, and the coil is slowly eroded. This approach requires that the coil be constructed of the same material as the magnetron cathode and that it be of the same high purity as the magnetron cathode, as atoms from the RF coil will certainly be included in the deposited films on the sample. This approach is also not really conducive to water cooling of the RF antenna/coil. High-purity tubing of most desirable materials is not available, and it would also be difficult to deal with water connections for some materials. Therefore, the approach of a always-eroded RF coil usually allows the coil to operate uncooled, which can result in coil temperature of 500°C. This can cause substantial radiant heating of the sample and may not be conducive to the ionized deposition of lower-melting-point materials such as Al.

Another aspect of the always-eroded, uncooled coil is that it no longer needs to have a round or hollow cross section. Commercial applications of this design use a flat plate to form the coil [8.7]. This leads to an inside-facing side that is heavily eroded and back-facing side that is lightly eroded. This design intrinsically removes concerns with buildup on the back-facing side and also reduces deposition near the vacuum feedthroughs, resulting in a long operating lifetime before cleaning.

The always-eroded coil can potentially lead to problems with directionality near the edge of the sample. Atoms sputtered from the coil differ from the atoms sputtered from the magnetron cathode in that (a) they have a much shorter trajectory from the source to the sample and as such as are less likely to be ionized in the discharge, and (b) since they are coming in from the perimeter coil electrode, their trajectory is mostly inward-facing as far as the wafer is concerned. These two issues may result in a lower step coverage on the bottom surfaces of features near the wafer edge. The local emission from the coil electrode may also lead to preferential deposition on the inner sidewalls of features near the wafer edge; the opposite effect of what is seen near the wafer edge in long-throw sputtering.

There has also been some experimentation with spiral coils instead of perimeter coils. The spiral coil is likely to lead to better plasma uniformity, and spiral coils are routinely used on etch-related inductively coupled plasma tools. The pitch of the spiral coil can be adjusted spatially to result in excellent uniformity for the plasma as well as allowing plasmas to be formed that are not purely circular but perhaps rectangular in cross section for use with flat panel display substrates.

However, spiral coils suffer from a very significant problem with an I-PVD tool: The deposition rate on the side of the coil that faces the magnetron sputtering cathode is much higher than the side that faces the wafer. This means that the top side of the coil is heavily deposited, and the underside (wafer side) of the coil is heavily etched, resulting in significant contamination of the film by coil material. This could only be used in the always-eroded operational mode described above.

The exact location of the RF coil has an effect on both the net deposition rate and the uniformity of the directional deposition. One obvious problem occurs is when the diameter of the RF coil is too small. This can geometrically block the edge regions of the sample from some fraction of the cathode emission surface. The inside diameter of the RF coil should be such that it does not intersect a line drawn from the edge of the sample to the edge of the cathode. Other geometrical problems are primarily found with the vertical location of the coil with respect to the sample. At distances less than 3–4 cm, emission from the coil can overwhelm deposition from the plasma and the edge regions of the sample may have poor or asymmetric directional deposition. If the coil is located less than 3–4 cm from the cathode surface, there can be a very strong coupling between the two nominally separate plasmas. This results in a much lower discharge voltage for the magnetron as well as increased tuning difficulties for the RF coil.

However, while it is critical that the RF coil not strongly interact with either the sample or the cathode, it is also important that the cathode-to-sample distance not be too large. Because the operating pressure (described below) will turn out to be several tens of mTorr, the overall deposition rate will fall strongly as a function of throw distance due to the high levels of gas scattering. At 40 mTorr, a throw distance of greater than 13–15 cm will result in over an order of magnitude reduction in the net deposition rate compared to a more typical PVD operating pressure of 1–2 mTorr. It is desirable that the throw distance be minimized as much as practical, and it appears that this minimum distance is on the order of 10 cm in a practical system.

The topics of RF coil design and operating frequency are likely to evolve in each successive generation of I-PVD tools. There are many different approaches for launching RF waves into plasmas, and subsequent designs may take advantage of the intrinsic magnetic field or some other geometry-dependent feature to generate a higher plasma density. Since the mean free path for ionization of the sputtered atoms is directly dependent on the plasma density and coupled to the electron temperature, it may be possible to increase the relative ionization of the atoms with denser or higher temperature plasmas.

The matchbox for an I-PVD tool differs from that of a conventional RF diode. The two primary differences are that the plasma coil antenna functions in place of the inductor in the conventional matchbox, and also that a third capacitor is included to help balance the potential on the coil. The matchbox is shown in Fig. 8.4. The power supply feeds both a shunt capacitor to ground and a series segment of a capacitor, the plasma electrode, and the third capacitor, which is then grounded. The capacitors before and after the plasma coil are typically of the same value (1000–2000 pf, variable air gap) and are tuned such that the voltage on each end of the coil is 180° out of phase. This limits the maximum voltage swing on the coil, reduces the amount of sputtering, and also makes it symmetric along the coil length. Depending on the operating conditions, the reflected power to the

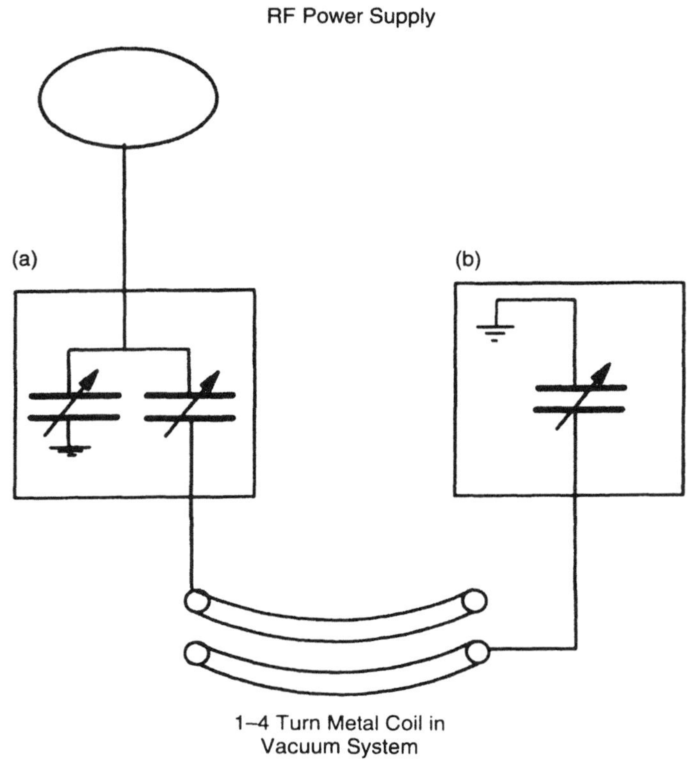

FIG. 8.4 RF matchbox design for I-PVD tool.

power supply may be as high as 20% of the incident power, although it is possible to reach very low reflected powers. The lowest reflected power may not be indicative of the best operating conditions, due to the interplay between inductive and capacitive coupling. In general, the higher the inductive coupling, the higher the plasma density, which might be seen as an increase in the ion saturation current to the sample. Higher coil voltages are consistent with a higher level of capacitive coupling, which would result in a lower plasma density as well as increased coil sputtering. Depending on the desired operating mode (from net deposition on the coil to no deposition to net erosion), the tuning of the circuit will vary — as will the reflected power. In some systems, the best inductive match (maximum plasma density, net deposition on the coil) is consistent with about

20% reflected power. The best matching conditions (minimum reflected power) tend to correlate with increased capacitive coupling and net erosion of the RF coil electrode.

Even though the initial manufacturing-scale tools for I-PVD appear to be based on inductively coupled plasma ionization of conventional magnetron-sputtered metal, several alternative designs have been developed and published. One is based on physical sputtering and ECR (electron-cyclotron resonance) plasma ionization [8.9]. This is shown schematically in Fig. 8.5. The ECR plasma in this case is set up by a permanent magnet assembly located at the top of the tool. Microwave power at 2.45 gHz is fed in near the top but from the side, creating a dense plasma in the central region of the system. A sputter cathode is located adjacent to the ECR region such that high rates of ion bombardment can occur on the cathode surface, resulting in sputtering of atoms into the ECR discharge region. The tool has been primarily used for Cu deposition, but the design would allow the use of other metals. The ECR tool typically reaches plasma densities that are 2 to 5 times higher than the inductively coupled RF plasmas, which allows some reduction in the operating pressure of the tool. This is important because of potential arcing and breakdown problems at higher pressures in the microwave-launcher region.

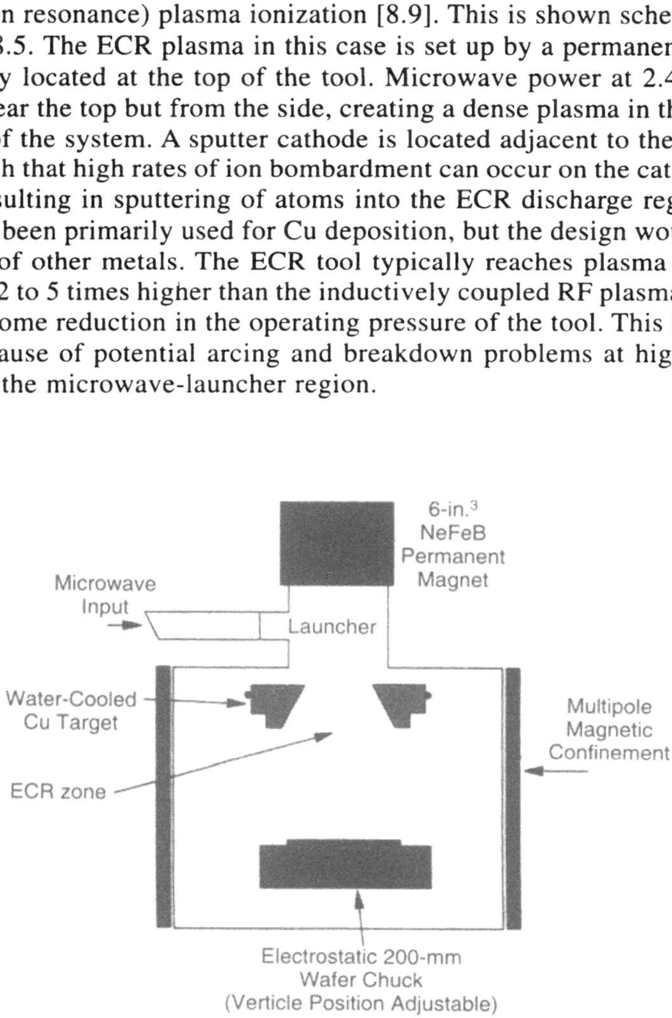

FIG. 8.5 ECR-based I-PVD tool [8.9].

A second, alternative design uses a magnetron sputtering source that has been configured as a hollow cylinder (Fig. 8.6) [8.10]. In this case, the magnetron discharge is located adjacent to the inner-facing walls of the cylinder and the magnetic field (not shown in the figure) is configured axially. Because of this geometry, an additional mechanism of electron confinement — geometrical in nature and known as the hollow cathode effect — leads to higher levels of plasma confinement and higher local densities. The sputtered metal in this case transits through the dense plasma region and has a much higher probability for ionization than in a conventional, planar discharge. The metal ions are then free to leave the source region in the form of a metal plasma beam, which diverges out toward a substrate location. This device differs from other I-PVD tools in that it has no additional active means of ionization for the metal: The ionization level is intrinsic to the hollow cathode design. It may also lead to rather high levels of ionization for the metal that leaves the source, as sputtered atoms that are not ionized and simply pass through the discharge are deposited and recycled on the opposite side of the cathode.

8.2 Plasma Aspects

The general operating process of an I-PVD system is as follows. (1) Metal atoms are sputtered from a conventional magnetron cathode and ejected as neutrals. (2) A second plasma is formed from the background or working

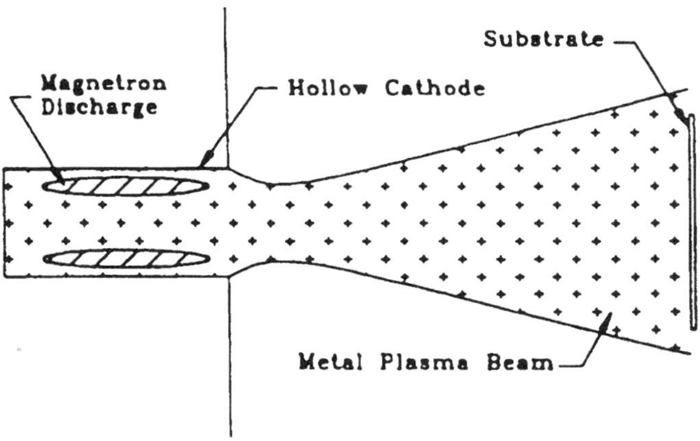

FIG. 8.6 Hollow cathode magnetron I-PVD system [8.10].

gas in the region between the cathode and the sample by the presence of the inductively coupled RF antenna. (3) Some of the metal atoms that enter this second plasma can be ionized under the right circumstances. (4) The metal ions (along with inert gas ions) can be accelerated from the second plasma to the sample or wafer surface by the presence of a bias on the sample.

The plasma characteristics in this type of tool can be explored with the use of a Langmuir Probe [8.11], which is typically just a fine wire tip exposed to the plasma. By biasing the wire both positively and negatively up to a few tens of volts, it is possible to measure the plasma density as well as the electron temperature at the probe's location. However, this technique does not give any indication about the composition of the plasma, i.e., it cannot differentiate between metal and inert gas ions. So its use is primarily to probe the electron dynamics and then to imply what might be going on with the various ion species.

For the purpose of examining the level of metal ionization, experiments have been done where a small, gridded energy analyzer was located at the sample position [8.5, 8.6]. The cathode of the energy analyzer was configured to be a quartz crystal microbalance that could measure both the mass of the depositing film and the net depositing ion current. Grids were configured above the energy analyzer cathode surface to suppress secondary electrons as well as to admit or repel positive ions. This type of analyzer is appropriate for low-to-moderate plasma densities where the Debye length of the plasma is still larger than the grid opening. Once the plasma becomes too dense (in this case at about 0.5×10^{12} cm^{3}), the grids no longer function and the analyzer becomes shorted. An optical emission monitor (monochromator) was also configured to examine the emission of metal ion and neutral lines at higher discharge powers.

Several parameters can have a significant impact on the level of metal ionization as well as the subsequent deposition of metal ions and neutrals. The most obvious is the density and electron temperature of the second plasma. Figure 8.7 shows the effect on relative metal ionization (of the depositing species) as a function of applied RF power to the RF coil electrode [8.6]. At zero RF applied power, the ionization level of the metal is quite low. This is consistent with conventional magnetron sputter deposition. As the RF power is increased, the relative ionization increases and then saturates. There is a slight advantage in relative ionization with the use of Ne instead or Ar, presumably due to the higher electron temperature expected with Ne.

The operating gas pressure of the I-PVD system is also quite critical due to two primary effects. First, the plasma density and temperature are a

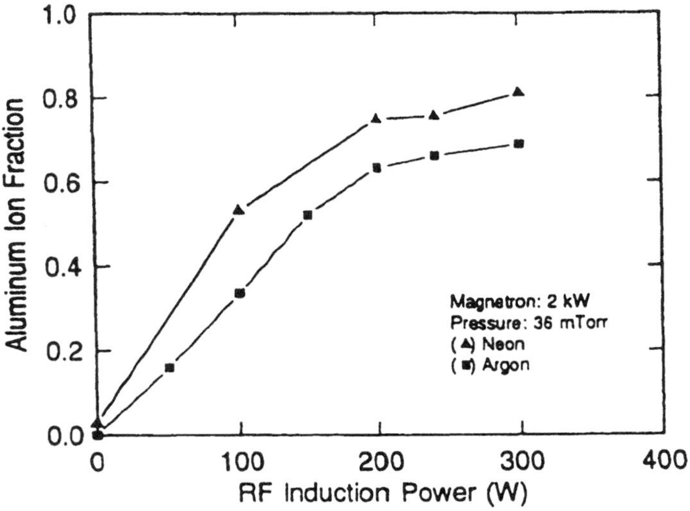

FIG. 8.7 Relative ionization of depositing metal species as a function of RF-inductive power for a fixed magnetron sputtering rate [8.6].

function of the gas density. Therefore, the level of metal ionization would be expected to be a strong function of pressure. Second, the background gas atoms tend to scatter the emitted, sputtered atoms due to collisions. This has been studied at great length for conventional sputter deposition processes. At pressures on the order of 1 mTorr, the amount of gas scattering is low and most of the sputtered atoms travel in a line-of-sight, ballistic mode from the target to the film surface. At pressures of several tens of mTorr, the path length for the sputtered atoms is much smaller than the cathode-to-film distance, and virtually all of the sputtered atoms lose their initial energy and direction due to gas atoms collisions.

The effect of pressure on the plasma or electron temperature is shown in Fig. 8.8 [8.11]. This figure shows a rapid fall in the electron temperature as the gas pressure goes from 1 to 10 mTorr, and then a much slower fall as the pressure is increased. Conversely, the electron density (not shown) would show a roughly inverse dependence to the temperature, with a gradual rise in density as the pressure is increased. This is a common dependence of a plasma powered at a constant power as a function of pressure: Low pressure correlates with fewer, hotter electrons, and higher pressures change the plasma into a higher density but cooler plasma.

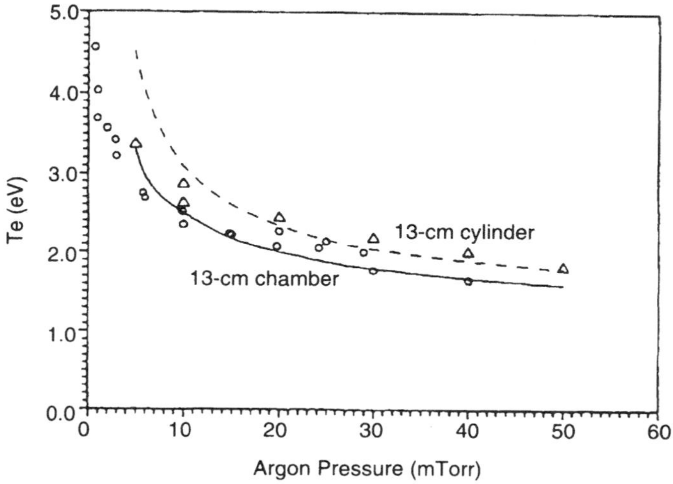

FIG. 8.8 Electron temperature as a function of chamber pressure [8.11].

The background gas pressure also alters the relative ionization of the depositing flux (Fig. 8.9) [8.6]. At low pressures (a few mTorr), the relative ionization is quite low. This is due to the relatively small cross section for ionization and the short path length. The ionization mean free path can be found as

$$L = \frac{v}{n<sv_e>} \tag{8.1}$$

where s is the average ionization cross section, v is the metal atom velocity, v_e is the electron velocity, and n is the electron density. For a plasma density of $10^{12}/cm^3$ — which is quite dense for a processing plasma — the mean free path is on the order of 30 cm, which exceeds the throw distance by 2 to 3 times. However, if the background gas density is increased, two things occur. First, the inert gas RF plasma becomes slightly denser, although with a lower electron temperature. Second, the sputtered atom can be slowed or even stopped by collisions with gas atoms or other sputtered metal atoms. This increases the net time that the metal ion spends in the second plasma, and as a result increases its chance of having an ionizing collision. In general, though, I-PVD will tend to be used at moderate system pressures in the tens of mTorr. The choice of operating pressure will depend on the required level of ionization, the system geometry, and the

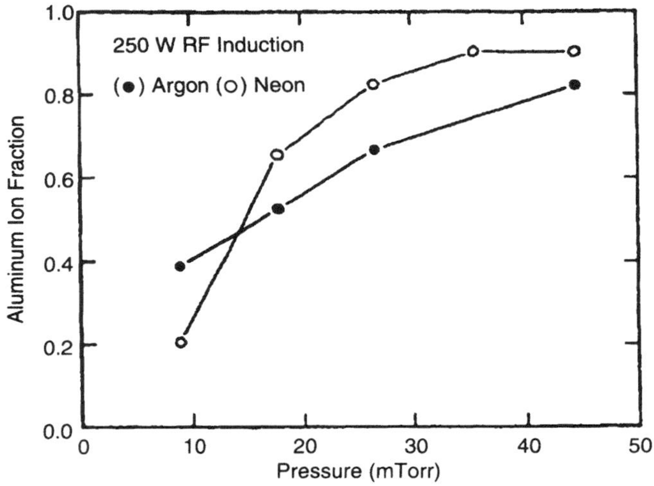

FIG. 8.9 Relative ionization of the depositing species as a function of chamber pressure [8.6].

necessary deposition rate. This final parameter falls rapidly as the operating pressure is increased.

A third parameter that can strongly alter the level of metal ionization is the number of metal atoms actually present in the discharge. The initial case is with no metal atoms present. Here the plasma is composed only of inert gas ions. Since the ionization potential of Ar is 15.7 eV and the electron temperature is typically 1 to 3 eV, only a small fraction of the Ar will be ionized. However, as metal atoms are sputtered into this inert gas plasma, their much lower ionization potential (5.9 eV for Al, for example) will mean that a much larger number of electrons have sufficient energy to ionize the metal atom and its ionization probability (compared to an Ar atom) will be much higher. Thus, with sufficient RF power coupled into the inert gas plasma, the relative ionization of the metal atoms can exceed 80%, even though the relative ionization of the Ar might be two orders of magnitude lower.

This situation is complicated, though, as large numbers of metal atoms are sputtered into the plasma and the RF power to the plasma is held constant [8.6]. An example of this is shown in Fig. 8.10, where the relative ionization of the metal species (as deposited into the energy analyzer) is measured as a function of RF power for three different magnetron power levels. As the magnetron power is increased, it corresponds roughly with a

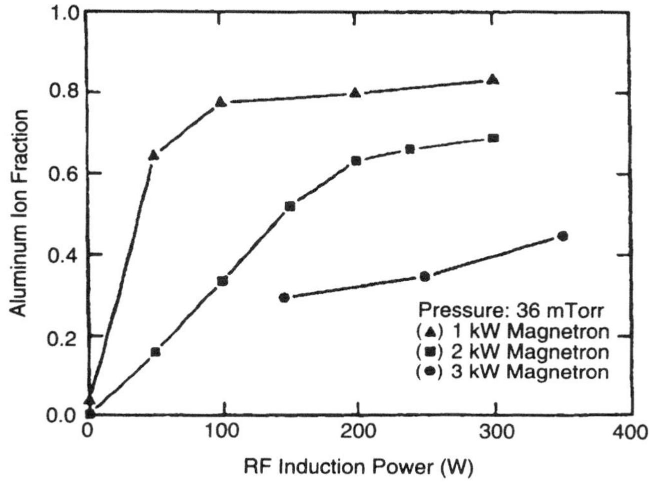

FIG. 8.10 Aluminum ion fraction as a function of RF induction power for three different magnetron powers [8.6].

linear increase in the number of metal atoms sputtered from the cathode. The figure shows a clear reduction in the relative ionization at all RF powers for the increased metal fluxes. Hopwood and Qian's I-PVD discharge model was used to describe the same situation, as shown in Fig. 8.11 [8.11]. In this case, the model uses a particle density of the sputtered Al atoms rather than a magnetron discharge power, but the general trends can be easily seen. The model was also used to predict the electron temperature and shows a significant cooling effect as a function of increased metal density (Fig. 8.12) [8.11].

An additional indication of the cooling of the discharge is the level of the ion saturation current to the sample. If a sample is biased sufficiently negatively to repel all electrons from the plasma, then the net ion flux to the sample is simply the Bohm presheath flux, which was listed as (see Section 3.4) [8.12]

$$F = 0.6(n_e)\left(\frac{KT_e}{M}\right)^{1/2} \tag{8.2}$$

where n_e is the electron density in electrons/cm^3, T_e is the electron temperature in eV, and M is the ion mass. The reduction of the ion saturation current as the level of metal atoms and ions in the discharge increases is

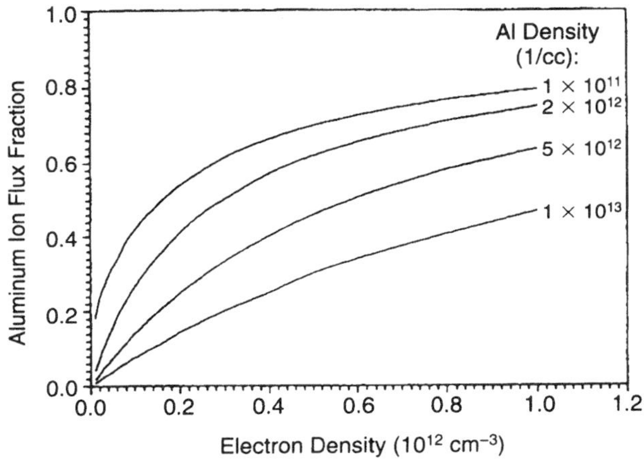

FIG. 8.11 Calculated Al ion fraction as a function of RF inductive power for several different Al atom densities [8.8].

another indication of the reduction in the electron temperature in the discharge.

There may also be a subtle effect of the chamber size on the level of metal ionization in an I-PVD tool. Hopwood and Qian have suggested that smaller chambers result in increased diffusional losses of the metal ions

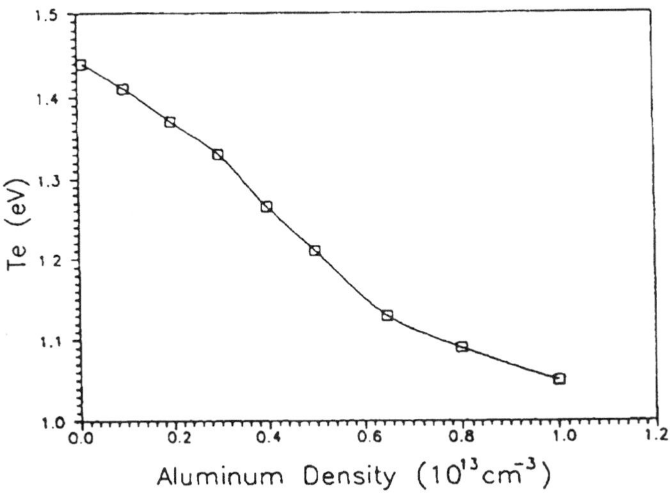

FIG. 8.12 Calculated electron temperature as a function of metal atom density [8.8].

[8.11]. He has computed the relative ion flux fraction as a function of the characteristic diffusion length (Λ) for the case of Al sputtered in Ne (Fig. 8.13) [8.11]. The implication is that if the diffusion length is on the order of or larger than the chamber dimensions, increased losses of metal ions will occur and the relative ionization will be reduced. Currently there are only two published reports of relative ionization in different chamber sizes to support this conclusion. In a relatively small chamber (8-cm-radius × 10-cm-length), Yamashita measured an Al ion fraction of 65% [8.13]. In the work of Rossnagel and Hopwood [8.5, 8.6], the measured ion flux fraction was over 80% in a chamber with a radius of 25 cm and a length of 30 cm.

The primary ionization process in an I-PVD tool was originally thought to be electron-impact ionization. However, Hopwood has published calculations of the relative levels of both electron-impact and Penning ionization. At low electron densities, Penning ionization is seen to dominate, particularly for Ar (Fig. 8.14) [8.11]. The experimental work of refs. 8.5 and 8.6 tend to suggest a dominant role for electron-impact ionization, particularly at high powers. This is consistent with the observation of a higher relative ionization with Ne as compared to Ar. A Ne plasma would be expected to have a higher electron temperature (2 times that of an Ar plasma).

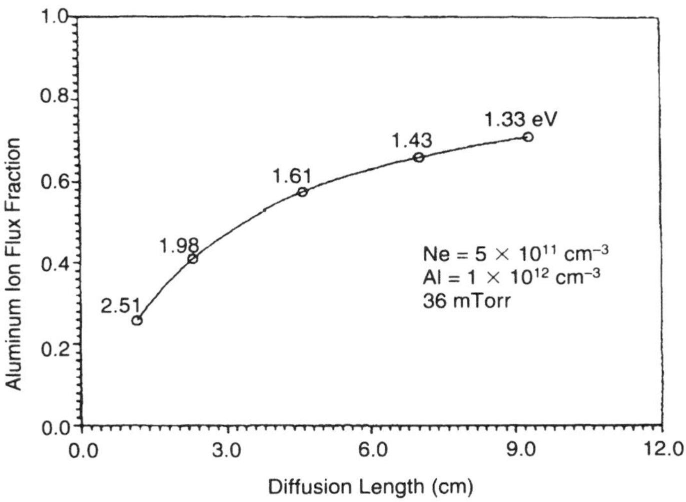

FIG. 8.13 Calculated Al ion flux fraction as a function of characteristic diffusion length for I-PVD of Al. The calculated electron temperatures are listed for each point [8.11].

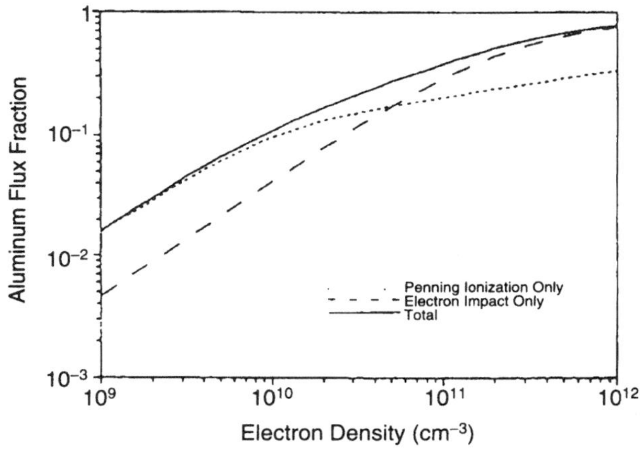

FIG. 8.14 The relative ionization level as a function of electron density showing the relative importance of Penning and electron-impact ionization processes [8.11].

Also, the Penning ionization cross section for Ne is smaller than that for Ar [8.11].

An interesting problem remains as a fundamental issue relating to I-PVD plasmas. The introduction of high levels of metal atoms into the dense, inductively coupled RF inert gas plasma appears to result in a decrease in both the electron temperature *and* the plasma density. Generally the electron temperature and the density would be expected to move in opposite directions; i.e., lower electron temperature is compensated for by higher plasma density. A recent paper has examined this dilemma [8.14] and has attributed the problem to gas rarefaction effects, which have been systematically observed during magnetron sputtering, particularly at higher pressures [8.15].

Gas rarefaction is a heating phenomenon, driven by the kinetic energy of the sputtered atoms. As they slow down due to gas-phase collisions, the sputtered atoms transfer some of their kinetic energy to the background gas, raising its effective temperature. Since all this occurs in an open container, the net result is a local reduction in the gas density in the region directly in front of the cathode [8.15]. The recent paper by Dickson *et al.* suggests that the gas rarefaction effect may be responsible for the density reduction as the sputtering level is increased. The electron temperature reduction is still driven by the much lower ionization levels of the metal atoms [8.14]. Experimental measurements of the local gas density in front of the I-PVD magnetron cathode show both a significant rarefaction due to the RF inductively coupled plasma and increasing rarefaction due to increases in the magnetron discharge power, the latter directly related to the amount of metal introduced into the discharge (Fig. 8.15).

The significance of this problem is actually quite important to the large-scale implementation of I-PVD. The addition of large amounts of metal atoms to the discharge is obviously necessary to attain a high deposition rate. Yet, as the metal flux is increased, the relative ionization decreases for two reasons: the cooling of the electron temperature and the increased transparency of the discharge region due to rarefaction. In effect, increased metal sputtering into the discharge is equivalent to reducing the operating pressure, which results in less relative ionization because of the shorter residence time for the sputtered atoms in the plasma.

As a result of this problem, the near-term implementations of I-PVD have tended to focus on liner and diffusion barrier applications rather than thick film deposition. This is convenient in that it supplants collimated sputtering, which is used on a wide scale in the industry, but is plagued by concerns over uniformity, rate, clogging, etc. However, I-PVD technology

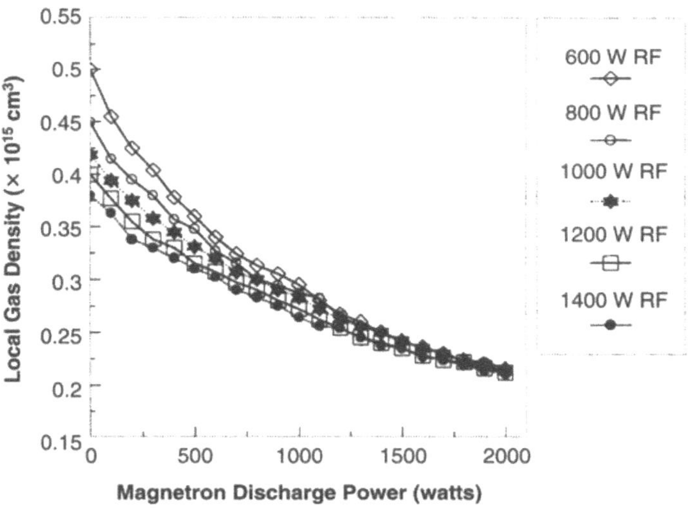

FIG. 8.15 The local gas density in the plasma region of an I-PVD tool (300-mm cathode) as a function of magnetron discharge power for various RF inductively coupled power levels. The starting chamber pressure was 30 mTorr, which is equivalent to $1 \times 10^{15}/cm^3$. Therefore, even at the lowest RF power used (600 W) the gas is already rarefied by 50% before the magnetron is turned on.

is still constrained by this reduced relative ionization problem, which may limit eventual applications at high rates.

8.3 Deposition and Experimental Results

The use of I-PVD for semiconductor applications encompasses two differing regimes: thin layers inside trenches or vias and the filling of trenches and vias. The thin layers may be conformal or may be selectively deposited only on the bottom or sides of features. They are not intended to be the primary current-carrying metallurgy, though. The fill metal's role is primarily low-resistance current flow.

The thin layers have numerous functions. For example, thin Ti layers often are used at the bottoms of vias or contact holes. The function of the Ti is to provide better electrical conductivity to the underlying metal. Ti is chemically reactive and can break down the surface oxide of the metal, in essence stealing the oxygen atoms from the underlying oxide in an attempt to form TiO_2. The Ti may also be used to form a silicide with underlying Si. Thin films within a via or a trench may also be used as a chemical diffusion barrier either between metal layers to reduce electromigration and interdiffusion problems or to protect an underlying layer from chemical attack. An example of this is the use of TiN layers on SiO_2 sidewalls, which protect the SiO_2 from chemical attack by the WF_6 working gas used for W CVD deposition. A third application is the use of Ti or TiN layers to promote growth of the desired (111) phase of AlCu on subsequently deposited films. This phase has been shown to be the most resistant to electromigration problems. It is not clear, however, whether buried, damascene-processed lines will have similar microstructure requirements.

At high levels of relative ionization of the metal species, the majority of the depositing species arrive at normal incidence to the sample surface. Three effects that might lead to a nonnormal arrival angle are (1) scattering within the sample sheath, (2) electrical deflection by the surface topography, and (3) the initial kinetic energy of the metal ion previously in the plasma. Gas scattering is an initial concern because the operating pressure is many tens of mTorr. However, the sample sheath thickness is on the order of a few hundred microns, which is 10 times smaller than the mean free path for gas scattering. Second, since the sample sheath is 100 or more times the scale of the surface topography (typically 1 micron or less), the surface appears flat to the incoming ions and they are not deflected electrically by ledges or via walls on the surface. Finally, even though the sputtered atoms that feed the metal-rich plasma can have several eV of kinetic energy when emitted from the cathode, gas scattering at 10–40 mTorr quickly damps out this energy and the sputtered atoms are nominally "thermalized" prior to ionization and acceleration. This means that at most the sputtered atoms have a few tenths of an eV in kinetic energy, which will cause only minor deviations from normal incidence at the typical ion acceleration voltages of several tens of volts. Therefore, none of these three problems will be significant to the ion's trajectory, and the metal ions will arrive at the sample surface at 90° (from the horizontal plane) with very little, if any, divergence.

8.4 Lining Trenches and Vias

Purely normal-incidence directional deposition within a via or trench would occur from the bottom up because there would be no depositing metal ions incident on the vertical sidewalls. However, since the relative ionization of the depositing species is always less than 100%, sidewall coverage is observed. An example is shown in Fig. 8.16, where the bottom step coverage is about 70% and the sidewall coverage is 15% or less. Similar work has been reported in commercially available tools, showing

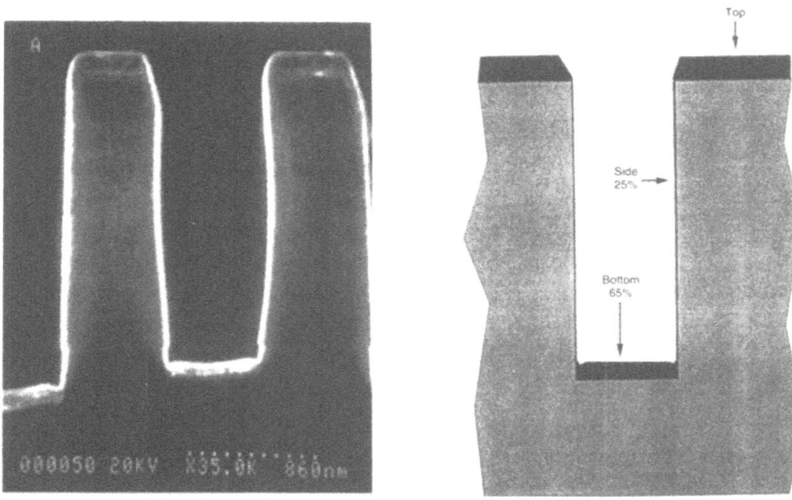

FIG. 8.16 (a) SEM of directional deposition of thin, 1000-Å film using I-PVD into a 3:1 aspect ratio trench, (b) sketch of the same structure.

bottom coverage of 50 to 70% as a weak function of the aspect ratio and feature width (Fig. 8.17 [8.16–8.19]).

The bottom-layer step coverage of the I-PVD films compares to a much lower value for collimated sputtering. Actually, this comparison can be misleading. The degree of collimation, effectively the aspect ratio of the collimator, directly correlates with the bottom surface step coverage. However, since the transmission of a high aspect ratio (5:1) collimator is less than 1%, it is effectively impractical to use such high levels of collimation even though the bottom-layer step coverage would be comparable to I-PVD.

Sidewall coverage is important primarily for diffusion barriers or for preferential growth surfaces. Diffusion barriers have the more stringent requirements: The films must be hermetic seals of the underlying surface. The ideal diffusion barrier would be amorphous and as such not have any grain boundaries that penetrate the film and allow for grain boundary diffusion. Preferential growth surfaces would be appropriate as seed layers for plating (e.g., Cu electroplating on Cu seed layers) and may be useful in helping to either define the orientation of the subsequently deposited film or facilitate the surface mobility of the next layer. As an example, on planar surfaces, TiN films tend to lead to a (111) orientation on subsequently deposited AlCu films. However, the extension of this effect into high aspect ratio features is unclear, and as such the presence of TiN on sidewalls may or may not be needed in high aspect ratio applications.

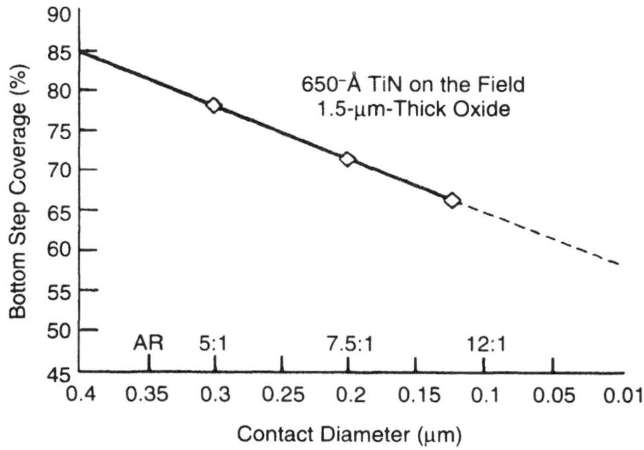

FIG. 8.17 Chart of step coverage versus aspect ratio for ionized Ti deposition [8.16] (courtesy of Applied Materials).

Directional deposition by itself would not lead to good sidewall coverage. However, three related effects can alter the sidewall coverage and the effective conformality of the thin films. The first effect is simply the degree of relative ionization of the deposit: Lower levels of directionality result in deposition on the upper sidewalls of the trench or via. Second, the re-emission of deposited atoms can also lead to better conformality. This remission can be due to either a reflection of the incident particle due to a highly grazing deposition angle or simply reemission consistent with a less-than-unity sticking coefficient for the depositing species. A third critical aspect of the directional deposition is the kinetic energy of the depositing ions. As the ion energy is increased, the possibility of self-sputtering of the deposited film increases. At ion energies of 10 eV or so, this effect is unimportant. However, as the ion's kinetic energy is raised (through a more negative bias voltage on the sample) to many tens of eV, physical sputtering becomes significant. The sputtering takes place primarily on the bottom surface of the trench or via, and the majority of the sputtered atoms land on the lower sidewalls of the feature. This is shown schematically in Fig. 8.18, where a deposition using only 50% relative ionization is changed significantly as the effective sputter yield of the depositing species is increased. A comparable experimental situation is shown in Fig. 8.19.

The degree of resputtering in these cases is expected to be only a mild function of aspect ratio because the primary sputtering is occurring on the bottom surface of the feature and the step coverage at this point is only weakly dependent on aspect ratio. Simulations by Hamaguchi and Rossnagel of this case show a general decrease in the bottom step coverage and a slow increase in the sidewall step coverage as a function of

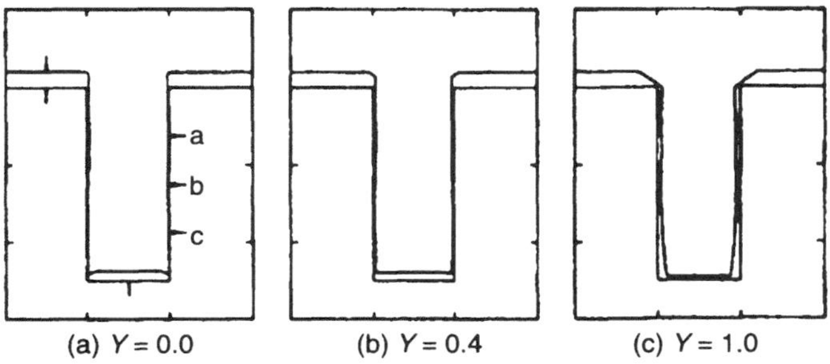

FIG. 8.18 Simulations of the effect of resputtering due to increases in the depositing ion kinetic energy as a function of sputter yield (Y). From left to right: a sputter yield of 0, 0.4, and 1.0 [8.20].

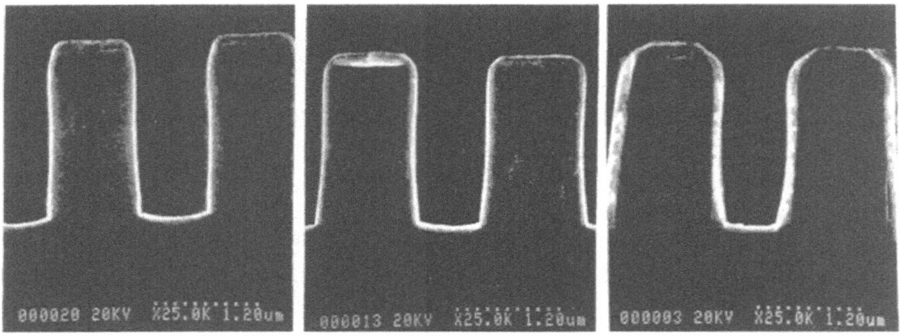

FIG. 8.19 SEM measurements of trenches deposited as a function of sample bias. The leftmost figure has a bias voltage of −10 V, the center SEM has a bias of −40 V, and the rightmost SEM has a bias of −100 V. The relative ionization of the depositing Cu species is 50% [8.20].

increasing sputter yield [8.20]. Figure 8.20 shows the case for an aspect ratio of 1.0, and Fig. 8.21 shows the case of an aspect ratio of 2.5:1. It should be noted that the ideal conformality in each case occurs at roughly the same sputter yield, which indicates that the overall conformality on a wafer with a variety of feature widths and dimensions should be uniform.

By tailoring the deposition and erosion processes, nearly ideal conformality can be obtained. Figure 8.22 is an example for a moderate aspect ratio feature, and shows a uniform 600-Å layer on both the sidewalls and bottom of the trench. The step coverage on the bottom surface can also be completely removed due to sputtering. In this case, the films would only be deposited on the wall surfaces. This may eliminate the need for the deposition of the bottom contact layer and may also prove advantageous for electromigration resistance in devices.

Another feature that begins to become important is the beveling of the upper corner of the deposit due to sputtering. Since the sputter yield is angle dependent and peaks somewhere near 45–50 degrees, a bevel can be formed in the deposited film at the upper corners. For thin liner films, this feature is reactively unimportant. However, at high levels of sputtering or with thicker films, the top bevel can become significant and deleterious to the film topography. An extreme example is shown in Fig. 8.23, where the top corner is completely etched away and a scalloped wall deposit is observed. The deposit on the upper sidewalls is now partly due to local redeposition from the opposing side and eventually could lead to closure of the feature. Normally for thin liner layers, this effect is unimportant or else very difficult to observe.

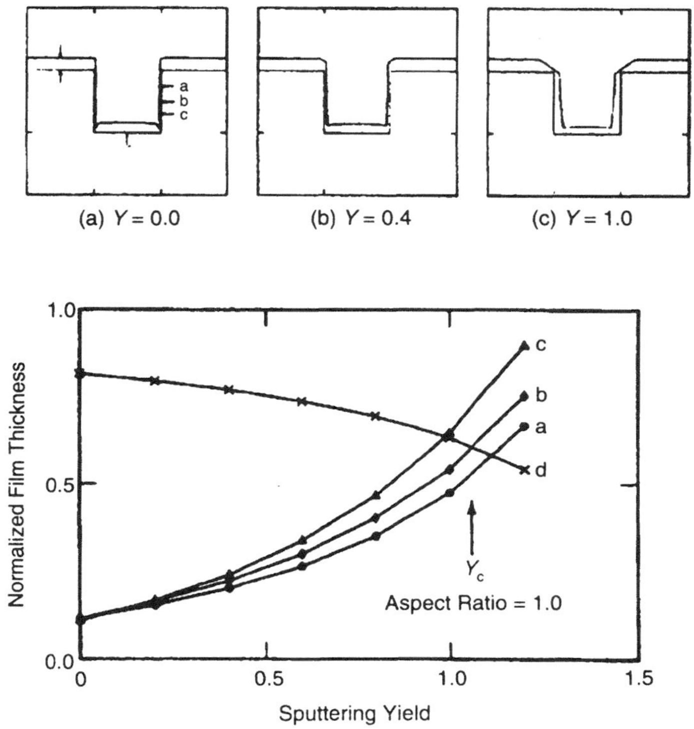

FIG. 8.20 Calculated bottom and sidewall step coverages for a feature of aspect ratio 1.0 as a function of the sputter yield of the depositing ions [8.20].

The sample bias for I-PVD can be attained in two ways. The easiest is by means of a DC potential on the wafer. However, because many wafers are configured with insulating layers or an oxided backside, it is usually necessary to use a front-surface, biased clamp ring to supply the potential. Once the film is continuous (50 Å), the current flow is across the wafer surface to the clamp ring. An alternate embodiment is to use an RF substrate bias that is not dependent on the oxide layers on the wafer. This allows bias to be from the backside of the wafer and no longer requires the use of front-surface clamps. The constraint with RF bias is that the potential and current flow to the wafer are more difficult to ascertain. In these cases, usually the RF power to the substrate chuck is measured (in watts), but this is only an indirect measure of the level of kinetic energy given to the depositing ions. However, once the effect of the RF bias is observed

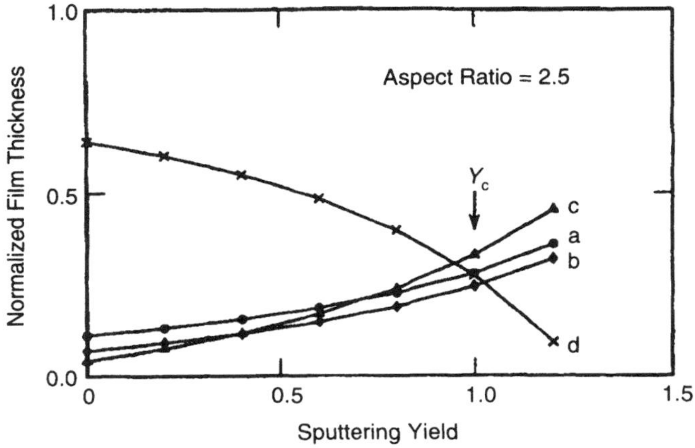

FIG. 8.21 Same as Fig. 8.19 for a 2.5:1 aspect ratio [8.20].

and calibrated, it is a more effective manufacturable process than the front-surface DC clamp ring.

At this point, it is worth a small digression to describe what might be the limits of PVD or I-PVD as applied to liner, diffusion barrier, or seed layer applications. An ideally conformal film has a uniform coverage on both the sidewalls and the bottom of the feature. The step coverage, however, is limited by the flux of ions and atoms from the top. For a trench, the best-case conformal step coverage, SC, is given as

$$SC_{trench} = \frac{1}{(1 + 2AR)} \tag{8.3}$$

where AR is the aspect ratio (depth/width) of the trench. For a via, the best-case conformality is

$$SC_{via} = \frac{1}{(1 + 4AR)} \tag{8.4}$$

Therefore, the maximum conformal step coverage for a trench of aspect ratio 4:1 is 11% and for a via of the same AR, 6%.

The ideal bottom coverage, though, is not limited by the aspect ratio. A good rule of thumb is that the bottom coverage scales directly with the relative ionization of the depositing flux: A 50% relative ionization leads to a bottom step coverage of 50%, independent of the aspect ratio at least for

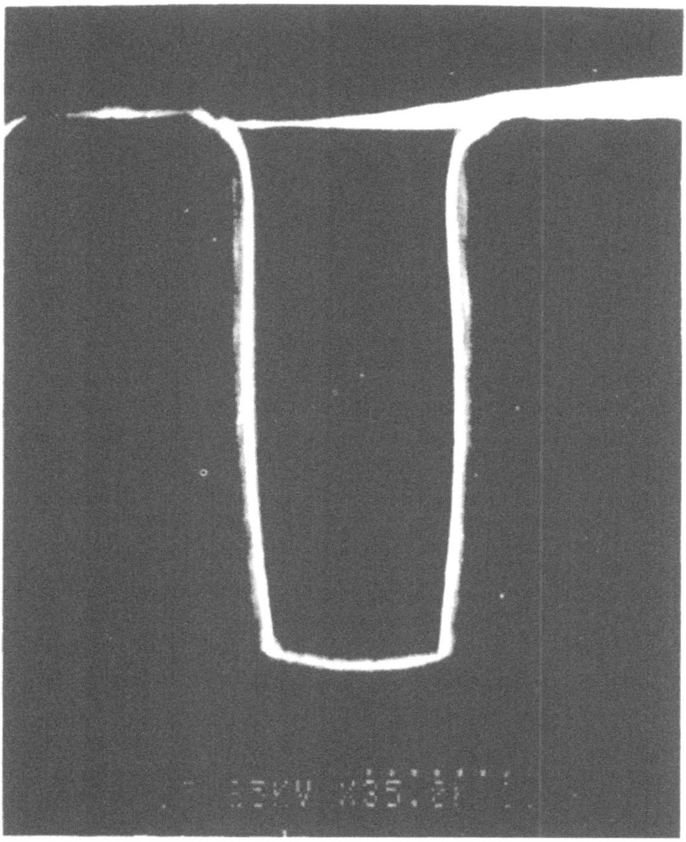

FIG. 8.22 SEM micrographs of a tailored thin film showing a highly conformal coverage inside a trench.

modest aspect ratios of to 7:1 or so. These numbers can be greatly altered due to the presence of sputtering, which can occur during I-PVD. This is primarily because the top or field thickness can be reduced due to sputtering. This reduces the denominator in the step coverage ratio (step coverage is just the ratio of the local thickness to the top or field thickness) and can result in step coverages greater than 100% in some cases. In effect, the sputtering process removes atoms from the field but atoms sputtered within a deep feature are unlikely to leave the feature, resulting in an apparently high step coverage.

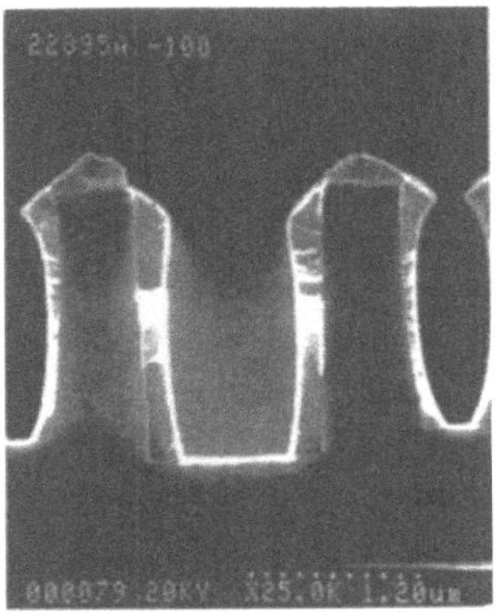

FIG. 8.23 An I-PVD deposit that had a high level of sample bias during deposition, resulting in a complete removal of the bottom layer as well as significant beveling and build-out near the top corners of the trench.

It is interesting to extend I-PVD toward extremely high aspect ratios, which might run from 7:1 to 10:1. From the formulas listed above, which are essentially just counts of the atoms incident onto the feature, the best-case (conformal) step coverage at 10:1 is 5% for a trench and 2.5% for a via. To form a viable film for either diffusion barrier or seed layer/surface diffusion would require at least 50 Å, implying field deposition thicknesses of 1000 Å for the trench and twice that for the via. Modeling of these very high aspect ratio features is ongoing, and preliminary results are shown in Fig. 8.24.

8.5 Trench and Via Filling

The filling of moderate to high aspect ratio features is more difficult than depositing a liner or bottom contact layer. This is driven primarily by the lack of 100% ionization of the depositing species. The neutral component of the deposit is directionally isotopic, and this results in eventual build-out of the sidewall deposits. An example of this is shown in Fig. 8.25 for a typical, partially ionized low-energy deposition.

At this point, the degree of filling is directly related to the directionality of the deposit, which is a simple function of the relative ionization of the depositing species. Low levels of relative ionization result in poor filling and the eventual close-off and void formation similar to nondirectional

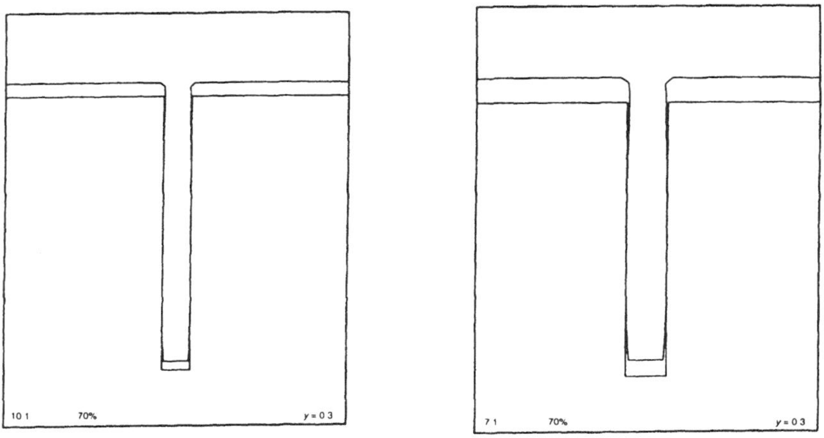

FIG. 8.24 Modeling results for 10:1 and 7:1 AR features using 75% ionization and a relative sputter yield of 0.4.

PVD. An example of this is shown in Fig. 8.26 for 30, 50, and 67% relative ionization. A simulation using the Hamaguchi model of the same experimental case is shown in Fig. 8.27 [8.21].

The step coverage of the bottom and sidewall surfaces has been characterized for Cu deposition at various aspect ratios and power levels. Figure 8.28 (see page 271) shows a representative feature. Figure 8.29

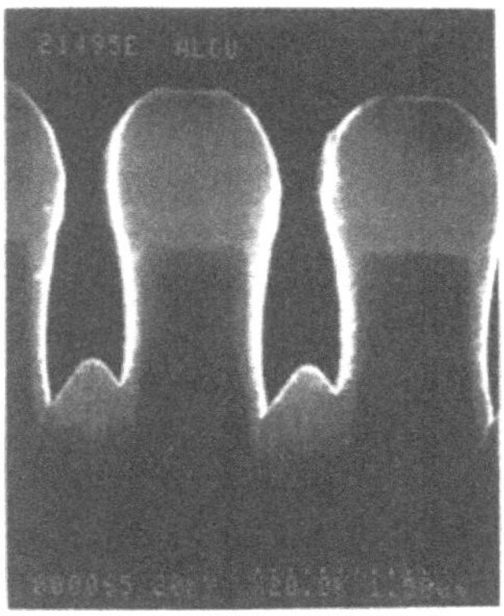

FIG. 8.25 I-PVD deposition of AlCu with approximately 60% relative ionization. The sample bias voltage is low to suppress self-sputtering.

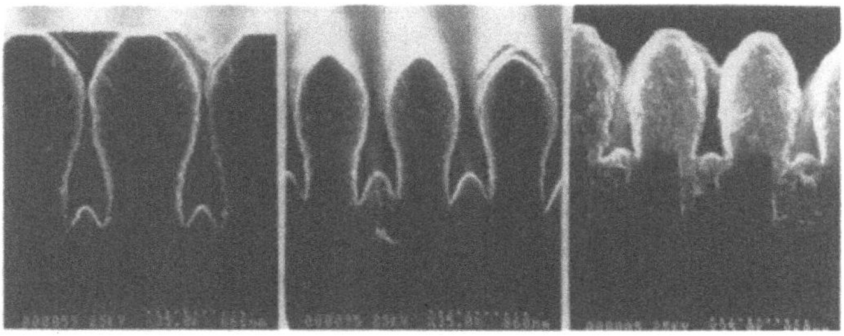

FIG. 8.26 Cross section SEMs of 4000-Å trenches deposited with various relative ionizations. From left to right: 33, 50, and 67% relative ionization [8.21].

shows the step coverages for the bottom and sides as the inductive RF power is increased, which results in increasing levels of metal ionization. Figure 8.30 shows similar data, now for the case of constant RF inductive power and increasing magnetron power. This shows that as the amount of metal is increased due to increased magnetron sputtering, the step coverage slowly degrades due to reduced relative ionization and directionality.

These results are all at room temperature and are characterized by two different regions within the feature: the bottom deposit and the sidewall deposit. Much like collimated sputtering, the bottom deposits are dense and fine-grained. The sidewall deposits, due to the intrinsic directionality

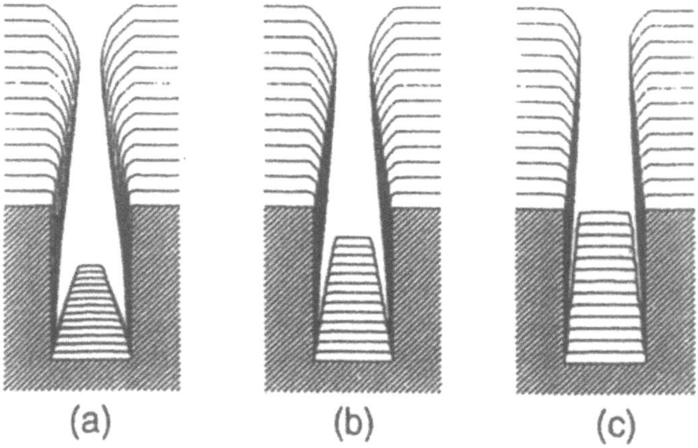

FIG. 8.27 Numerical model of the same deposition case as in Fig. 8.26 [8.21].

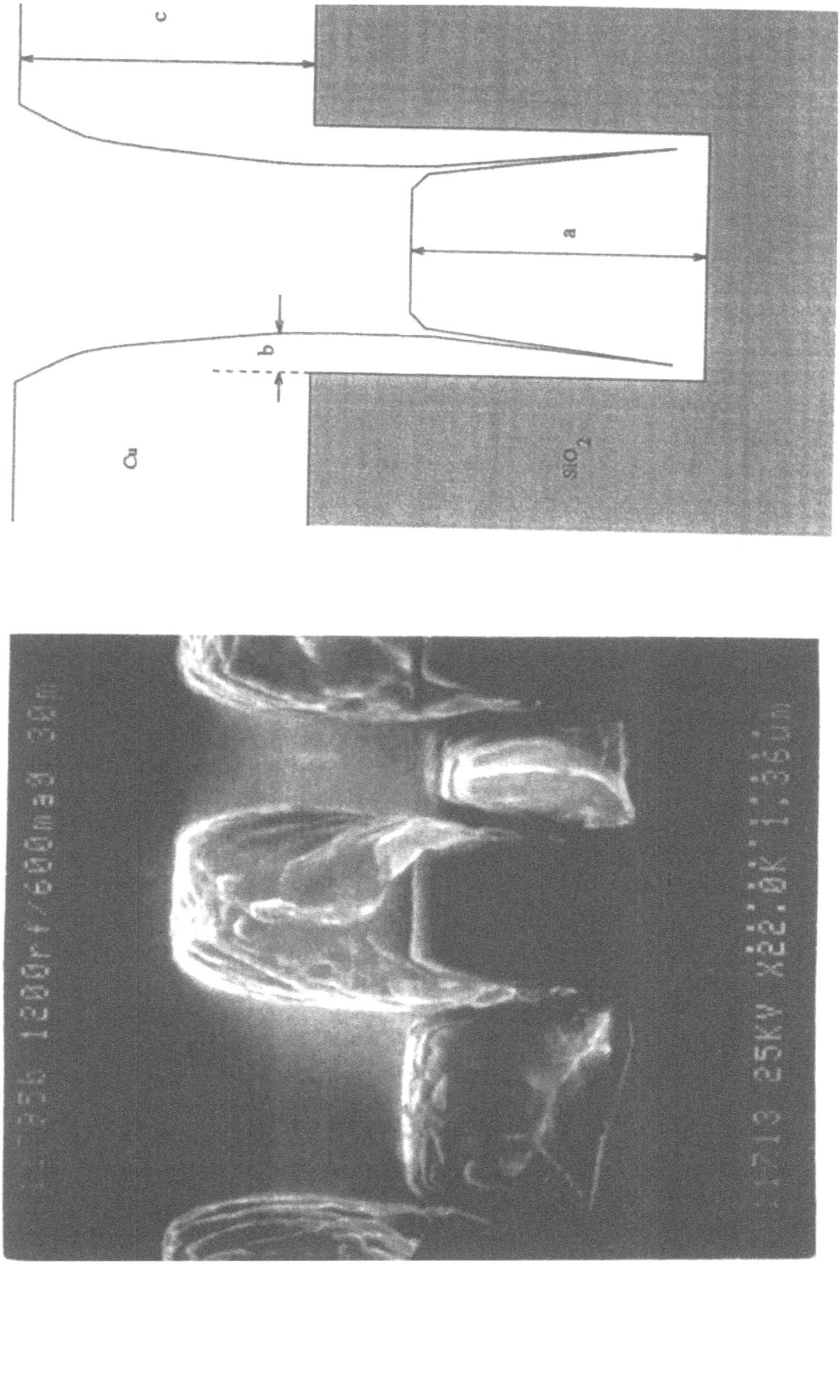

FIG. 8.28 Sketch and SEM of bottom and sidewall depositions [8.22].

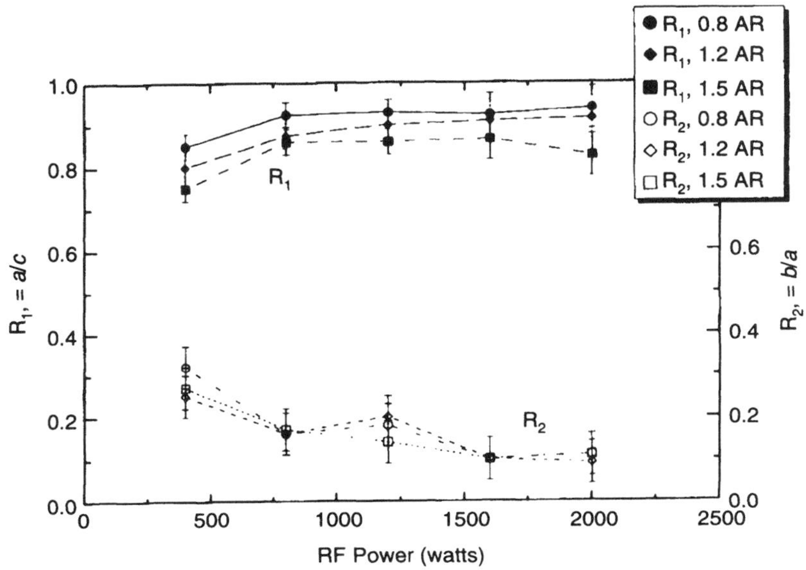

FIG. 8.29 Step coverage ratios, $R_1 = a/c$ and $R_2 = b/a$, shown as applied RF power is increased. The magnetron was kept at 600 W, pressure = 45 mTorr, and wafer bias = −10 V [8.22].

of the deposition, tend to be columnar and underdense. In addition, there is usually a very noticeable seam between these two deposits, which limits the rearrangement of atoms on the surface by diffusion. Figure 8.27 shows these characteristic regions as well as the seams between the sides and the bottom. Figure 8.31 shows the results of this effect after the polishing of the top or "overburden" film, as would normally be done in the chemical mechanical polishing (CMP) step during wafer processing. The seams result in a characteristic boundary visible from the top. This seam is clearly undesirable in that it is indicative of an underdense deposit, which will have higher-than-bulk electrical resistivity and will also be quite susceptible to electromigration due to the elongated void in the direction of current flow.

Room-temperature I-PVD deposition for filling is a direct function of the relative ionization of the depositing flux. For moderately high levels of ionization (70%), filling appears possible for aspect ratios of up to about 2:1 and feature sizes down to about 3500 Å. At higher aspect ratios or narrower feature size, the lateral, side deposits close off the feature prior to

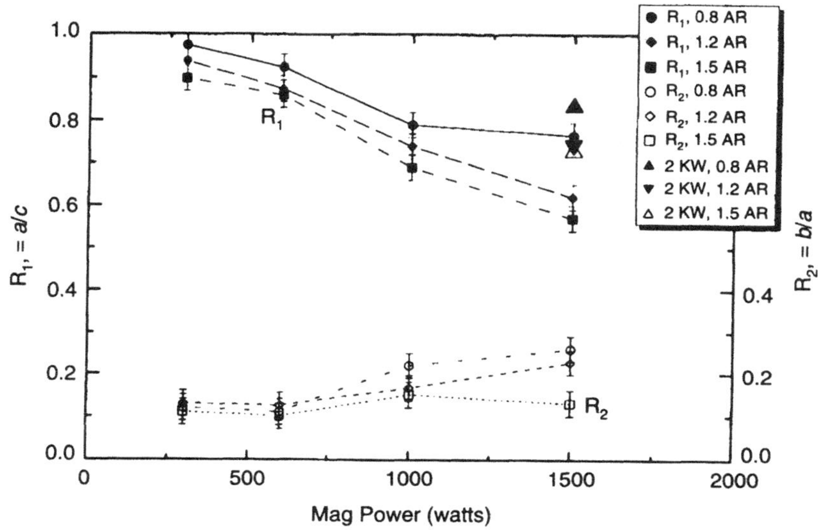

FIG. 8.30 Step coverage as a function of magnetron power level at constant RF inductive power (800 W) [8.22].

complete filling. It is clear that some degree of temperature-enhanced mobility is necessary to overcome this intrinsic structure caused by the I-PVD deposition.

It is intriguing to consider what might occur if the depositing ion energy were increased sufficiently to cause self-sputtering of the film during deposition. It might be expected that the incoming ions would sputter-back the overhanging sides of the deposit and keep the trench or contact hole open. In addition, it might be anticipated that the sputtering process would result in a forward motion of atoms sputtered from the sidewalls, which would push them farther down into the feature. The answer to these expectations is both positive and negative, depending on both the feature width and the aspect ratio.

If the ion energy of the depositing atoms is increased sufficiently to cause sputtering of the deposited film, bevels are formed on the top edges of the trench or via. This is a good feature in that it tapers back the sidewall. However, the atoms that are "missing" from the bevel due to sputtering are redeposited on nearby surfaces. If the feature width is large, these redeposited atoms are spread over a large area and are unimportant. This

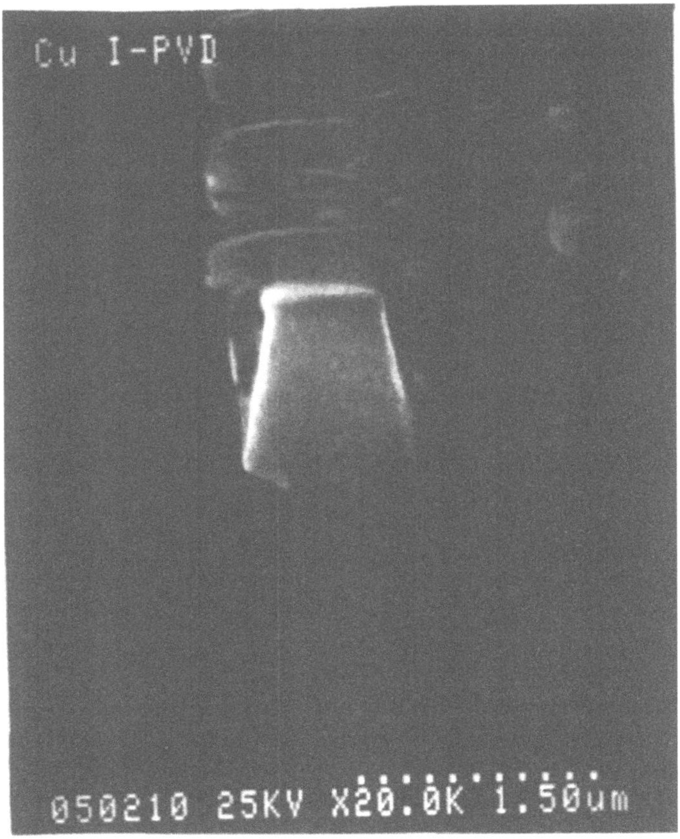

FIG. 8.31 Polished-back features showing seam formation.

is shown in Fig. 8.32 for an aspect ratio of < 1 and in Fig. 8.33 for simulations of a low aspect ratio feature.

However, as the feature width is lowered the redeposition process starts to become quite significant. This is because the redeposition can occur mostly on the opposing sidewall, which results in the build-out of that sidewall. This effect is quite deleterious to subsequent filling and can rapidly lead to a pinching off of the feature and void formation. This is shown in Fig. 8.34 for experimental results and in Fig. 8.35 for simulations [8.21]. The amount of sputtering necessary to see this effect for narrow features is quite low, and even yields of 0.3 will result in rapid closure and void formation for features below 0.4 microns in width.

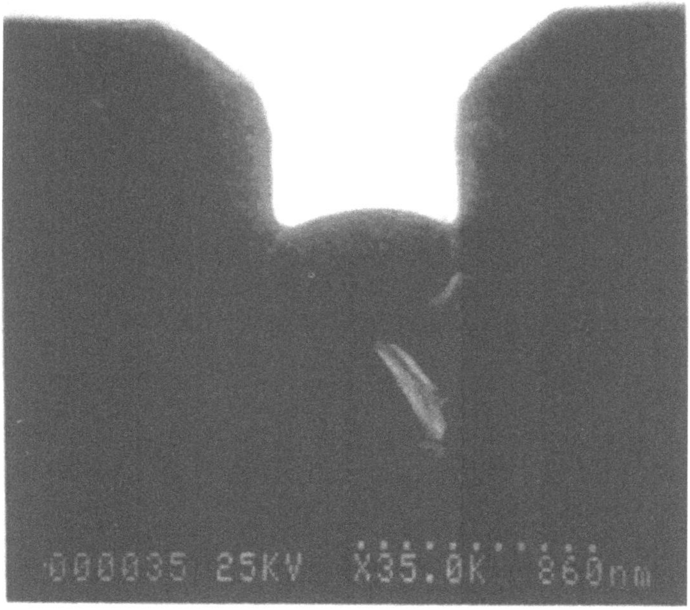

FIG. 8.32 An I-PVD deposit for a low aspect ratio feature showing the effect of significant resputtering of the deposit.

One potential solution to the redeposition problem is to use sufficient sputtering so that the entire top or field layer is completely removed. The reasoning here is that if there is no bevel formation because there is no film to bevel, then cross-trench redeposition will be reduced. An example of this is shown in Fig. 8.36 for a very high level of resputtering during deposition. The top or field layer is completely gone and the upper sidewalls are tapered back significantly. There is often a small seam formed at the centerline of the feature, but it does not lead to void formation. The negative aspect of this high level of sputtering during deposition is that the oxide sidewalls are exposed by the sputtering process. They are etched (and beveled) and can also lead to impurity incorporation in the deposited

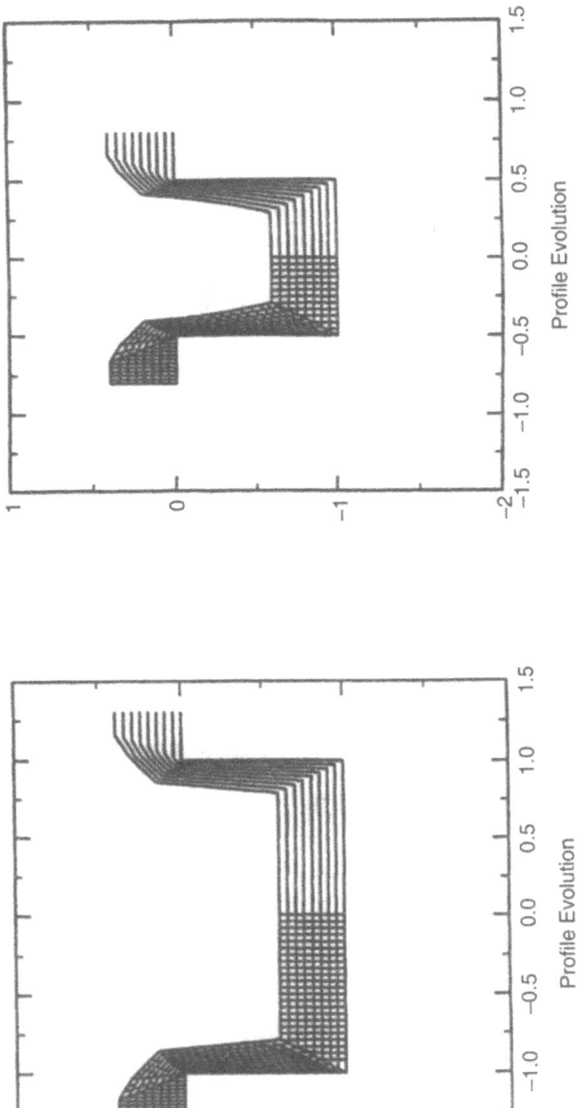

FIG. 8.33 Modeling studies of low aspect ratio features with significant sputtering of the deposit.

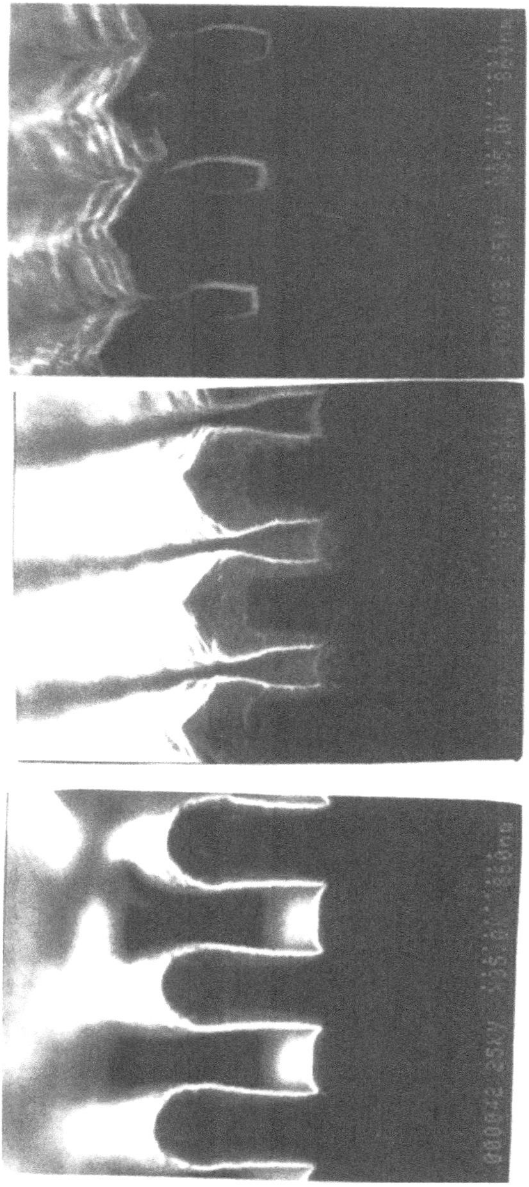

FIG. 8.34 The effect of increasing ion energy during I-PVD deposition (left to right) [8.21].

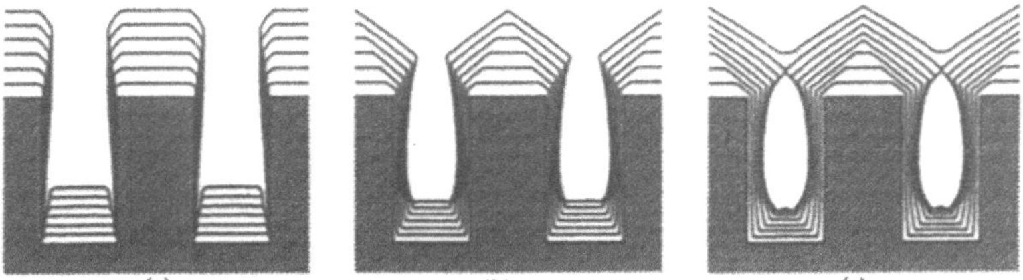

FIG. 8.35 Simulation of I-PVD deposition with high levels of ion energy in the depositing ions [8.21].

metal line. In addition, the energy density on the sample is quite high in these cases, typically several watts per square centimeter, and this results in significant wafer heating and potential problems.

8.6 Electrical Measurements

The use of I-PVD for wafer applications has intrinsic advantages over collimation and some CVD technologies. For contact resistance, it is critical that a dense Ti layer be deposited primarily at the very bottom surface of a via or contact hole. I-PVD of Ti should be superior to collimation in this application because of the increased directionality and kinetic energy of the depositing atoms. Initial studies, exploring the difference between I-PVD of Ti and collimated Ti have shown both lower contact resistance and a lowered Kelvin resistance distribution (Figs. 8.37 and 8.38 [8.24] and Fig. 8.39 [8.18]). These studies, as well as others, have not shown any increase in damage levels due to the more energetic process [8.17].

Initial results with I-PVD of TiN are also encouraging. The film structure is dense and large-grained. The sidewall coverage in trenches is also slightly smoother than collimation. One of the principle suppliers, Applied Materials, has claimed that this less porous surface is more conducive to surface diffusion of the subsequent AlCu layers than collimated TiN [8.16]. Other reports indicate that the functional resistivity of the I-PVD TiN can be significantly lower than conventionally deposited TiN [8.10]. In general, the resistivity of the TiN is related to the deposition conditions.

For the I-PVD TiN, the kinetic energy of the Ti^{+} depositing particles (set by the sample bias) has been shown to strongly alter the resistivity of the resulting thin films (Fig. 8.40). One clear advantage of I-PVD TiN is that the material is easily made at low temperatures, unlike TiN sputtered conventionally. The I-PVD TiN, as a function of sample bias voltage,

FIG. 8.36 I-PVD deposition with very high levels of ion bombardment of the sample.

shows a strong transition into stoichiometric TiN at voltages lower than −20 V [8.25]. The best-case resistivity in this case appears to be about 35 micro-Ohm-cm, which is about 2 times the bulk, but significantly lower than reactively sputtered, conventional TiN deposited at temperatures of 300°C or more.

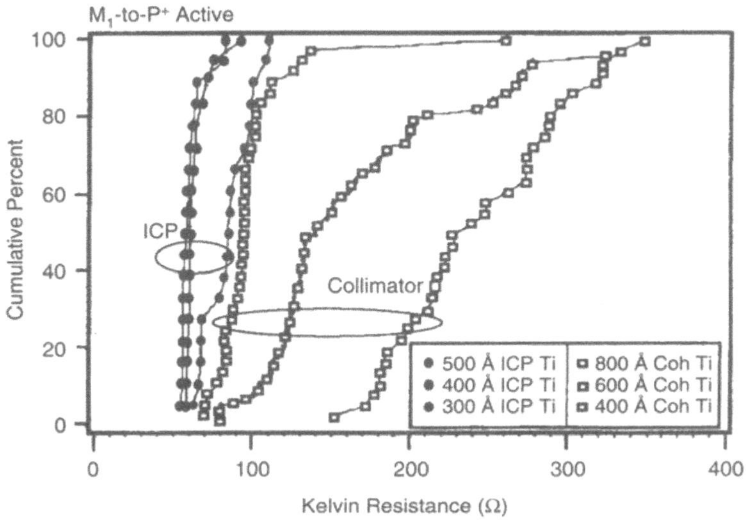

FIG. 8.37 Resistance in contact chains, comparing I-PVD Ti (described in figure as ICP) and collimated Ti (conventional) [8.24].

8.7 Materials Properties

In addition to the case of TiN described above, Al deposited with I-PVD techniques is significantly different than conventionally sputtered or collimated-sputtered Al. Experiments have both examined the crystallinity of I-PVD Al and AlCu and compared results to ionized-beam deposition techniques. In general, several results all point to a similar conclusion. The (111) texture of the Al films (on silicon dioxide) is strongly enhanced if either the relative ionization of the depositing species (at constant energy) or the kinetic energy (at constant relative ionization) is increased (Fig. 8.41) [8.26]. Another measure of this is to examine the dispersion or

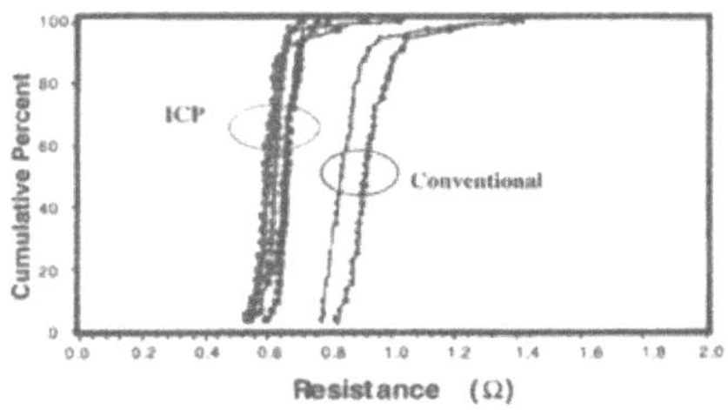

FIG. 8.38 Kelvin resistance distribution comparing I-PVD (IMP) to collimated Ti deposition [8.24].

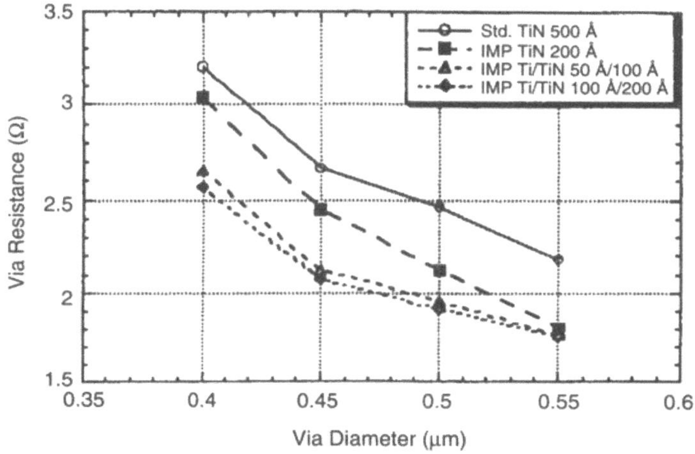

FIG. 8.39 Via resistance as a function of via diameter for standard TiN and I-PVD Ti/TiN layers [8.18].

degree of orientation of the (111) grains. As a function of increased Al ion energy, this also shows a strong effect, resulting in very highly oriented films at higher ion energies (Fig. 8.42) [8.26]. Similar results have been confirmed recently [8.10].

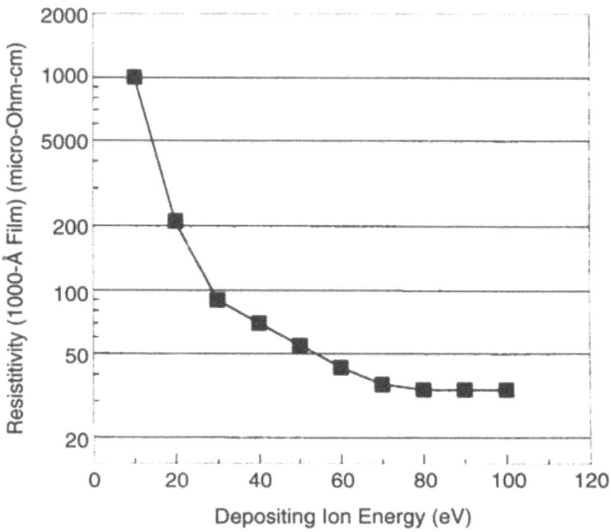

FIG. 8.40 The resistivity of 1000-Å TiN as a function of sample bias voltage for I-PVD reactive deposition to TiN [8.25].

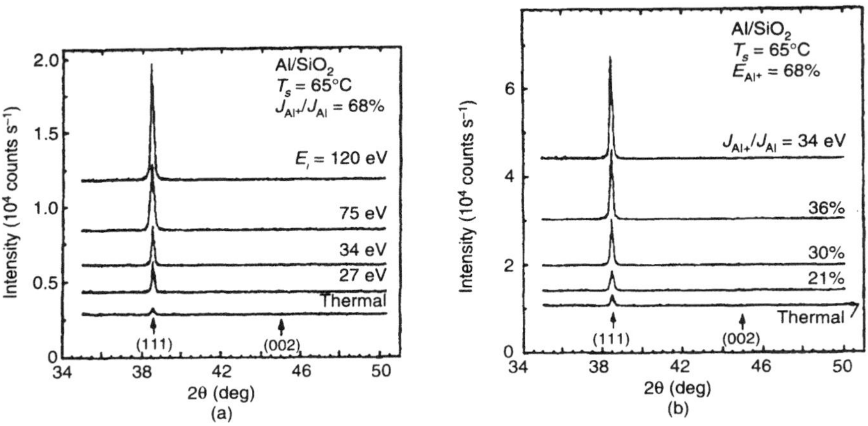

FIG. 8.41 Intensity of (111) peaks for Al films deposited on SiO_2 samples as a function of (a) increased ion energy at constant relative ionization and (b) increased relative ionization at constant ion energy [8.26].

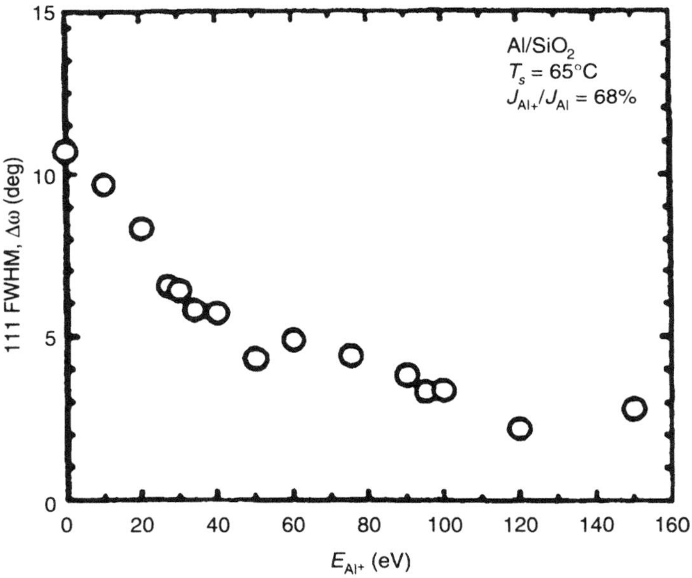

FIG. 8.42 Full width at half maximum intensity of w-rocking curves obtained from 3000-Å Al films grown at 65°C using a partially ionized Al beam [8.26].

References

8.1. J. J. Cuomo, R. J. Gambino, J. M. E. Harper, J. D. Kuptsis, and J. Webber, "Significance of negative ion formation in sputtering and SIMS analysis," *J. Vac. Sci. & Tech.* **15**: 281 (1978).

8.2. P. Kidd, "A magnetically confined and electron cyclotron resonance heated plasma machine for coating and ion surface modification use," *J. Vac. Sci. & Tech.* **A9**: 466 (1991).

8.3. W. M. Holber, J. S. Logan, H. J. Grabarz, J. T. C. Yeh, J. B. O. Caughman, A. Sugarman, and F. Turene, "Copper deposition by ECR plasma," *J. Vac. Sci. & Tech.* **A11**: 2903 (1993).

8.4. M. Barnes, J. C. Forster, and J. H. Keller, "Apparatus for depositing material into high aspect ratio holes," U.S. Patent No. 5,178,739 (Jan. 12, 1993).

8.5. S. M. Rossnagel and J. Hopwood, "Magnetron sputter deposition with high levels of metal ionization," *Appl. Phys. Lett.* **63**: 3285 (1993).

8.6. S. M. Rossnagel and J. Hopwood, "Metal ion deposition from ionized magnetron sputtering discharge," *J. Vac. Sci. & Tech.* **B12**: 449 (1994).

8.7. Applied Materials "Vectra Source," Santa Clara, CA.

8.8. J. Drewery, F. Cerio, K. F. Lai, Q. Lu, G. Reynolds, and M. Vukovic, "Ionized physical vapor deposition for next-generation integrated circuit manufacturing," in *Proc. VMIC*, Santa Clara, CA, 1997, pp. 274–276 (unpublished).

8.9. S. M. Gorbatkin, D. B. Poker, R. L. Rhodes, C. Doughty, L. A. Berry, and S. M. Rossnagel, "Cu metallization using a permanent magnet electron cyclotron resonance plasma/sputtering hybrid system," *J. Vac. Sci. & Tech.* **B14**: 1853–1859 (1996).

8.10. K. F. Lai, Q. Lu, G. J. Reynolds, L. M. Tam, C. J. Case, C. B. Case, M. A. Marcus, and J. E. Bower, "Ultra low resistivity Ti/TiN diffusion barriers deposited by hollow cathode magnetron sputtering," in *Proc. VMIC*, p. 274 Santa Clara, CA, 1997, (unpublished).

8.11. J. Hopwood and F. Qian, "Mechanisms for highly ionized magnetron sputtering," *J. Appl. Phys.* **78**: 758 (1995).

8.12. Brian Chapman, *Glow Discharge Processes*, John Wiley and Sons, NY 1980.

8.13. M. Yamashita, "Fundamental characteristics of a built-in high-frequency coil-type sputtering apparatus," *J. Vac. Sci. & Tech.* **A7**: 151 (1989).

8.14. M. Dickson, F. Qian, and J. Hopwood, "Quenching of electron temperature and electron density in ionized physical vapor deposition," *J. Vac. Sci. & Tech.* **A15**: 340–344 (1997).

8.15. S. M. Rossnagel, "Gas density reduction effects in magnetrons," *J. Vac. Sci. & Tech.*, **A6**: 19–24 (1988).

8.16. Y. Tanaka, T. Tanimoto, P. Gopalraja, J. Forster, and Z. Xu, "Ionized metal plasma deposition of titanium and titanium nitride," in *Proc. VMIC* Santa Clara, CA, 1997, pp. 437–439 (unpublished).

8.17. Z. Wang, W. Catabay, J. Yuan, J. Ku, N. Krishna, V. Pavate, A. Sundararajan, D. Saigal, B. Chang, M. Narasimhan, J. Egermeier, and S. Ramaswami, "IMP Ti/IMP TiN and IMP Ti/CVD TiN liners for W-plug metallization schemes," in *Proc. VMIC* Santa Clara, CA, 1997, pp. 258–261 (unpublished).

8.18. S. Bothra, S. S. Sengupta, B. Chang, M. Narasimham, and S. Ramaswami, "Extending PVD TiN to sub-0.25 micron technologies using ionized metal plasmas," in *Proc. VMIC*, Santa Clara, CA, 1997, pp. 240–245 (unpublished).

8.19. H. J. Barth, H. Helneder, D. Piscevic, M. Schneegans, G. Birkmaier, G. Crowley, H. Kieu, S. Ramaswami, and U. Richter, "Integration of a novel IMP Ti/TiN barrier with W-plug fill for contact and via applications," in *Proc. VMIC*, Santa Clara, CA, 1997, pp. 225–230 (unpublished).

8.20. S. Hamaguchi and S. M. Rossnagel, "Liner conformality in ionized magnetron metal sputter deposition processes," *J. Vac. Sci.& Tech.* **B14**: 2603 (1996).

8.21. S. Hamaguchi and S. M. Rossnagel, "Simulations of trench-filling profiles under ionized magnetron sputter metal deposition," *J. Vac. Sci. & Tech.* **B13**: 183–191 (1995).

8.22. C. Nichols, S. M. Rossnagel, and S. Hamaguchi, "Ionized physical vapor deposition of Cu for high aspect ratio damascene trench fill applications," *J. Vac. Sci. & Tech.* **B14**: 3270 (1996).

8.23. P. F. Cheng, S. M. Rossnagel, and D. N. Ruzic, "Directional deposition of Cu into semiconductor trenches using ionized magnetron sputtering," *J. Vac. Sci. & Tech.* **13**: 203 (1995).

8.24. T. P. Hong, R. Fiordalice, S. Garcia, H. Chuang, M. Thompson, V. Pol, B. Chu, J. Klein, F. Pintchovski, and R. Marsh, "Ionized metal deposition for ULSI interconnect," presented at 1996 Advanced Metallization and Interconnect Systems for ULSI Applications, Boston, MA, pp. 10–96, to be published by *MRS Proceedings*.

8.25. S. M. Rossnagel, "Directional and preferential sputtering-based physical vapor deposition," *Thin Solid Films* **263**: 1 (1995).

8.26. Y.-W. Kim, J. Moser, I. Petrov, J. E. Greene, and S. M. Rossnagel, "Directed sputter deposition of AlCu: Film microstructure and microchemistry," *J. Vac. Sci. & Tech.* **A12**: 3169 (1994), Y.-W. Kim, I. Petrov, J. E. Greene, and S. M. Rossnagel, "Development of (111) texture on Al films grown on SiO2/Si (001) by ultrahigh vacuum primary ion deposition," *J. Vac. Sci. & Tech.* **A14**: 346 (1996).

Chapter 9 PVD Materials and Processes

9.1 Introduction

There are many excellent monographs, conference proceedings, and articles that treat microelectronic thin film metrology, properties, and processing (see Section 1.3). This chapter presents a brief overview of the key microelectronic materials and process sequences that are utilized in PVD for multilevel metallization (MLM), including contact, via, and interconnect applications. Since this book is a tutorial about PVD and not thin films per se, this chapter will emphasize PVD-specific aspects of thin film processing. By way of introduction, Fig. 9.1 summarizes material properties of the key metallic elements encountered in PVD for microelectronics.

Figure 9.2 illustrates schematically how PVD films are utilized for MLM applications in a 0.5-μm VLSI device, and Fig. 9.3 illustrates how PVD metals might be incorporated in a 0.18-μm ULSI device (see also Fig. 1.3, Chapter 1). For VLSI devices, as discussed in Chapter 1, the metallurgy of PVD is dominated by relatively thick ($\approx$ 1 μm) Al alloy interconnect lines — clad with Ti and/or TiN for improved performance — and relatively thin ($\approx$ 500 Å) Ti/TiN barriers, liners, and adhesion layers to deal with CVD W plugs at contact and via levels. As a result, much of this chapter deals with PVD process conditions and issues for Ti, TiN, and Al alloys. On the other hand, the desire to replace or augment Al alloys with a lower-resistance interconnect for ULSI devices has focused considerable attention on Cu and Cu-compatible barriers such as Ta and TaN. In the future, as shown in Fig. 9.3, it is likely that these films will join the list of key PVD materials, so this chapter also includes a discussion of PVD Cu issues.

Both Al and Cu are also of interest with regard to contact and via plugs. Namely, although CVD W has been the plug fill material of choice for VLSI devices, MLM technology roadmaps indicate that W via plugs, and possibly W contact plugs, will be replaced for ULSI devices by the more conducting Al and/or Cu. Another reason for moving away from W is that it is a diffusion barrier for the critical diffusing Al and Cu species in electromigration and prevents their being replenished. As a result, void damage can occur at the via once the limited source of migrating species is depleted. Therefore, interconnect roadmaps show W plugs being replaced by Al and/or Cu for deep submicron devices ($\leq$ 0.18 μm). Replacing W with Al or Cu will allow one to take advantage of planarized processes (e.g., the two-step process for Al and dual-damascene process for Cu) so that the same metal can be used in both the plug and the line. In spite of

	Al	Co	Ti	Cu	Ta	W	Au
Atomic number (Z)	13	27	22	29	73	74	79
Atomic mass (AMU)	26.98	58.9	47.90	63.54	180.95	183.85	196.97
Density (gm/cm^3)	2.7	8.71	4.51	8.96	16.6	19.3	19.32
Melting point (°C)	660	1495	1668	1083	2996	3410	1063
Resistivity (μΩ-cm)	2.7	6.2	42	1.7	13.5	5.6	2.35
Thermal conductivity (cal-cm/cm^2-sec-K)	0.53	0.17	0.22	0.94	0.13	0.4	0.71
Coefficient of thermal expansion (x 10^{-6}/K)	23.6	13.8	8.4	16.5	6.5	4.6	14.2
Binding energy (eV)	3.39	4.46	4.85	3.49	8.10	8.90	3.81
First ionization potential (eV) [IP_{Ar} = 15.7 eV]	5.98	7.86	6.83	7.72	7.7	7.98	9.22
Yield (atoms/ion) for 600 eV Ar^+ [Y_{Si} ~ 0.5, Y_{SiO2} ~ 1.3, Y_{Al2O3} ~ 0.2]	0.83	0.99	0.45	2.00	0.3	0.32	1.18

FIG. 9.1 Selected material properties of key elements encountered in PVD for microelectronic applications.

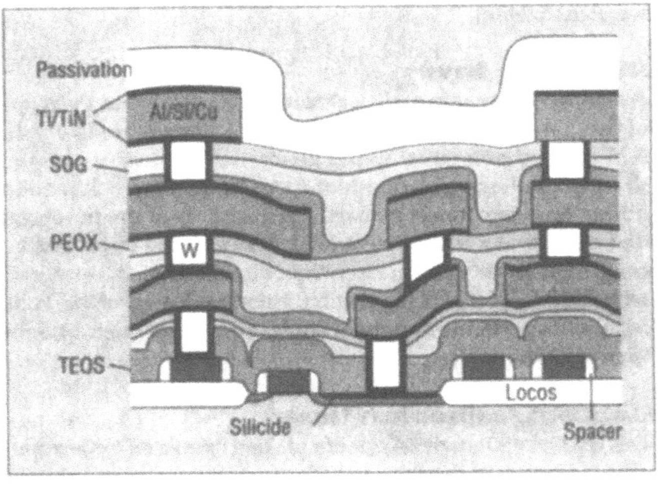

FIG. 9.2 Illustration of the application of PVD films in a three-level metallization scheme of a 0.5-μm very large scale integrated (VLSI) device [9.1]. The wiring is characterized by CVD W plugs with PVD Ti/TiN liners and PVD slab Al lines (i.e., PVD AlSiCu or AlCu clad with Ti/TiN). The interlayer dielectrics are based on a combination of spin-on glass (SOG) and thermal or plasma CVD oxides.

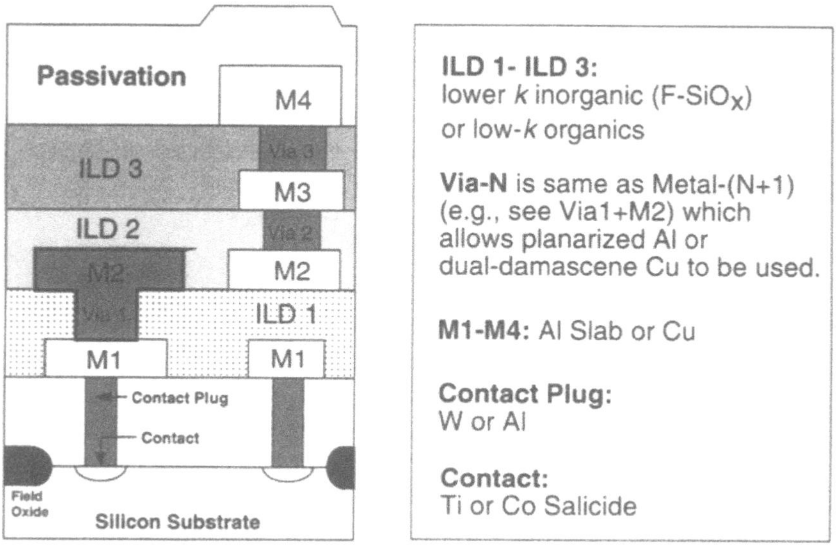

FIG. 9.3 Illustration of the application of PVD films in an ultra-large scale integrated (ULSI) device.

encouraging results for hot and/or high-pressure PVD processing (see Chapter 7) and ionized PVD (see Chapter 8), it is an open question whether PVD, CVD, or a combined CVD/PVD approach will be used for filling these high aspect ratio plugs. Therefore, some comments are included on the integration of PVD and CVD. Finally, this chapter includes a brief discussion of PVD issues of refractory alloys (TiW), refractory metal silicides (e.g., $MoSi_2$, WSi_2, $TiSi_2$ and $CoSi_2$), and the use of PVD in back-end-of-line bonding applications.

9.2 Metrology

The science and technology of thin film metrology for microelectronics has developed enormously over the past 20 years, driven by the need of research scientists for increasingly sensitive surface analytical tools and by the need of IC technologists to measure key thin film properties with device-scale spatial resolution and to map these properties over large-diameter wafers [9.2–9.6]. As a result, commercial equipment or analytical services are now available to measure and map critical electrical, mechanical, and optical properties of PVD films, including film thickness, chemical composition and purity, surface roughness, grain size distribution and orientation, step coverage, electrical resistivity, optical reflectivity, stress, and the size distribution and composition of fine particles added by the process or by mechanical handling. In addition, noncontact, nondestructive methods are being developed to measure film properties on actual product wafers, thereby reducing costs associated with test wafers — a particular issue for 300-mm technology due to the excessive cost per wafer (> $1000). For example, a novel "laser sonar" method based on picosecond ultrasonic laser (PULSE) technology has been developed that can simultaneously measure the thickness of a multilayer metal film stack (e.g., TiN/Ti/AlCu/TiN/Ti/SiO_2/Si) with high spatial resolution (20 μm) and sub-Angstrom precision over a wide range of film thickness of $\approx$ 20 Å-5 μm [9.7].

There are literally hundreds of analytical methods that can be used to characterize PVD films used in microelectronics; however, in the context of IC production only about a dozen are routinely used for process qualification or failure analysis: (1) full-wafer mapping of electrical sheet resistance (R_s) with a four-point probe or noncontact eddy current method; (2–4) thickness mapping by physical profilometry and, more recently, by X-ray fluorescence (XRF) and thermal-wave methods; (5) microscopic cross-sectional imaging of contacts, vias, and interconnects obtained by a combination of sample cleaving/polishing and secondary electron microscopy (SEM) — often with high-resolution field emission (FE) electron sources and elemental information provided by energy dispersive X-ray (EDX) analysis; (6–9) surface and in-depth elemental and chemical analysis by a complementary combination of the "Big Four" methods — auger electron spectroscopy (AES), secondary ion mass spectrometry (SIMS), X-ray photoemission spectroscopy (XPS or ESCA), and Rutherford backscattering spectroscopy (RBS); (10) particle detection and mapping by laser light scattering; (11) optical properties by reflectometry; (12) crystal structure and orientation by X-ray diffraction (XRD); (13) film stress by laser reflection; and (14) surface roughness by atomic force microscopy (AFM). Ultimately, of course, it is device performance and reliability that determines the quality of a PVD film for a microelectronic application. Figure 9.4 shows representative probing areas and detection sensitivity for several of the principal surface-sensitive analytical tools.

In the context of IC production, the most routinely measured PVD metal film properties are probably resistivity, thickness, and morphology (e.g., step coverage of a via). In this regard, a widely used measurement for PVD equipment qualification is the uniformity of sheet resistance for an unpatterned (i.e., a blanket) PVD film. This uniformity is often specified by the

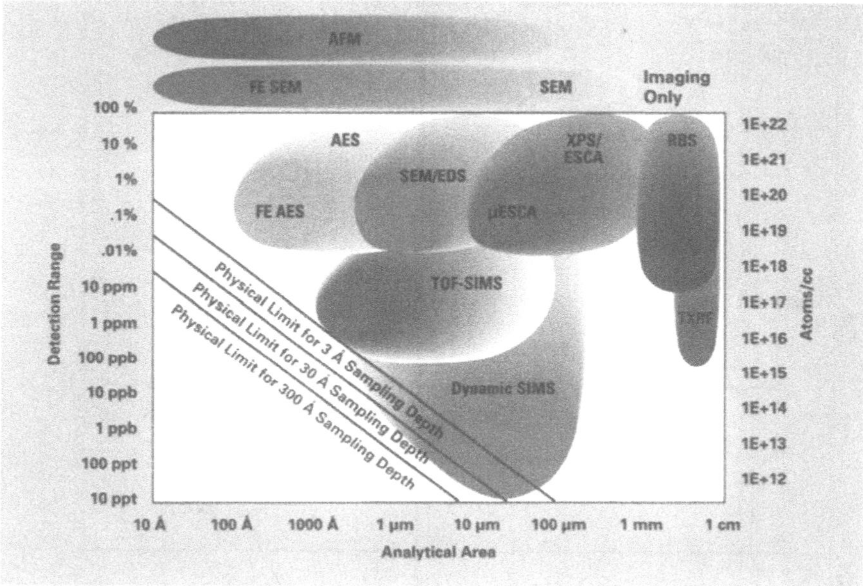

FIG. 9.4 Analytical sensitivity and probing depth of common surface-sensitive tools used in PVD metrology (courtesy of Charles Evans & Associates, Redwood City, CA).

supplier in the statistically based unit of standard deviation, sigma or σ. For example, the R_s uniformity of a 1-μm Al alloy film on a 200-mm wafer with edge exclusion of 3 mm might be given as $3\sigma = 5\%$ or, equivalently, as $3\sigma = \pm 5\%$. For a statistically normal, bell-shaped distribution, this would imply that about 99.7% of the R_s data points lie within a range extending from 5% below the mean to 5% above (see Fig. 9.5). Unfortunately, the distribution of thickness R_s of a PVD film over a wafer is not dominated by random events like the tossing of a coin, but often has obvious patterns (e.g., a W-shaped profile) that reflect the design of the magnetron source and chamber geometry. In addition, rarely are more than 100 points collected in routine sheet resistance mapping, so that talking about 99.7% of the data points is meaningless unless more than 1000 points were collected. As a result, a more realistic way of reporting PVD film uniformity is to take the maximum (M) and minimum (m) values from the data set and report the ratio of the data range (M − m) to its sum (M + m). Using this "max-min" notation, a film might be stated to have $(M - m)/(M + m) = 5\%$ or, equivalently, $(M - m)/(M + m) = \pm 5\%$,

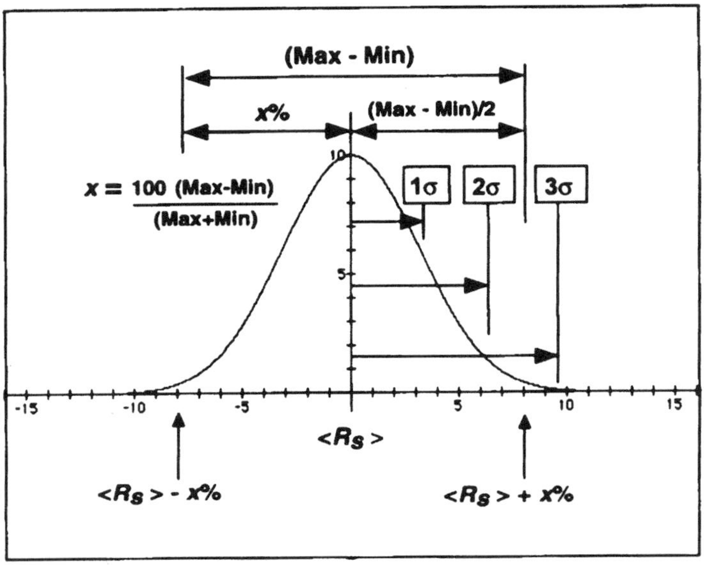

FIG. 9.5 Comparison of a gaussian distribution with the measured statistical distribution of sheet resistance for a PVD Al film.

which means that all of the data points lie within 5% below the mean value of (M + m)/2 to 5% above. As an empirical rule of thumb, and assuming that one could in fact treat the PVD distribution as normal, the max/min ratio of PVD films typically has a value between 2σ and 3σ. Even though the distribution of PVD film thickness R_s over the wafer is not normal, random statistical processes may be the determining effect on the *repeatability* of uniformity. In this case, normal statistical notation is appropriate.

> The sheet resistance R_s of a thin film of thickness t is often referred to as "sheet rho," and sometimes improperly called "sheet resistivity." While the Greek symbol rho (ρ) is used to denote resistivity that has CGS units of ohm-cm, sheet resistance $R_s = \rho/t$ has units of ohm/square, also written as Ω/sq. or Ω/□. So "sheet rho" is an oxymoron. By way of illustration, if an 8000-Å AlCu film of bulk resistivity $\rho = 3.0\ \mu\Omega$-cm is deposited onto SiO_2, the measured sheet resistance of the film is given by $R_s = \rho/t = (3.0 \times 10^{-6}\ \Omega\text{-cm})/(8 \times 10^{-5}\ \text{cm}) = 3.8 \times 10^{-2}\ \Omega/\text{sq}$.

The repeatability of a statistically variable parameter is often expressed in terms of a dimensionless quantity C_p — or C_{pk} in the most general case — which is referred to as the process capability index or manufacturability. C_p is of particular importance to the production use of semiconductor hardware (including PVD tools) and gives information about the relationship between design tolerance and process width. In particular, C_p is defined as the ratio of design tolerance (i.e., the spread between upper and lower specified control limits) to the process width (max-min, 6σ, etc.). A high value of C_p means that the process is tight and that statistical variations are unlikely to produce a defective, out-of-spec product. For example, assume that the thickness of a desired PVD Al film is targeted to be 1.0 μm, but might be acceptable if its thickness were no greater than 1.1 μm and no less than 0.9 μm — a specified control limit of $\pm 10\%$. If the wafer-to-wafer repeatability of this deposition on a given PVD tool has a standard deviation of $1\sigma = 3\% = 30$ nm, then the process capability is calculated to be $C_p = (1.1\ \mu\text{m} - 0.9\ \mu\text{m})/(6 \times 30\ \text{nm}) = 1.1$, where a process width of 6σ was chosen. For state-of-the-art PVD tools and processes, one desires $C_p \geq 2$, in which case only about 1 film in 10^6 will be outside of the specified control limits. Part-per-million levels of defective parts in semiconductor fabrication was a concept pioneered by Motorola and is referred to as a "six-sigma" or "zero-defects" quality control methodology.

While the uniformity of blanket PVD film thickness or sheet resistance are often used for process or equipment qualification, it is important to note that there are a number of other PVD "uniformities" that impact device performance and whose distribution can be quite different from that of blanket thickness; these include bottom coverage, sidewall coverage, and film composition. For example, early-generation magnetrons (e.g., the Con-Mag™ from Varian) gave extremely high blanket uniformity, but their sidewall coverage in high aspect ratio vias was not nearly as uniform. Also, differences in the sputtered angular distributions and gas-phase scattering of an AlCu(1%) alloy's component elements may lead to highly nonuniform Cu distribution from center to edge, even though the film thickness and resistivity may be much more uniform.

Another metrology issue relates to the sheet resistance of extremely thin PVD films ($\ll 500$ Å) such as Ti and TiN used for contacts, barriers, and adhesion layers. If the film thickness and/or polycrystalline grain size are smaller than the electron mean free path, then scattering of conduction electrons at free surfaces and grain boundaries adds to the intrinsic resistivity. The resistivity ρ is found to have a dependence on film thickness t of the form $\rho = \rho_0 (1 + a\lambda/t)$, where ρ_0 is the resistivity inside the crystal grains, λ is the mean free path of conduction electrons in the film, and a depends on the grain boundary scattering cross section and the density of grain boundaries [9.8]. Therefore, even though the intrinsic bulk resistivity of a thick polycrystalline film of TiN may be $\rho_0 \approx 50\ \mu\Omega$-cm, the actual resistivity can be greater for very thin films, leading one to underestimate their thickness from a sheet resistance measurement. For example, Fig. 9.6 shows how the electrical resistivity of both PVD Ti and PVD TiN films sputtered onto SiO_2 increases greatly for thickness below about 200 Å. Regarding thin film effects, it is worth noting that even though ULSI interconnect lines are hundreds of times thicker than their barriers and liners, thin film thinking is still appropriate. For example, nearly 50% of the Al atoms in a 0.75-μm × 0.25-μm interconnect line are located within 500 Å of a surface or interface.

9.3 Al Alloys

9.3.1 Metallurgical Considerations for PVD

Aluminum alloys (with a few weight percent of Si and/or Cu to prevent junction spiking and enhance electromigration resistance, respectively) used in combination with Ti and TiN cladding layers are likely to remain

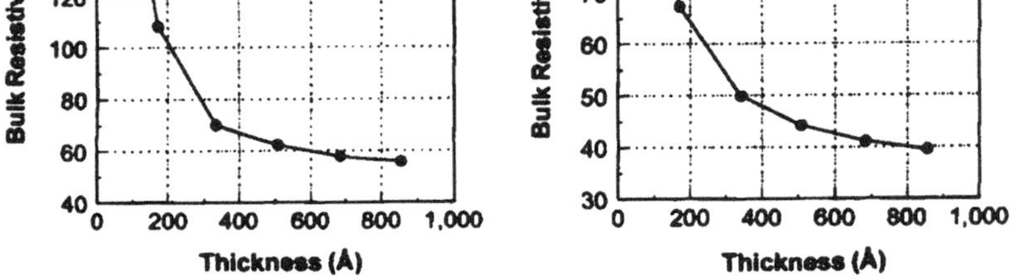

FIG. 9.6 Dependence of resistivity on film thickness for collimated PVD Ti and PVD TiN.

the dominant horizontal interconnect for 0.25-μm (and possibly 0.18-μm) devices. In fact, the primary application of PVD cluster tools in production today is to deposit planar films of "slab Al" interconnect as opposed to the more aspect-ratio-challenging applications of contacts and barriers. The advantages of Al in microelectronics are numerous [9.9], with the following points being of special relevance to its use as a PVD interconnect.

1. The room temperature electrical resistivity of pure Al ($\rho = 2.7$ $\mu\Omega$-cm), although not as low as Cu, Ag, or Au, is one of the lowest among all the metals. This resistivity is only slightly increased to ≈ 3 $\mu\Omega$-cm when the Al is alloyed with a few weight percent of Cu to improve electromigration resistance ($\Delta\rho \approx 0.3$ $\mu\Omega$-cm for 1 wt % Cu) and/or with a few weight percent Si to limit void formation at the metal–Si interface ($\Delta\rho \approx$ 0.7 $\mu\Omega$-cm for 1 wt % Si). In any case, PVD Al resistivity is many times lower than that of the other films used in the MLM stack (e.g., $\rho \approx 70$ $\mu\Omega$-cm for PVD Ti and TiN; $\rho \approx 10$ $\mu\Omega$-cm for CVD W).
2. Al sputter targets can readily be obtained with ultrahigh purity ($\geq$ 5N5) in either elemental or alloy composition.

> PVD alloy film and target compositions are often given in weight percent (e.g., an Al-Si(1%)-Cu(0.5%) film or a Ti(10%)-W target) which, depending on the relative masses and concentration of the elemental constituents, can differ significantly from atomic percent (see Chapter 11). For example, 1 wt % of Si or Cu in Al is equivalent to $\approx$ 1.0 and 0.4 at %, respectively, while 10 wt % of Ti in W is equivalent to $\approx$ 30 at %. Therefore, in comparing results it is important that the same type of percents are being reported. This is particularly relevant for surface analytical results, which are often given in atomic concentrations.

3. The DC magnetron sputter rate of Al is high enough ($>$ 1 μm/min) that blanket 1-μm Al films can be deposited with production-worthy throughput of $>$ 40 wafers per hour.
4. Although Al is highly reactive with SiO_2 and reduces it to Si (heat of formation of Al_2O_3 is 399 Kcal/mol vs 205 Kcal/mol for SiO_2), the reaction is self-limiting and stops when a sufficiently thick Al_2O_3 layer has formed. This ensures that when Al is sputtered onto field oxide regions the reaction does not compromise the integrity of either the Al wiring or the interlayer dielectric. On the other hand, the limited reactivity of Al films toward SiO_2 is very important since this ensures good adherence of the PVD Al film to the field oxide surface and to the sidewalls of a via cut

through the oxide, obviating the need for a separate glue layer such as the TiN that is used between oxide and CVD W.

5. Al has a relatively low melting point (T_{mp} = 660°C) with high self-diffusion rates at moderate process temperatures (≈ 400–550°C). This has allowed a variety of elevated-temperature PVD processes such as reflowed Al and the cold-hot Al, two-step process (TSP) to be used to improve the step coverage and filling of Al in high aspect ratio features (see Chapter 7).

6. Al films and Al alloy films with moderate weight percents of Cu can be readily patterned into interconnect lines using plasma-assisted, dry etching methods. This ability to use subtractive metal patterning (i.e., etching of a photoresist-patterned Al overlayer on oxide) means that one does not have to resort to a single- or dual-damascene approach in forming the multilevel metal interconnect stack, such as is the case with Cu (see Section 9.8). Damascene processing not only removes the need for plasma etching of the metal lines but also the need to fill the gaps between the lines with insulator. Since plasma etching of metals and dielectric gap fill are two of the most difficult processes in ULSI device fabrication, this is a considerable simplification. Damascene processing can require filling of higher aspect ratio structures such as simultaneous filling of a via and trench; however, the potential cost savings has led to its being applied to Al as well as Cu even though a subtractive method of patterning the Al could be used. On the other hand, the chemical-mechanical polishing (CMP) step used to planarize the metal layer involves creation of an anodized metal surface. When damascene processing is applied to Al, the CMP step then requires polishing back a layer of alumina (Al_2O_3) whose hardness is greater than either CuO or SiO_2.

The two major concerns about PVD Al interconnect lines are (1) their relatively poor electromigration (EM) resistance and (2) the effects of stress that can result in the formation of voids within the lines (stress voiding) or the formation of protruding bumps on their surface (hillock formation). Electromigration refers to the migration of matter due to momentum exchange between the conduction electrons and Al atoms of the interconnect line. Even though the total current flow in a thin film interconnect is small, its microscopic cross-sectional area leads to an enormous current density ($10^{6\text{-}7}$ A/cm^2 for advanced devices), which can lower device reliability and even result in catastrophic open-circuit line failure. Historically, PVD has addressed concerns about EM by depositing Al alloys with a few weight percent of Cu and by choosing deposition conditions favoring a strongly (111) oriented film. Also, since thinning down of the metal along the vertical sidewalls of via holes can lead to local heating and EM failure,

PVD processes with improved step coverage are preferred. Concerns about PVD film stress have been addressed by reducing process temperature and using cladding layers such as Ti and/or TiN on the Al to "harden" it against stress voiding and hillock formation as well as to provide a low-resistance shunt should the Al line start to void. A useful summary of film issues associated with either thermal stress or electromigration is provided in ref. 1.15 (Chapter 8 on "Electro- and Stress-Migration in MLM Interconnect Structures," M. L. Dreyer and P. S. Ho).

9.3.2 Deposition Rate

Advanced DC magnetrons are capable of depositing the 1-μm-thick Al alloys used in a slab Al interconnect with a uniformity of $3\sigma < 5\%$ over 200-mm wafers. A high rate of sputtering (> 1 μm/min) is also needed for production-worthy cluster tool throughput of $\approx$ 40–60 wafers/hour. As a practical matter, it is not deposition rate that matters but the deposition rate normalized to the sputter cathode power, or *specific deposition rate* (SDR).

Figure 9.7 shows SDR values (Å/sec-kW) as a function of power to an Al planar magnetron cathode showing a flattening above $\approx$ 5 kW. As the power applied to the cathode increases, the number of sputtered Al atoms in the volume between target and wafer also increases, and the increased Al-Al gas-phase collisions scatter Al away from the wafer, limiting the gain in Al deposition rate below that expected from the increased sputter erosion rate of the target. Using the data in Fig. 9.7, we see that while 3 kW gives an Al deposition rate of 5220 Å/min (SDR = 29 Å/sec-kW), one must go to 9 kW (SDR = 21 Å/sec-kW) to double the deposition rate to $\approx$ 1.1 μm/min. With regard to cathode size, the SDR tends to decrease linearly with increasing target area. For example, the SDR values in Fig. 9.7 taken with a 12-inch-diameter cathode (Varian Quantum™ source) were empirically found to be $\approx$ 50% higher when an 8-inch-diameter magnetron was used (Varian Mini-Quantum™ source). On the other hand, coating uniformity of 200-mm wafers was not nearly as good when the smaller cathode was used.

9.3.3 Deposition Temperature and Microstructure

Since chemical reaction rates and physical diffusion phenomena depend strongly (often exponentially) on temperature, it is not surprising that the deposition temperature of PVD Al has a strong influence on its

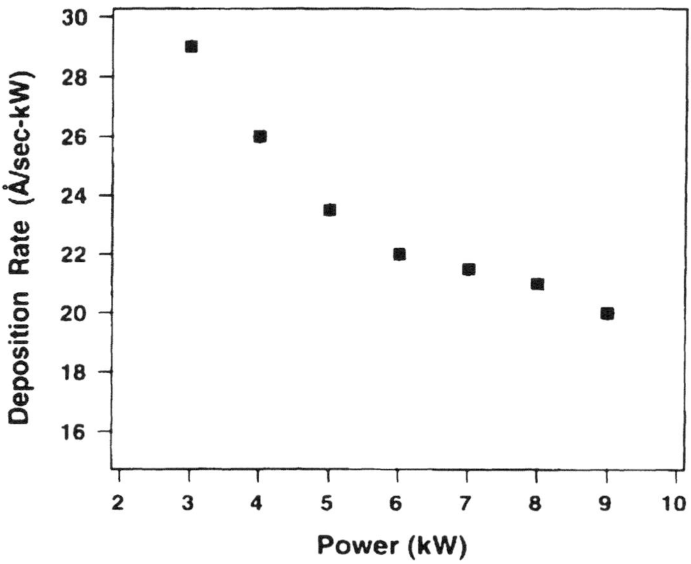

FIG. 9.7 Deposition rate of Al as a function of magnetron cathode power.

microstructure and therefore on its electrical, optical, and mechanical properties. In addition, as discussed in Chapter 5, outgassing from the heated wafer and the indirectly heated chamber walls and fixtures can release oxidants (H_2O, O_2) and other contamination that degrade film properties. For example, high specular reflectivity is often used as a measure of film quality, with milky-looking, rough Al films indicative of oxidation during sputter deposition.

Grain Size PVD Al and Al alloy films are polycrystalline in nature with the dominant orientation, grain size, and grain size distribution dependent on a variety of process conditions, but strongly influenced by temperature. Figure 9.8 shows the dramatic increase in average grain size for a PVD Al-Si-Cu film as a function of wafer temperature (20–400°C) during deposition onto a thermal oxide-coated Si wafer. The grain size distribution can be visualized from Fig. 9.9, where a dark field optical micrograph (≈ 1000×) of the Al alloy film surface is shown at the low- (20°C) and high-end (400°C) temperatures.

Reflectivity Even though the use of Al in microelectronics is driven by its electrical properties, its optical properties are routinely measured be-

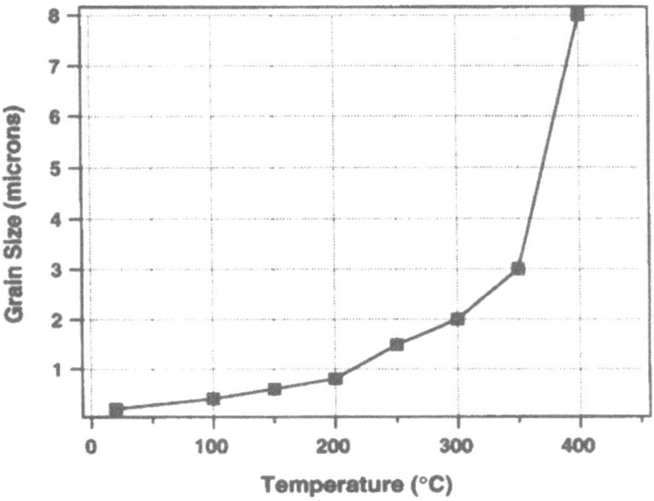

FIG. 9.8 Grain size of PVD Al-Si-Cu alloy as a function of wafer temperature during deposition onto SiO_2.

cause they directly impact subsequent lithographic patterning steps and indirectly indicate film purity and microstructure. The reflectivity of Al is probably the most common optical property measured even though the real and imaginary parts of the complex dielectric constant $N = n + ik$ are the more fundamental physical parameters. (*Note*: The refractive index n is

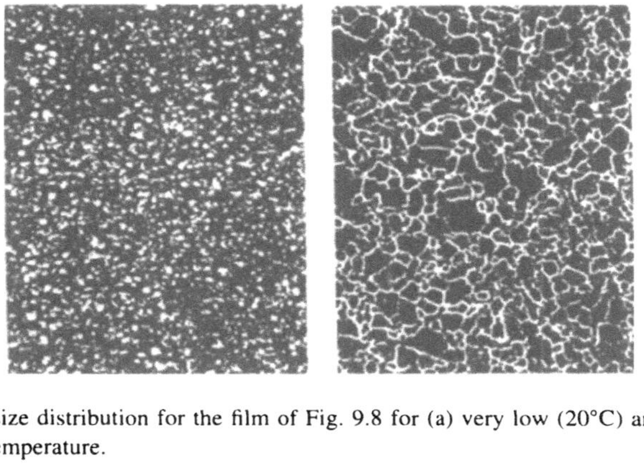

FIG. 9.9 Grain size distribution for the film of Fig. 9.8 for (a) very low (20°C) and (b) very high (400°C) process temperature.

sometimes called out on thin film spec data sheets as "RI".) The specular reflectivity of Al is typically measured at a wavelength used for optical lithography (such as 440 nm) and is given in absolute units or relative to that of Si. As with grain size, reflectivity also depends on temperature but in a rather complicated way that is related to changes in both grain size and film morphology (see Fig. 9.10). The effect of temperature on Al film morphology is conveniently summarized in Fig. 9.11 using the structure zone model first proposed by Movchan and Demchishin [9.10] whereby the structure of a film deposited on a substrate at temperature T depends universally on the normalized temperature ratio T/T_m, where T_m is the melting point of the film in degrees Kelvin (this ratio is also referred to as the homologous temperature). The initial work of Movchan and Demchishin was based on e-beam evaporated films and did not consider the structure of PVD films per se. The model was later amended by John Thornton for application to sputter deposition by addition of another independent variable — the pressure of the inert sputter gas in the deposition chamber. Thornton then introduced the amended model to the semiconductor industry in the early 1970s [9.11, 9.12]. As a result, the three-dimensional pictogram shown in Fig. 9.11 relating zones of PVD film morphology to both sputter gas pressure (x-axis) and normalized temperature T/T_m (y-axis) is popularly referred to as a *Thornton diagram*.

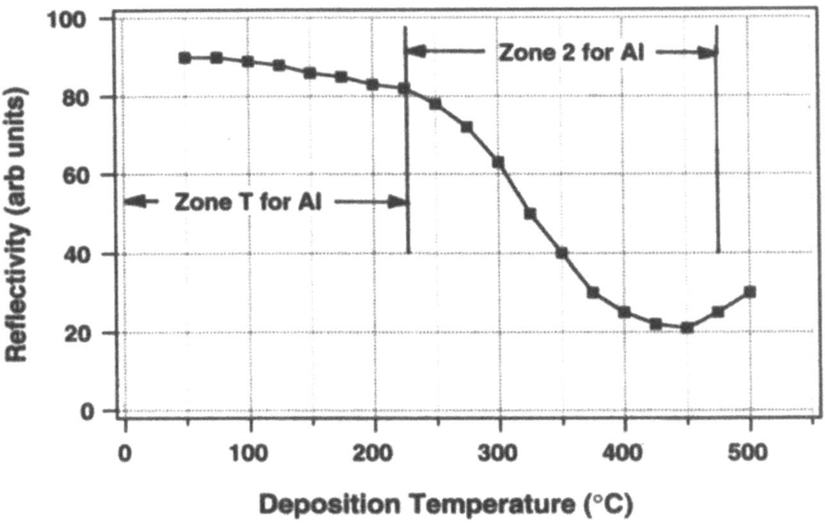

FIG. 9.10 Reflectivity of a PVD Al film as a function of deposition temperature.

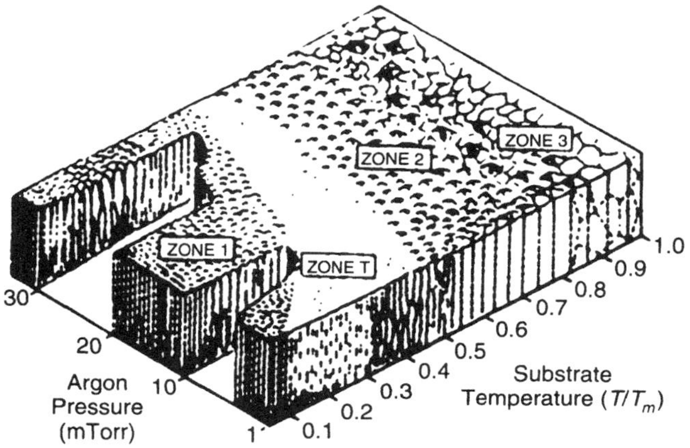

FIG. 9.11 Visualization of PVD film morphology versus process pressure and temperature can be made using a Movchan-Demchishin diagram [9.10], also referred to as a "Thornton" diagram [9.11]. (Reprinted with permission from J. A. Thornton, *J. Vac. Sci. & Tech.* **A11(4):** 666 (1974). Copyright 1974 American Institute of Physics.)

The structure zone model graphically shows how PVD film microstructure evolves with increasing deposition temperature from highly porous and columnar (Zone 1), to densely columnar (Zone 2), and finally to a recrystallized dense grain structure (Zone 3). Given the range of PVD Al sputter pressure ($\approx$ 3–5 mTorr) and deposition temperature (20–550°C) used for microelectronics, the relevant regions of the diagram are the "transition" Zone T and Zones 2 and 3. Zone T films are characterized by small grains. The surface is flat relative to the wavelength of the incident radiation so that the entire film surface acts as one large reflector, and film reflectivity is high — $R \approx 90\%$ absolute at 440 nm. As temperature increases, the film morphology moves into Zone 2, where the grains are larger and comparable to the wavelength of incident light. The surface angle of the individual grains differ from those surrounding a random grain and the reflectivity is reduced. At sufficiently high temperature ($>$ 450°C), a Zone 3 film with recrystallized, larger grains is formed. The individual grains are now large enough to act as individual reflectors and R increases slightly.

Resistivity The bulk resistivity of a PVD Al film typically decreases slightly with deposition temperature because the larger individual grains lead to a reduced number of grain boundaries per unit length, leading to reduced grain boundary scattering of the conduction electrons. This is illustrated in Fig. 9.12, which shows the bulk room-temperature resistivity for a t = 1-μm-thick PVD Al-1%Si-2%Cu film as a function of deposition temperature. The measured sheet resistance of the film would have been $R_s = \rho/t = (3\text{–}4\ \mu\Omega\text{-cm})/(1\ \mu\text{m}) = 0.03\text{–}0.04\ \Omega/\text{sq}$.

Thermal Stress Highly stressed films are not desirable in IC processing since this can lead to reliability problems, particle generation, and even the possibility of delamination of the film from the substrate or underlayer. In general, the total stress in a PVD film results from the sum of three components: (1) external stress, (2) intrinsic stress, and (3) thermal stress. External stress is usually not important given the small weight of a Si

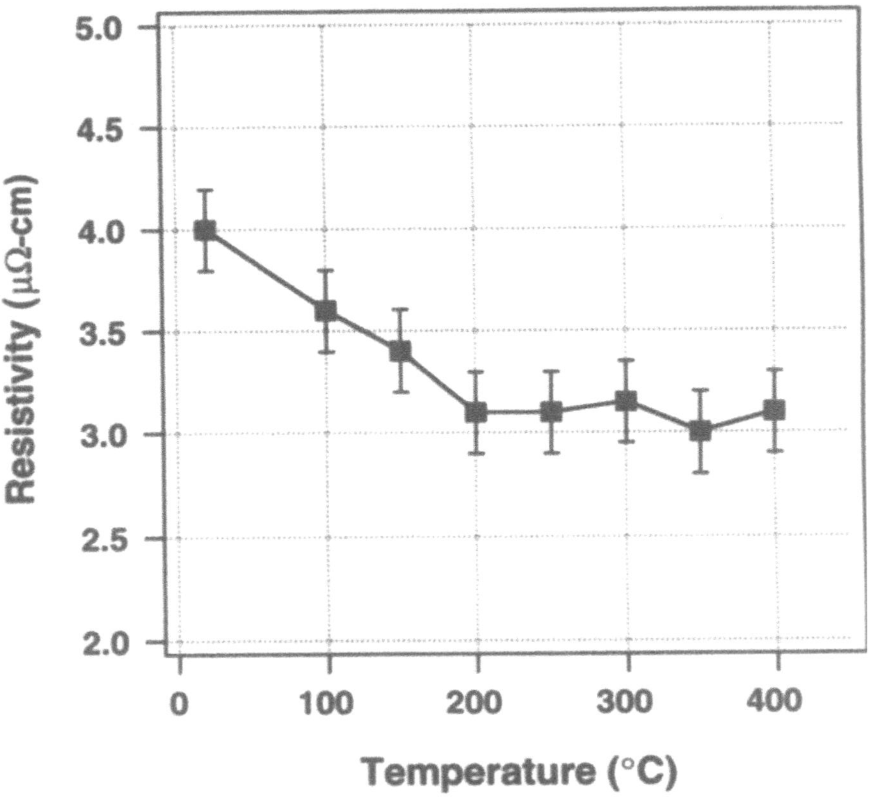

FIG. 9.12 Bulk resistivity of PVD Al-1%Si-2%Cu film as a function of deposition temperature.

wafer and the subatmospheric pressure of PVD processing (one notable exception is the Forcefill™ method described in Chapter 7 in which extremely high external pressure (> 600 atm) is applied to cause an Al film to flow into fine structures.) Intrinsic stress is related to the detailed microstructure of the film (e.g., lattice defects and impurities) and by the mismatch in lattice spacing between film and substrate. Intrinsic film stress depends on a number of deposition and film parameters (e.g., deposition rate, temperature, ion bombardment during deposition, argon incorporation, and film thickness) and can usually be controlled by choosing appropriate process conditions. Thermal stress results when the film and substrate expand or contract at different rates during thermal cycling. For a blanket two-dimensional film on a substrate, the thermal stress σ_{th} is given by

$$\sigma_{\text{th}} = \frac{E_f(\Delta\alpha_{\text{fs}})(\Delta T)}{(1 - \nu)} \tag{9.1}$$

where $\Delta\alpha_{\text{fs}} = \alpha_{\text{f}} - \alpha_{\text{s}}$ is the difference between the coefficients of thermal expansion (CTE) of the film and substrate, ΔT is the difference between the deposition temperature and measurement temperature (i.e., room temperature), E_f is the Young's modulus of the film, and ν is the Poisson's ratio. In general, the CTE of film and substrate are different. Hence, following PVD at elevated temperature, the film and substrate shrink by different amounts during cooling — resulting in a thermal stress. If the CTE of the film is greater than the CTE of the substrate, then during cooling the confining substrate will prevent the film from shrinking, leaving it under tension. If the CTE of the substrate is greater than that of the film, then the film will be pulled into compression.

Due to the large difference in the linear thermal expansion coefficient between Al ($\alpha = 23.2$ ppm per °K $= 23.2 \times 10^{-6}$ per °K) and that of Si ($\alpha \approx 2.6$ ppm per °K) or SiO_2 ($\alpha \approx 4$ ppm per °K), the film and underlayer shrink dimensionally by quite different amounts as they cool from the elevated temperature of deposition down to room temperature. The result is that a thermally induced stress develops in the PVD Al film. To estimate the magnitude of the stress, we assume that PVD Al was deposited on a Si wafer at 300°C (573 K) so that $\Delta\alpha_{\text{fs}} \approx 23.2$ ppm $-$ 2.6 ppm $=$ 20.6 ppm and $\Delta T = 573$ K $- 293$ K $= 280$ K. Young's modulus for Al is 9×10^6 psi $= 6.2 \times 10^{11}$ dyne/cm², and Poisson's ratio is ≈ 0.34. Therefore, using Eq. (9.1) we calculate that $\sigma_{\text{th}} \approx (6.2 \times 10^{11}$ dynes/cm²$)(20.6 \times 10^{-6}$ per K) (280 K)/(0.66) $= 5.3 \times 10^9$ dynes/cm² $=$ 530 MPa. This tensile stress would then add to the intrinsic film stress, which, if it were compressive, would then serve to reduce the net stress in the film.

Stress is measured in megapascals (MPa) or dynes/cm^2, where 1 MPa = 10^7 dynes/cm^2. By convention, values are written positive for tensile stress and negative for compressive stress.

One mechanism for relieving this stress is mass transport to the surface, which manifests itself as surface bumps or hillocks. This topography can induce interlayer short circuits and changes in metal reflectivity leading to difficulties with photolithography. The onset of hillock formation occurs around half the melting point in degrees Kelvin, which for Al is ≈ 190°C and unfortunately well within the typical PVD process window. The use of Cu-containing Al alloys, reduced process temperature, and PVD conditions that lead to small Al grain size can all be exploited to minimize hillock formation.

It is important to note that the thermal stress of an unpatterned, two-dimensional PVD film on a substrate (Eq. (9.1)) does not accurately represent the thermal stress of a real metal line. This is because in multilevel metallization, the metal films are embedded in dielectric layers and are either patterned into narrow lines or confined within three-dimensional contacts and vias. For example, simply confining an Al line within an oxide can double the stress that it would experience due to thermal cycling between room temperature and 400°C. Also, a large stress concentration exists at the corners of lines and at interfaces, which is where one usually observes void formation.

9.3.4 Crystal Orientation

The typical orientation of PVD Al films deposited on Si or SiO_2 is predominantly (111) with a small amount of (200), which has important consequences for PVD. In particular, it has been found that the mean-time-to-failure (MTTF) of an Al line due to electromigration can be correlated with a microstructural parameter η, which is a function of median grain size (s), standard deviation of grain size distribution (σ), and the peak intensities I of the (111) and (200) reflections in the X-ray diffraction pattern. This parameter is given by [9.13]

$$\eta = \left(\frac{s}{\sigma^2}\right)\log\left(\frac{I_{(111)}}{I_{(200)}}\right)^3 \qquad (9.2)$$

Large values of MTTF have been found to correlate with large values of η, and therefore a PVD Al interconnect film should have a narrow distribution of large grains with a strong (111) texture.

Al films are typically deposited over amorphous films such as the exposed oxide sidewalls in a via; however, they are also deposited onto films such as Ti and TiN that can have their own preferred crystal orientation. These underlayers can in turn effect the resulting Al orientation. Figure 9.13 shows data on the texture of PVD Al films deposited onto thermal SiO_2 (a thermally oxidized Si wafer) or onto a PVD Ti film that had been deposited onto the SiO_2 at temperatures in the range of 300–500°C. The Ti deposition temperature was found to influence its own crystal orientation, being (002) at 300°C and becoming (1010) above ≈ 400°C. Figure 9.13 shows that the preferred orientation of the Al was (111) in all cases, but the texture was much weaker when deposition was directly on oxide. Surprisingly, the crystal orientation of the Ti underlayer had little effect on the texture of the Al film, nor did the deposition temperature of the Al (100°C or 300°C).

9.3.5 Interaction of Al with Ti

Both Al and Ti are highly reactive metals and are often used in combination. For example, a Ti wetting layer is used to promote the flow of Al in high-temperature applications such as reflow Al and the cold-hot Al process (see Section 7.2). Therefore, the interaction of Al and Ti needs to

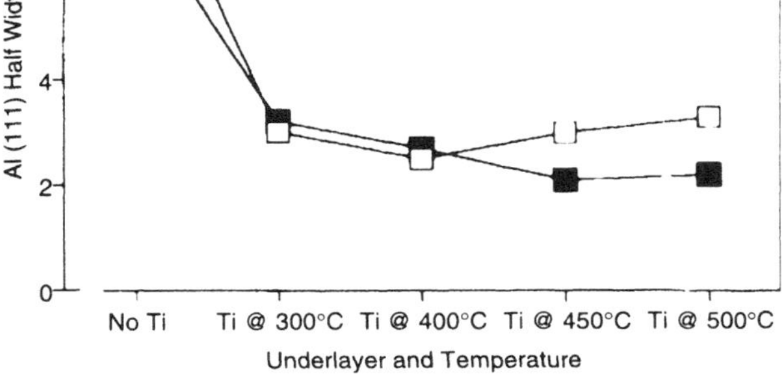

FIG. 9.13 Crystal orientation of PVD Al films deposited on bare SiO_2 and on SiO_2 with a PVD Ti overlayer.

be considered. When Al and Ti are in contact at elevated temperature, the Ti and Al interdiffuse and react to form the intermetallic compound $TiAl_3$, or Ti aluminide, in a layer-by-layer fashion. It has been found that the growth of the aluminide proceeds with a rate constant $K = K_0 \exp(-E_a/kT)$, where $E_a \approx 1.85$ eV and $K_0 \approx 0.15$ cm²/sec [9.14]. The rate constant is the same whether the Ti/Al is deposited on Si or SiO_2; however, if an Al alloy is used, the activation energy needs to be modified from the 1.85 eV value for pure Al. For example, a value of $E_a = 2.4$ eV has been found for Al alloys with 3 at % Cu, leading to a slower rate of growth [9.15]. In all cases, the thickness of the $TiAl_3$ formed after time t is given by the expression

$$d = (Kt)^{1/2} = (K_0 t)^{1/2} \exp\left(\frac{-E_a}{2kT}\right) \tag{9.3}$$

Using Eq. (9.3), we estimate that in 1 minute at a temperature of 450°C, a $TiAl_3$ film of thickness $d = 270$ Å will form. Given that Ti films used in interconnect or barrier applications are on the order of 100 Å while the Al films are on the order of 10,000 Å, we see that all of the Ti layer can be quickly consumed by the Al. Since the resistivity of $TiAl_3$ ($\approx 20\ \mu\Omega$-cm) is many times greater than that of Al ($\approx 3\ \mu\Omega$-cm), the overall resistance of the Al–Ti stack increases. On the other hand, the mechanical properties of the aluminide act to "harden" the slab interconnect against stress migration and electromigration in the overlying Al conductor, leading to use of sandwiched structures with the Ti deposited directly under the Al (e.g., TiN/Ti/Al/TiN) or above it (e.g., TiN/Al/Ti/TiN). In both cases, having an ultraclean PVD chamber is desired to prevent oxidation of the Al–Ti interface that would poison $TiAl_3$ formation.

Finally, we note that the Al–Ti reaction has been used with monitor wafers to measure the temperature and/or temperature uniformity of heater tables used in PVD [9.16, 9.17]. In this case, the monitor wafer might be an oxidized Si wafer on which a relatively thick Ti film (e.g., 1000 Å) and Al overlayer (e.g., 6000 Å) have been sputter deposited. Annealing of such a wafer in a calibrated furnace for different times and/or temperatures would yield sheet resistance curves like those shown in Fig. 9.14. The layer-by-layer formation of high-resistivity $TiAl_3$ at the Ti–Al interface consumes Al, and the thickness of the conducting Al layer measured by the sheet resistance probe decreases. Since $R_s = \rho/t$, a corresponding increase in sheet resistance is observed. Using this calibration data, the decrease in R_s of a monitor wafer can then be used to compare one PVD heater against another (Fig. 9.15) or to assess the uniformity of

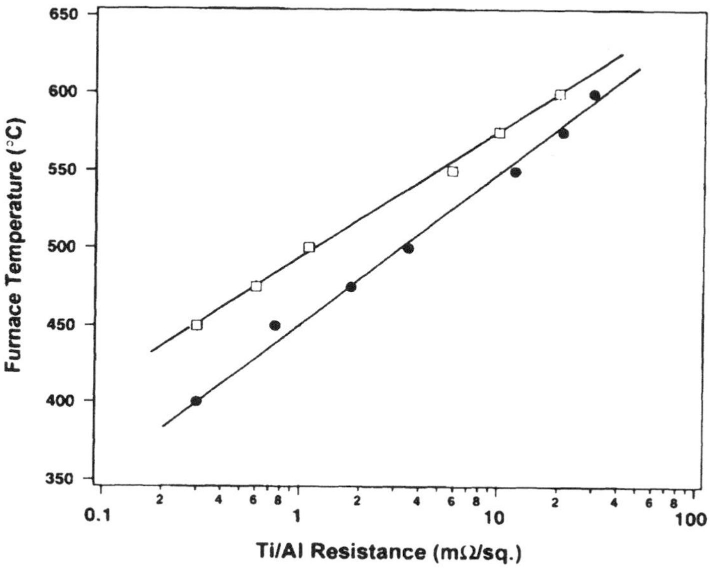

FIG. 9.14 Calibration curves of sheet resistance of an Al (6 kÅ)-Ti(1 kÅ) bilayer on SiO_2 after furnace annealing at 128 sec (boxes) and 180 sec (dots).

a given heater table by mapping monitor wafer sheet resistance before and after annealing on the table.

9.3.6 Uniformity of Alloy Composition

Although Al-Cu alloys typically contain only a small weight percentage of Cu ($\approx$ 0.5–1%), the distribution of Cu over the wafer surface can affect the uniformity of resistivity, electromigration performance, and interconnect line definition during plasma etching. While the target may contain a uniform distribution of Cu, differences in the emission and transport properties of the alloy constituents can give rise to compositional variations across the wafer [9.18, 9.19]. Figure 9.16 shows experimental and measured radial thickness profiles for elemental Al and Cu targets sputtered at 5 mTorr and 20 mTorr from a 5-cm-diameter magnetron with a 6-cm source-to-substrate spacing [9.19]. At the lower pressure, both targets produce a thickness uniformity with a pronounced off-axis peak at $\approx$ 4.5 mm that is associated with the annular erosion groove produced by

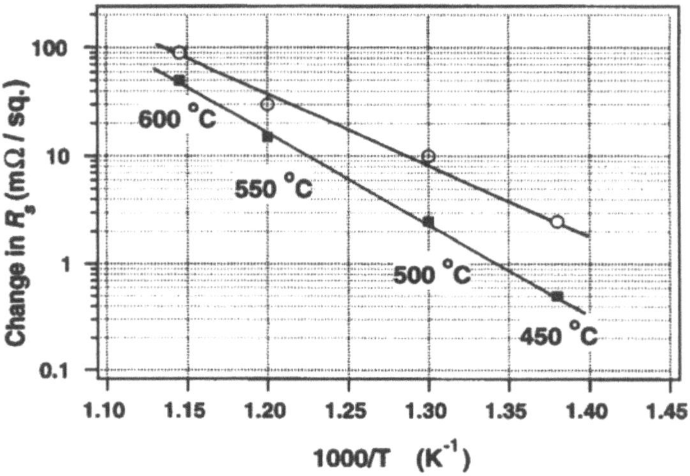

FIG. 9.15 Heaters from different PVD vendors are compared using the $Ti + 3Al = TiAl_3$ reaction method described in section 9.3.5. At the same nominal setpoint of 500°C, wafer temperature differed by 45°C. (Reprinted with permission from R. Wilson et al, *J. Vac. Sci. & Tech.* **B15(1):** 122–126 (1997). Copyright 1997 American Institute of Physics.)

the particular magnetron source used. In effect, the erosion profile of the target is imaged into the thin film. Although the curves are similar, they are not identical but reflect differences in the angular emission distribution of Cu and Al. At higher pressure, gas-phase scattering smears out the peak and broadens each profile, although the effect is less pronounced for the ^{64}Cu, whose scattering angle with the ^{40}Ar is lower than that of ^{27}Al. Thus, at higher pressure, the memory of the sputtering distribution at the target is lost through randomizing collisions with the sputtering gas.

To the extent that the separate Al and Cu sputter distributions can be superpositioned to describe a compound Al-Cu target, we would expect the composition of an Al-Cu film deposited from the magnetron of Fig. 9.16 to depend on sputter pressure and be relatively Cu-rich at the center versus the edge, as shown in Fig. 9.17. In practice, such effects may be smaller than predicted due to such things as the surface diffusion of Cu at elevated deposition temperature. Also, it is worth noting that very little Cu is involved in an absolute sense. For example, if all of the Cu in the bulk of a 8000-Å Al-0.5%Cu alloy film segregated to the surface, the thickness of the resulting Cu layer would only be about 15 Å.

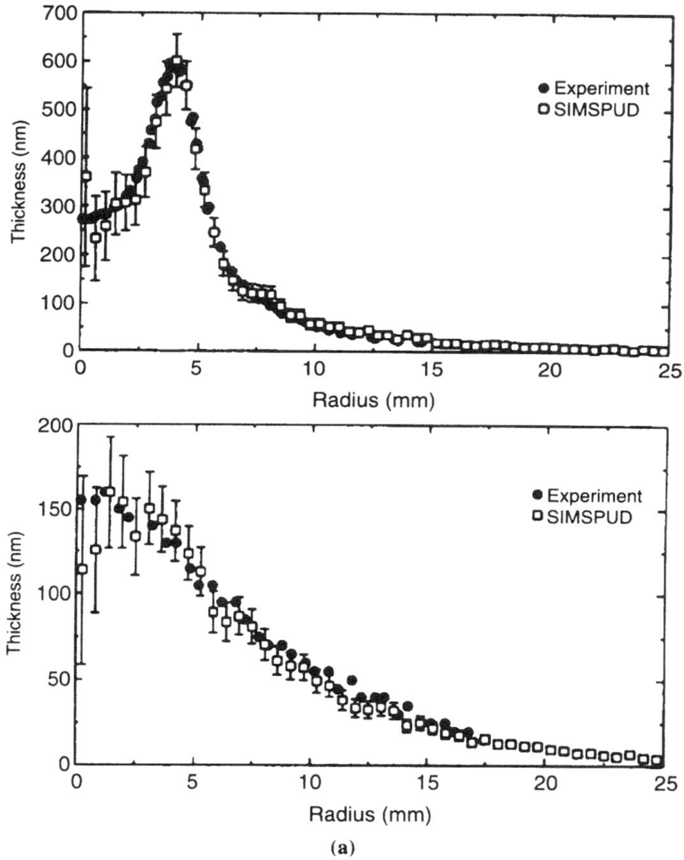

FIG. 9.16 Experimental and measured radial thickness profiles for elemental (a) Al and (b) Cu targets sputtered at 5 mTorr and 20 mTorr from a 5-cm-diameter magnetron with a 6-cm source-to-substrate spacing [9.19].

9.4 Titanium

9.4.1 Metallurgical Issues for PVD

While PVD Ti is used for a variety of purposes in multilevel interconnect schemes (e.g., its role as a wetting layer to enhance hot Al PVD processing is discussed in Chapter 7), its critical application is to reduce interfacial

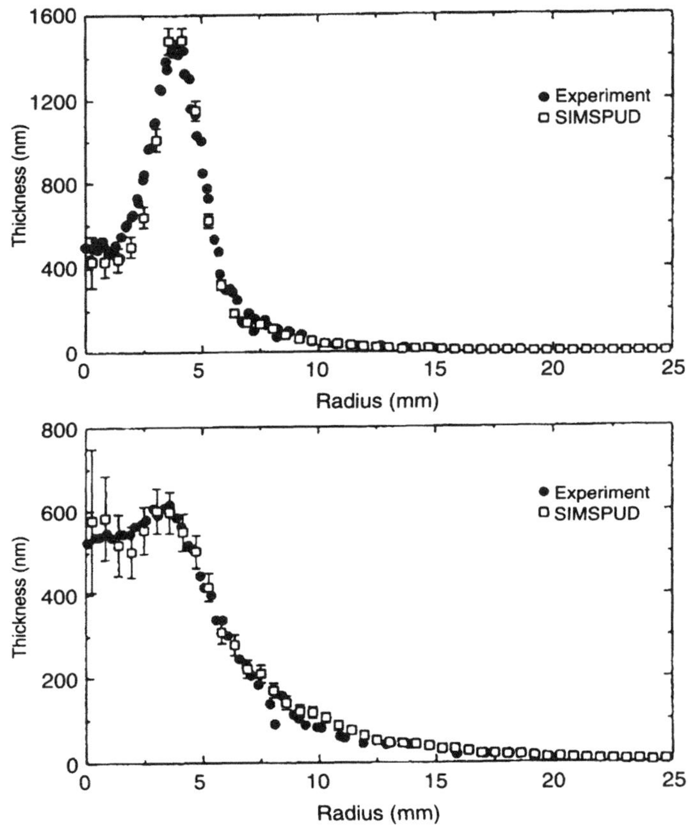

FIG. 9.16 **(b)**

oxide impurities and thereby improve adhesion and reduce contact resistance between a via plug and an interconnect line or between a contact plug and silicon. The key attribute of Ti that makes this possible is its ability to reduce native silicon oxide through the formation of TiO or TiO_2 (e.g., $SiO_2 + Ti = TiO_2 + Si$) as well as to reduce other insulating metal oxides such as Al_2O_3 whose formation cannot always be prevented and whose in-situ removal by sputter etching can be problematic (see Section 5.3.3).

The key challenge for PVD Ti is getting sufficient bottom coverage in high aspect ratio features — in both an absolute and a percentage sense (see Fig. 9.18). Consider the case of a contact to Si. In this case, there should be enough Ti at the bottom to completely reduce the native sili-

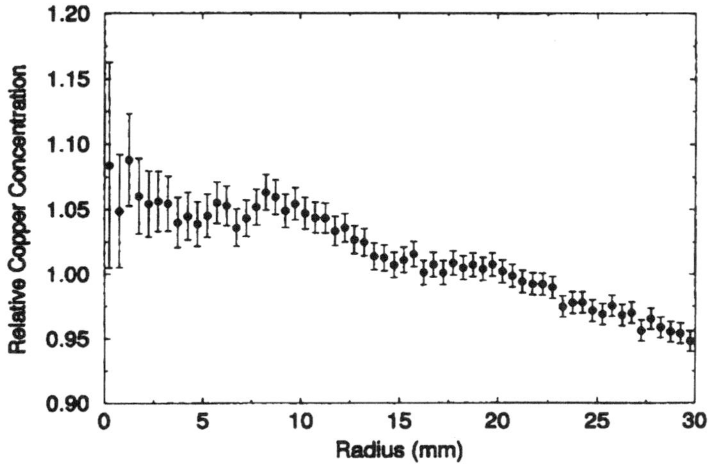

FIG. 9.17 Cu concentration variation expected for an Al-Cu alloy sputtered deposited under the conditions of Fig. 9.16 [9.19].

con oxide. Even though the native oxide is ultrathin ($\approx$ 20–30 Å), the PVD Ti initially forms discontinuous islands so that a rather thick Ti film ($\approx$ 100 Å) is required to reduce the native oxide over the entire surface area of the contact. Subsequent high-temperature annealing is often used to convert the unreacted Ti into $TiSi_2$, which has relatively low resistivity and can further reduce contact resistance by consuming interfacial contamination. On the other hand, if too thick a Ti film is deposited, so much of the underlying, active Si region may be consumed during silicidation as to compromise junction integrity.

Even if an optimal absolute amount of Ti reaches the bottom of the contact (e.g., 100 Å), the percentage of bottom coverage needs to be high enough to avoid bread-loafing that could restrict the top of the hole. Also, it is desirable to prevent depositing very thick Ti films on the field regions since this could lead to the formation of even thicker and more resistive $TiAl_3$ films after PVD Al. Since the bottom coverage of conventional PVD Ti in a 4:1 aspect ratio contact hole is $< 5\%$, 100 Å of Ti at the bottom translates into 2000 Å or more on the field. For this reason, advanced device applications of PVD Ti in contact or via holes involve some directional enhancement (such as low pressure) to reduce gas-phase scattering, physical collimation (variously called coherent sputtering, filtered sputtering, and controlled divergence sputtering or cds) and, more recently,

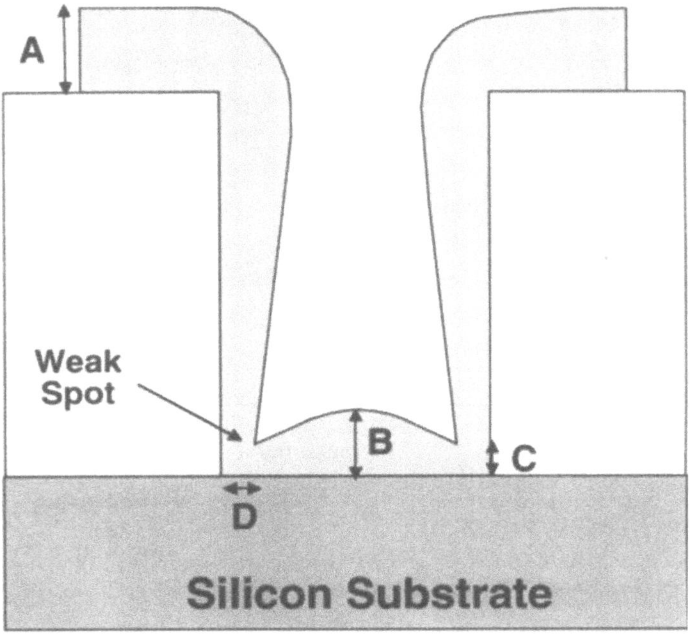

FIG. 9.18 For PVD Ti and TiN barriers and liners, one generally desires a high-percentage bottom (B/A) and sidewall (D/A) coverage, with robust corner thickness (large C). A flat-bottomed profile (B ≈ C) is also preferred for such applications as a Ti contact layer.

ionized metal PVD. Deposition of two-dimensional Ti films for a planar, slab Al interconnect do not typically use such directional enhancements — which may involve an unwanted trade-off of blanket uniformity against bottom coverage.

Finally, we note that a flat profile for the PVD Ti at the bottom of the hole is in general preferred over the domed shape that can result from applying PVD to high aspect ratio features. A domed profile of Ti in a contact hole would lead to an unwanted variation in Ti-silicide thickness over the contact area. Also, in a via hole, the thinning of Ti at the edge could be replicated in a barrier overlayer (e.g., TiN) and compromise its ability to perform as intended. Unfortunately, Ti is a refractory metal whose melting point is sufficiently high ($T_{mp} \approx 1670°C$) that hot PVD processes cannot easily be exploited to flatten surface topography as with Al. Therefore, other methods of redistributing Ti mass at the bottom (such as resputtering) must be considered.

9.4.2 PVD Ti Process Results

Figure 9.19 presents representative PVD Ti process conditions and film properties, and Fig. 9.20 shows the bottom and sidewall coverage of PVD Ti films deposited with the directional enhancement of a 1.5:1 aspect ratio collimator (1.5:1 cds Ti), which allows moderate coverage ($\approx$ 25%) in high aspect ratio topography. Ironically, even though the percent coverage of PVD Ti films in steep structures is relatively low, the films can appear very conformal. From Fig. 9.20 we see that the bottom and sidewall coverage of steeper features (AR $>$ 4:1) are comparable, so we would expect collimated Ti (and TiN) films to uniformly coat such structures. This is seen in the SEM micrographs in Fig. 9.21, where 1.5:1 cds TiN was deposited onto a very high aspect ratio (AR $\approx$ 8:1) sub-0.25-μm hole. While

Process conditions and film attributes	Typical values
Process pressure	2 mTorr Ar
Magnetron power	12 kW
Deposition rate on field	580 Å/min
Specific deposition rate	0.8 Å/kW-sec
Thickness nonuniformity	(M-m)/(M+m) = 5%
Sheet resistance (1000 Å film) R_s non-uniformity R_s repeatability	R_s = 5.8 Ω/sq 3σ = 9% 3σ = 5%
Bulk resistivity	ρ = 58 μΩ-cm
Absolute reflectivity (@ 440 nm)	R = 53.8%
Bulk density	4.7 gm/cc
Mechanical stress	1.6×10^9 dyne/cm^2

FIG. 9.19 Representative PVD Ti process conditions and film properties (1.5:1 collimator, 200-mm wafer).

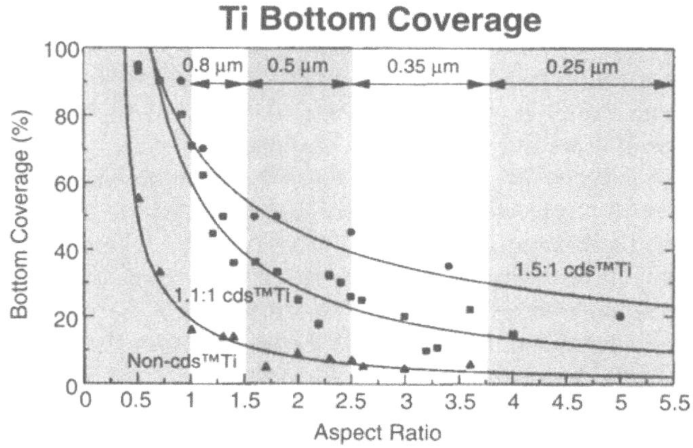

FIG. 9.20 Bottom and sidewall coverage (i.e., step coverage) of PVD Ti film in a contact or via hole as a function of hole aspect ratio and collimation.

collimation greatly improves bottom coverage, its use may degrade blanket uniformity somewhat. This relates to the fact that obtaining uniformly thick PVD films generally involves tailoring the target erosion profile to compensate for the finite geometric size of the PVD source. Unfortunately,

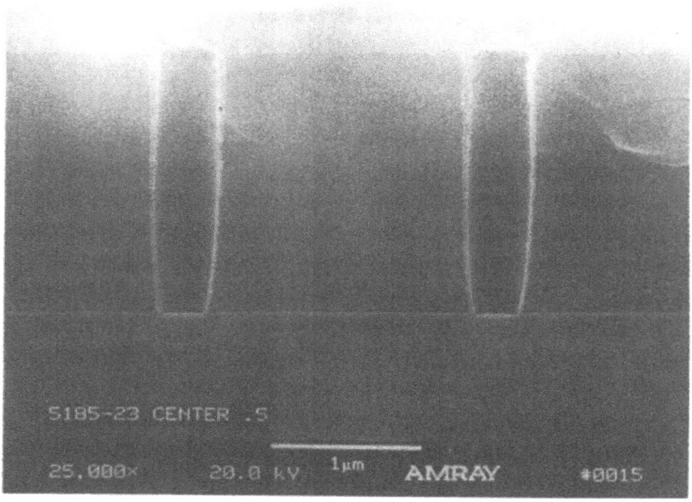

FIG. 9.21 SEM micrograph showing the step coverage of a collimated PVD TiN film in a steep contact hole (aspect ratio of collimator = 1.5:1; aspect ratio of hole = 8:1).

high aspect ratio collimator cells tend to image the nonuniform erosion profile of the target onto the wafer, and sources with extremely uniform erosion are difficult to design. Therefore, although blanket Ti nonuniformity of $3\sigma = 3–5\%$ over 200 mm is typical of state-of-the-art magnetrons used without collimation, 3σ values of $\approx 10\%$ are more typical of highly collimated Ti processing. On the other hand, since Ti films in microelectronics are usually thinner than 300 Å, a $3\sigma = 10\%$ value represents a variation of only about 30 Å, or about 10 Ti atoms.

While collimation improves directionality, it also reduces the specific deposition rate of Ti at the wafer by removing low-angle material from the sputtered flux; to compensate for this effect, higher magnetron power is used. For example, while noncollimated Ti might be deposited at $\approx$ 1–2 kW, a 1:1 or 1.5:1 collimated deposition might require 5–10 kW for equivalent throughput. Figure 9.22a shows the field thickness obtained when trying to obtain a 65-Å film of Ti at the bottom of a contact with noncollimated, 1:1, and 1.5:1 aspect ratio collimation, and Fig. 9.22b shows the number of such films that can be deposited before having to change the collimator or target (the collimator is changed when buildup of Ti on the cell walls reduces transmission by 50%). Collimation clearly reduces deposition on the field and, in spite of impact on absolute deposition rate, still allows a rather large number of wafers to be processed.

Collimation also can impact the microstructure developed in the Ti film. In general, columnar growth arises in PVD films due to limited surface diffusion and competition or shadowing between columns. The surface diffusion length depends on both substrate temperature and the presence of contamination, while the shadowing is a result of the surface topolology. Given the limited surface mobility of Ti at PVD temperatures and the fact that collimation removes obliquely incident adatoms from the incident flux, it is not surprising that collimated Ti films have a dense columnar microstructure on the field regions. On the other hand, collimated coatings on vertical sidewalls of high aspect ratio features can have reduced density and increased porosity due to shadowing of the highly directional Ti flux by the growing Ti grains [9.20, 9.21].

9.5 Titanium Nitride

9.5.1 Metallurgical Issues for PVD

While the applications for PVD Ti and TiN can be quite different (e.g., Ti contact layers and TiN ARC layers), it is difficult to separate the two materials in a PVD context since TiN is deposited by reactive sputtering of a

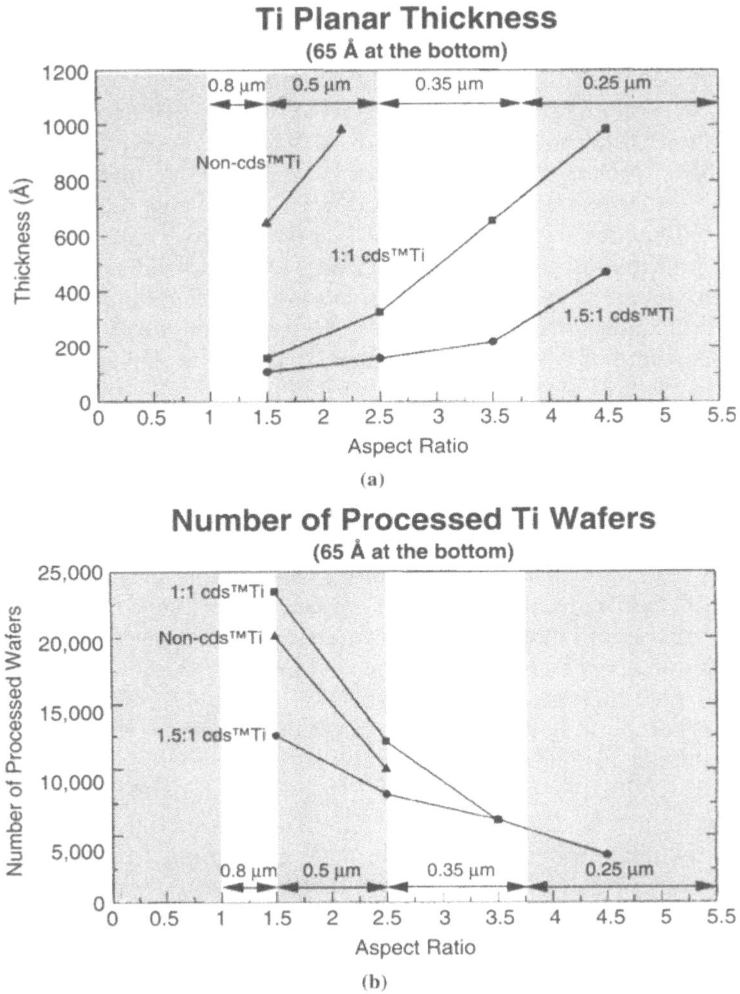

FIG. 9.22 (a) Thickness of Ti that must be deposited on the field to obtain 65 Å of Ti on the bottom of a contact hole for different hole and collimator aspect ratios.(b) Number of Ti films obtained before the end of collimator life (defined as point where transmission of Ti flux through the collimator has dropped to 50% of the value when new).

Ti target in a nitrogen-containing ambient, typically Ar/N_2. Also, there are many cases where the complementary cleaning properties of Ti and barrier properties of TiN favor their use as a Ti/TiN bilayer. Therefore, much of the data used in this section will include both Ti and TiN.

The major use of PVD TiN is as a barrier layer, e.g., to prevent diffusion of an Al or W contact plug metallurgy into the underlying Si substrate. For example, TiN prevents the interdiffusion of Si and Al at the contact level, which could lead to junction spiking. With regard to CVD W contact and via plugs, TiN is also widely used as a "glue" layer to promote the adhesion of W to the oxide walls and to help it nucleate — although in many cases, the TiN is deposited over an intermediate Ti layer that makes the actual bond to the oxide (this glue layer is not required in the case of Al, which adheres well to SiO_2). Since the WF_6 precursor commonly used for CVD W reacts strongly with Ti, the TiN also serves as a protective coating for the underlying Ti. However, if this TiN coating has any breaks or delaminations, the volatile reaction of WF_6 + Ti to form TiF_4 combined with the deposition of CVD W on the peeled-back TiN can give rise to a dramatic defect resembling a miniature volcano (Fig. 9.23). Attack of the underlying Si by the WF_6 is also possible, leading to the subsurface migration of W into the Si and giving rise to a wormhole-shaped structural defect. Whether used for barrier, adhesion, or protection purposes, PVD TiN should be pinhole-free and as conformal as possible, particularly at sharp bottom corners where PVD coverage can be reduced and give rise to weak spots as shown in Fig. 9.18. It should also be noted that different applications require different thickness. For example, while a 50–100-Å-thick

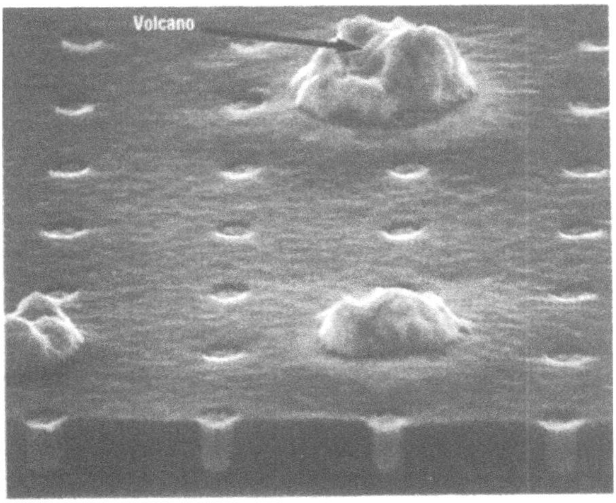

FIG. 9.23 "Volcano" defect that is formed by chemical reaction of Ti with the WF_6 chemistry used in CVD of W. (Reprinted from S. Bothra et al in the February 1997 edition of Solid State Technology, copyright 1997 by PennWell.)

TiN film might suffice as a glue layer for CVD W, a 250–400-Å film may be required as the contact diffusion barrier for a high-temperature Al-Cu alloy reflow or two-step process.

As the atomic concentration of N in Ti is increased, the resulting material evolves from pure Ti, to a solid solution of N in Ti, to the compound Ti_2N (33% N), and finally to TiN (50% N). At concentrations above 50 at %, the excess N exists in solid solution with stoichiometric TiN. TiN can accept large vacancy fractions on both the anion and cation sublattices, and over-stoichiometric TiN_x ($x > 1$) remains single phase in the NaCl structure with excess nitrogen fractions up to about $x = 1.2$. But it is the 1:1 stoichiometric TiN phase that is preferred over other compositions due to its superior barrier properties and that is readily deposited by reactive sputtering of Ti in Ar/N_2. However, treatments to enhance as-deposited TiN barrier performance — such as air exposure or in-situ or ex-situ annealing in an oxidizing ambient to "stuff" the grain boundaries — are often done following PVD. While PVD Ti is silver colored, stoichiometric TiN has a characteristic gold or brownish-gold color under reflected light. This has led to "gold TiN" sometimes being used as an indicator of 1:1 film composition, although in reality the perceived color depends on both stoichiometry and other film properties in a complex way [9.22, 9.23].

> It is a popular misconception that TiN is a metal. However, even though TiN films are gold and shiny with electrical conductivity comparable to that of titanium, TiN is not a metal. The high conductivity is associated with a strong overlap of N 2p and Ti 3d bands, while the gold color arises from interband transitions combined with a high reflectance in the red and infrared. [9.22].

9.5.2 Reactive PVD of TiN

TiN is deposited by reactive sputter deposition of a Ti target in the presence of nitrogen, typically by using an Ar/N_2 admixture. The kinetics of the resulting PVD TiN film formation depend on process and hardware parameters in an interactive way (e.g., N_2 partial pressure, magnetron power, collimator, and PVD shield design), which has important practical consequences. The basic issue relates to nitridation of the Ti target — in particular, one wants to minimize nitridation of the target surface to increase the sputtered flux of Ti atoms, yet at the same time maximize nitridation of Ti at the wafer surface to produce a stoichiometric TiN film (see also Chapter 3 for a discussion of reactive PVD).

The steady-state condition of the Ti target surface during PVD can range from fully metallic to fully nitrided, with the exact ratio of exposed Ti and TiN areas reflecting the detailed consumption and liberation of nitrogen at the surface — e.g., gas-phase nitrogen is consumed by the reaction of molecular N_2 with Ti ($2Ti + N_2 = 2TiN$) while nitrogen bound as TiN is liberated by Ar^+ or N_2^+ ion bombardment, etc. Regardless of the state of target nitridation and contrary to what one would expect, the primary sputter-ejected particles from a Ti target in an Ar/N_2 discharge are always Ti and N [9.24]. That is, even when the target surface is fully nitrided, sputter ejection of molecular TiN is not significant. On the other hand, the sputter yield of Ti from TiN is several ($\approx$ 3) times less than from Ti, so the ejected flux of Ti atoms is much less from the nitrided target. Gas-phase recombination of the sputtered Ti with nitrogen via a two-body collision does not occur since the heat liberated in the formation of a molecule of TiN cannot be dissipated while simultaneously conserving energy and momentum. Instead, this occurs at the wafer surface where the sputter-deposited Ti adatoms are nitrided to TiN_x primarily by heterogeneous reactions such as the dissociative chemisorption of N_2. In effect, both the Ti target and the wafer act as solid state "pumps" of nitrogen, whose relative pumping speeds reflect their state of nitridation and affect the overall TiN_x deposition rate. In addition, the nitridation state of target and wafer are influenced by PVD shields and collimators, which become coated with Ti and themselves behave as dynamic getter pumps for nitrogen. Since the surface area of a high aspect ratio collimator ($\approx$ 3500 cm^2 for an AR = 1.5:1 hexagonal cell collimator) is much greater than either a 200-mm wafer ($\approx$ 315 cm^2) or DC magnetron target ($A \approx$ 700–1000 cm^2), we see that the collimator can have a major influence on the consumption of N_2 in the chamber. The overall situation is schematically illustrated in Fig. 9.24.

The net result of these competing processes typically leads to experimental data such as that shown in Fig. 9.25, where the deposition rate and sheet resistance of reactive sputtering of Ti in Ar/N_2 is plotted as a function of N_2 mass flow. At low flows of nitrogen into the process chamber, the deposition rate is high and characteristic of sputtering from an elemental Ti target, and the deposited film is Ti-rich TiN_x. The N/Ti ratio in the film increases with nitrogen fraction in the Ar/N_2 admixture. As nitrogen flow continues to increase, the curve exhibits a sharp fall off in deposition rate, reflecting the greatly reduced sputter yield of the nitrided target and the lower ionization cross section and sputter efficiency of N_2^+ versus Ar^+. If DC magnetron power is increased, the onset of this abrupt fall off occurs at a higher N_2/Ar fraction because the additional Ar^+ bombardment of the target sputter etches away the TiN that is forming. The deposition rate

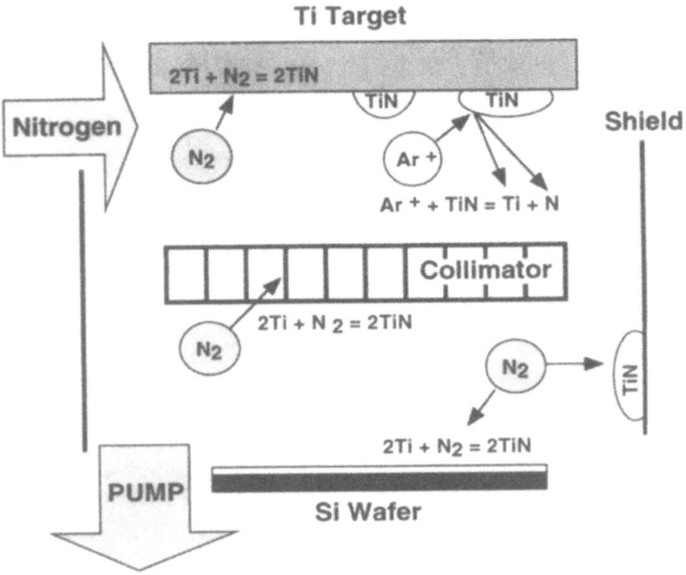

FIG. 9.24 Schematic illustration of N_2 generation and consumption in a PVD chamber during reactive PVD of TiN.

finally stabilizes at the lower value characteristic of sputtering from TiN, a target condition that workers sometimes describe as "poisoned" in that the target sputter yield has been degraded by the nitride surface layer. This terminology is somewhat hypocritical though, since one rarely hears that the desired TiN film produced from the "poisoned" target is "toxic"! In any event, the overall behavior seen in Fig. 9.25 has been modeled by several workers based on mass balance considerations, with similar phenomena observed in reactive PVD of TiO_2 in Ar/O_2 discharges [9.25–9.28]. We also note that good control of target temperature is desired for process repeatability, since the rate of target nitridation involves temperature-dependent steps such as dissociative N_2 chemisorption. A large increase in target temperature could, for example, change the Ti target state from metallic to nitrided for a given N_2/Ar ratio [9.29].

Deposition of PVD TiN with the target in the nitrided mode (NM) raises concerns about the deposition of sequential Ti/TiN bilayers since the nitrided target would contaminate with nitrogen the Ti film of the next Ti/TiN bilayer. This can be avoided either by depositing the Ti and TiN in

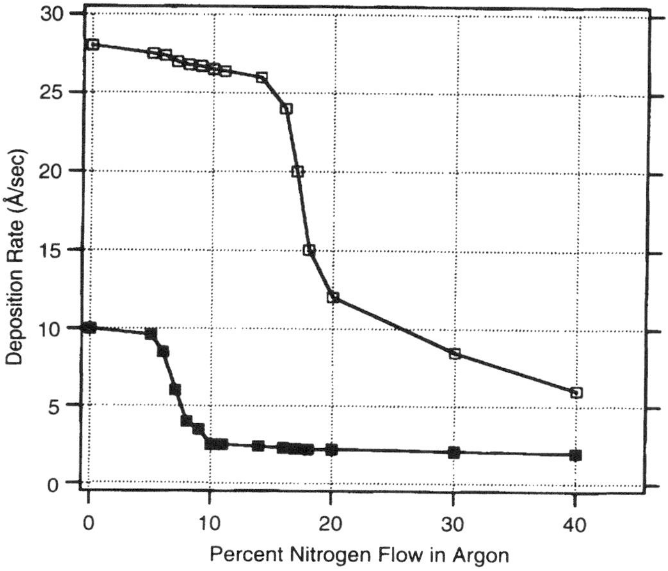

(a)

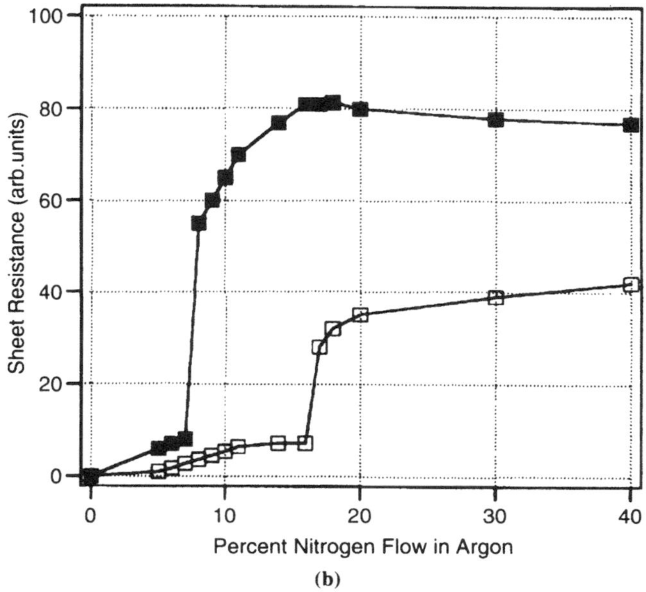

(b)

FIG. 9.25 Representative data on deposition rate and sheet resistance of reactive PVD of Ti in Ar/N_2 (after ref. 9.27). Reproduced by permission of The Electrochemical Society, Inc.

two separate chambers or by using a single chamber with a mechanical shutter that allows the nitrided Ti target to be sputter-cleaned in pure Ar between successive wafers. It is sometimes possible to avoid the shutter when collimated Ti/TiN is deposited since N sputtered from the temporarily nitrided target is pumped by unreacted Ti on the collimator surface, leading to deposition of an acceptably clean Ti film [9.30]. Finally, we note that it has been possible to operate in the high-deposition-rate, non-nitrided mode (NNM) in which the Ti target is not saturated with N_2 yet there is a sufficient partial pressure of N_2 at the wafer to ensure that TiN_x with $x = 1$ is grown. This has been done by exploiting the nitrogen-pumping action of a high aspect ratio collimator and carefully controlling nitrogen flow and partial pressure to achieve a stable target nitridation state [9.30].

An interesting aspect of collimated TiN PVD is that the chemical composition of the film can, under some process conditions, vary over topography. In particular, TiN films deposited into high aspect ratio contact holes have been observed to be substantially nitrogen-deficient ($TiN_{0.75}$) at the bottom relative to the stoichiometric TiN on the field [9.31]. This is a result of the flux of Ti atoms coming through the collimator being highly directional whereas the nitrogen flux is diffuse and characteristic of molecular N_2 in the gas phase. Hence, deep enough in topography, conditions can be reached where there is insufficient nitrogen to fully nitride the Ti. This is less of a problem for a nitrided target since in this case the relative N flux is initially much higher at the wafer. Also, a postdeposition thermal anneal in N_2 (e.g. 30 min at 450°C) has been found sufficient to restore the composition of the in-depth depleted films to near-stoichiometric TiN [9.31].

Figure 9.26 summarizes selected PVD film properties for Ti, noncollimated, nitrided-mode TiN and collimated, nonnitrided-mode TiN. It is worth noting that the resistivity of 1.5:1 collimated TiN ($< 45\ \mu\Omega$-cm) is considerably lower than that of the uncollimated TiN (80–200 $\mu\Omega$-cm). In part this reflects the excess nitrogen incorporated in the noncollimated $TiN_{x=1.2}$ films that were deposited in the nitrided mode. However, the elimination of low-angle Ti atoms by collimation reduces the TiN film's lateral growth, resulting in more densely packed columnar grains with a more bulk-like conductivity — although not as low as the bulk resistivity of single-crystal, stoichiometric TiN, which has been reported to be $\approx 15\ \mu\Omega$-cm at room temperature for either (111) or (110) orientation [9.22]. The mechanical film stress in PVD TiN films is generally compressive and much greater in magnitude than that of PVD Ti deposited under similar conditions. In addition, the stress depends on temperature of deposition, degree of collimation, and underlying substrate (Si, SiO_2). A major concern

	All Titanium	Standard non-collimated TiN (nitrided mode)	1:1 collimated TiN (nitrided mode)	1.5:1 collimated TiN (non-nitrided mode)
XRF thickness NU (M-m)/(M+m) (%)	< 5	< 5	< 5	< 5
Sheet resistance NU 3σ (%)	< 10	< 10	< 10	< 10
Bulk resistivity (μΩ-cm)	< 60	80 to 200	< 60	< 45
Density (g/cm³)	4.40	4.75	4.90	5.10
Mechanical stress (MPa)	300	3000	2500	2500
Absolute reflectivity measured at 440 nm (%)	55	19	N/A	19
Grain size (nm)	33	12	N/A	14
Stoichiometry	N/A	1.2:1	N/A	1:1
Film color	Silver	Brown Gold	(Light) Brown Gold	Light Gold

FIG. 9.26 Selected properties of uncollimated and collimated (i.e., cds) Ti films on 200-mm wafers (after ref. 9.30).

about high TiN film stress relates to particle generation from films deposited on shields and other chamber fixtures. Fortunately, this is much less of a concern for PVD Ti/TiN bilayers since the film stress of the composite structure — particularly for higher deposition temperatures — can be quite low (see Fig. 9.27). In a similar way, when a PVD chamber is used almost exclusively for TiN deposition (e.g., a dedicated chamber for TiN ARC layers), one can periodically sputter a layer of Ti onto the TiN-covered surfaces, which serves to reduce overall stress and additionally functions as a paste to prevent the flaking of thick TiN layers. On the other hand, if this pasting is done too frequently (e.g., more than once per cassette of wafers), the cost per wafer will increase due to the unproductive consumption of the Ti target (see Section 11.8).

9.5.3 Antireflection Coating (ARC)

High specular reflectivity is one indication of PVD Al film quality (e.g., $R \approx 90\%$ from 200–800 nm). However, this high reflectivity can adversely affect subsequent photolithographic patterning and etching of the blanket

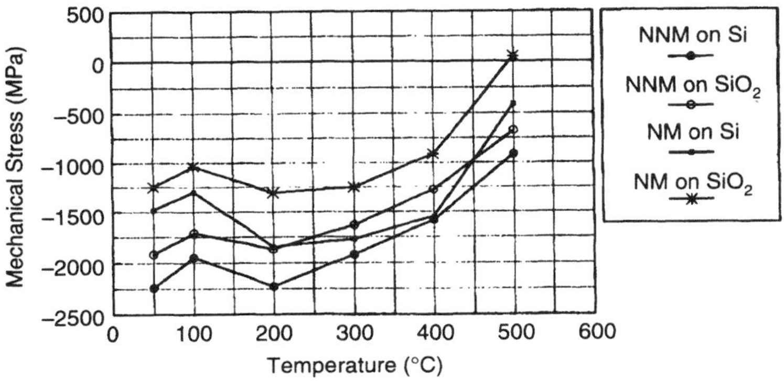

FIG. 9.27 Stress of composite Ti(300 Å)/TiN(500 Å) bilayers on Si and SiO_2 as a function of deposition temperature. The TiN was deposited in either a nitrided mode (NM) or non-nitrided mode (NNM).

PVD Al film into separate metal interconnect lines [9.32, 9.33]. In particular, as shown in Fig. 9.28, light reflections can degrade pattern resolution by three mechanisms: (1) off-normal incidence light can reflect back through and expose regions of the resist that were intended to be masked; (2) thin film interference effects can produce linewidth variations in areas where the resist thickness varies; and (3) "reflective notching" can occur in which the resist pattern is undercut at the metal–resist interface due to extraneous backscattered reflection of light from nearby regions of topography. This can lead to notching of the metal line after pattern definition etching with resulting reliability problems. For example, each nanometer of gate width change in an advanced MOSFET can reduce chip speed by 1 MHz. As a result, a chip designer might want to keep these dimensional variations in gate width across the chip below 1%, which can require keeping reflected light levels $< 0.5\%$ during lithographic patterning.

To minimize these problems, a PVD TiN antireflection coating (ARC) layer is often deposited on top of the Al prior to photoresist application [9.34, 9.35]. One typically thinks of an antireflection layer as a transparent thin film whose index of refraction and thickness are engineered to give phase-shift cancellation of specific reflected wavelengths. On the other hand, TiN strongly absorbs visible light. It turns out that a sufficiently thin film of TiN transmits enough light to be used for the ARC application. For example, the absorption coefficient of TiN at the 436-nm wavelength used in g-line optical lithography is about 3×10^5 cm^{-1}, lead-

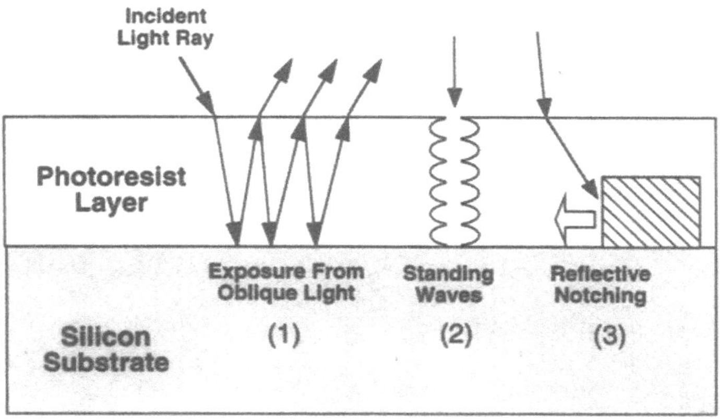

FIG. 9.28 Light scattering and reflection through a photoresist layer can degrade lithographic patterning and lead to artifacts such as metal line notching.

ing to a transmission of about 30% for a 400-Å-thick film. As shown in Fig. 9.29a, the reflectivity of an Al-Si alloy at 436 nm could be reduced by nearly 90% by deposition of a 350-Å film of TiN. Moreover, the reduction in reflectivity held over a very wide range of wavelength (Fig. 9.29b). Reducing the reflectivity at the metal–film interface also reduces the sensitivity of the lithographic process to variations in resist thickness, which improves across-wafer uniformity of line width. While sputtered films of amorphous Si have been similarly used to control lithographic variations over reflective topography, TiN is preferred due its more robust adhesion to photoresist and the fact that it can be left in place after metal patterning as part of the Al slab interconnect. The TiN ARC layer also protects the underlying Al layer from corrosion by the chemically basic photoresist developer, and for this reason a pinhole-free ARC layer is required.

9.6 Titanium-Tungsten (Ti-W) Alloys

9.6.1 Metallurgical Issues for PVD

TiW is actually a pseudo alloy of Ti in W, with the Ti added to improve the adhesion of the W to oxide and its oxidation resistance. While TiW is commonly sputtered from targets with 10% Ti by weight (equivalent to a target of $Ti_{0.3}W_{0.7}$ atomic composition), the deposited film is depleted in

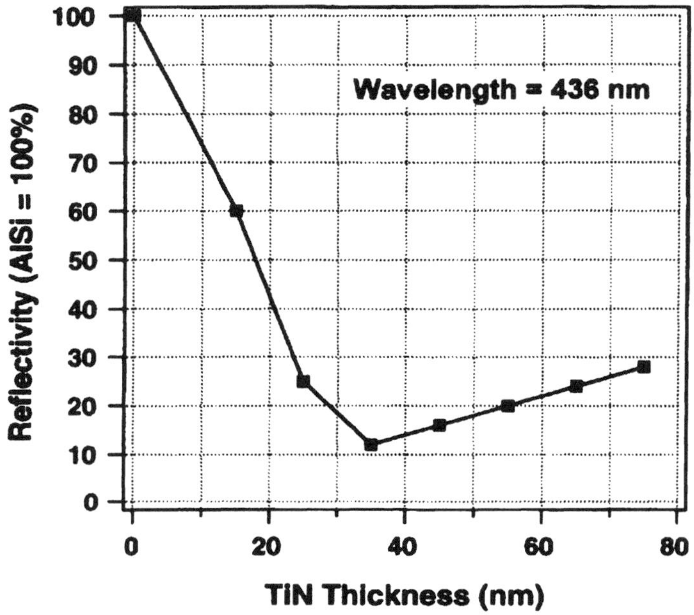

(a)

FIG. 9.29 Reflectivity of an Al-Si alloy with a PVD TiN overlayer as a function of (a) TiN layer thickness for a wavelength of 436 nm, and (b) wavelength for a TiN film thickness of 35 nm (after ref. 9.34). (Reprinted with permission from M. Rocke and M. Schneegans, *J. Vac. Sci. & Tech.* **B6(4):** 113–115 (1988). Copyright 1988 American Institute of Physics.)

Ti such that its Ti content is only ≈ 5–7% by weight. Since the films are ≈ 95% tungsten by weight, one should really refer to them as "tungsten-tie" as opposed to the conventional "tie-tungsten." In any event, the resistivity of such TiW films (≈ 50–80 μΩ-cm) is comparable to that of PVD TiN, and the films are similarly applied in IC processing, e.g., as a diffusion barrier between Al and Si. Unlike TiN, however, TiW can be sputtered easily in a nonreactive magnetron process but is generally not considered as good a barrier as TiN, so its use in advanced devices is becoming less common. The step coverage of TiW films over topography can be quite good, relative to, say, Al-Cu, because the large mass of W (184 amu) minimizes any loss of directionality due to gas-phase scattering with Ar (40 AMU).

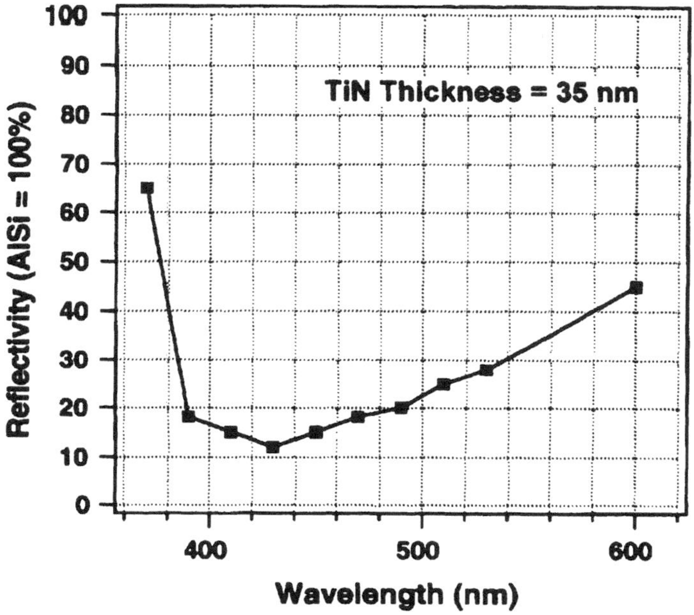

FIG. 9.29 **(b)**

9.6.2 PVD OF Ti_xW_{1-x}

As mentioned above, PVD TiW films are Ti-deficient with respect to the target composition, and this can affect both their electrical resistivity and barrier properties. In addition, TiW film composition in topography is affected by differences in the angular distribution of incident Ti and W flux at the wafer [9.36, 9.37]. While the angular emission of Ti and W from an alloy target is similar, the lower-mass Ti has a higher average scattering angle than W with the Ar sputter gas. Therefore, a wider incident flux distribution is expected for Ti than for W at the wafer, although this difference will be more pronounced at greater source-to-substrate spacing due to the additional gas-phase scattering. This is illustrated in Fig. 9.30, where angular distributions are calculated for sputtering at 7 mTorr at source-to-substrate spacing (sss) of sss = 2 cm (Fig. 9.30a) and 7 cm (Fig. 9.30b) from a commercial 5-cm-diameter magnetron. The two well-defined peaks correspond to the annular erosion track designed into the sputter source, and the integrated area under each curve is proportional to the total flux of

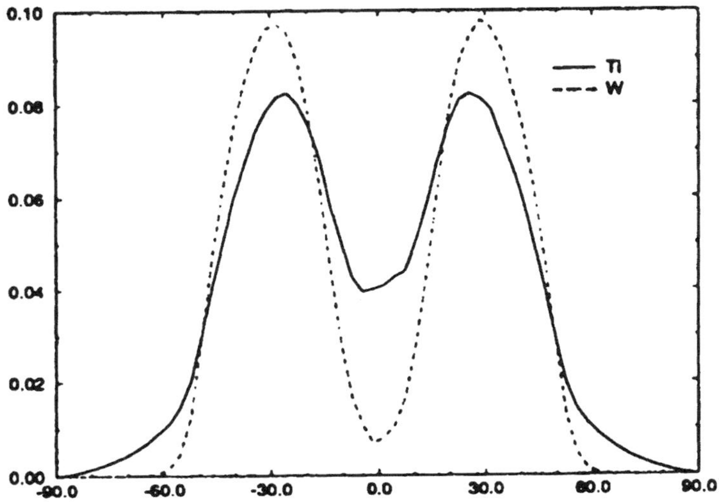

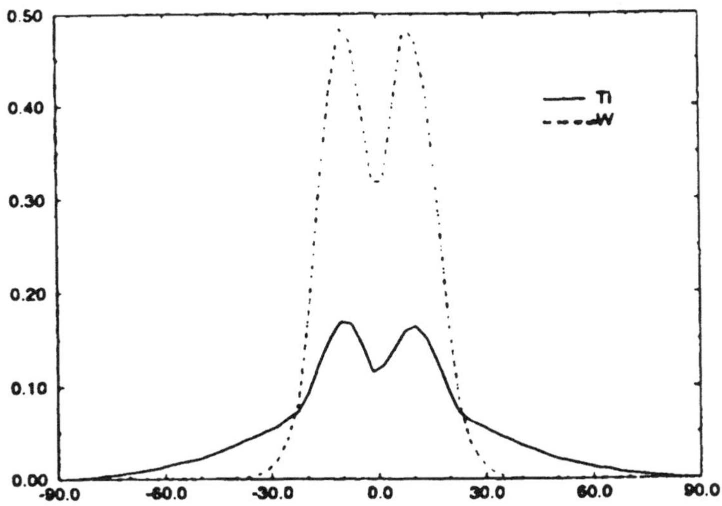

FIG. 9.30 Calculated Ti and W incident flux distributions for magnetron sputtering of a 5-cm-diameter TiW target at 7 mTorr and a source-to-substrate spacing of 2 cm (upper) and 7 cm (lower) [9.37].

Ti or W reaching the wafer surface. It is clear from Fig. 9.30 that the stoichiometry of a blanket TiW film deposited at sss = 7 cm will be heavily depleted in Ti compared to the film deposited at sss = 2 cm. In addition, it has been shown that preferential resputtering of the Ti by energetic Ar (40 AMU) that is backscattered from the W target atoms can have a dominant effect on further depleting the Ti content of the deposited TiW film [9.38].

With regard to Fig. 9.30b, we see that the W flux is much more directional than the Ti flux. Hence, we would expect the W component to penetrate deeper into a via and exhibit relatively better bottom coverage than the Ti. On the other hand, the Ti component should exhibit relatively better sidewall coverage due to a higher fraction of Ti atoms arriving at oblique incidence. The net result is that sputter deposition of TiW over topography is expected to lead to Ti_xW_{1-x} on the field, an increased concentration of Ti on via sidewalls (Ti_yW_{1-y} with $y > x$), and a decreased concentration of Ti on the via bottom (Ti_yW_{1-y} with $y < x$). This is seen in the simulated film composition profile of Fig. 9.31, which show W-enriched TiW at the bottom and Ti-enriched TiW on the sidewalls — an effect supported by experimental data of Liu *et al.* [9.36].

9.7 Refractory Metal Silicides

The four basic applications for refractory metal silicides in advanced IC processing are as (1) contacts to Si, (2) polycide gate electrodes in which a layer of silicide is deposited on top of polysilicon to produce a lower-resistance gate stack, (3) short-length local interconnects, and (4) self-aligned silicides (so-called salicides) in which the source, drain, and polysilicon gate regions of an MOS transistor are selectively shunted with a low-resistance silicide by a process of metal deposition and a two-step annealing that leaves the oxide regions free of both unreacted metal and silicide. PVD is widely used for all of these applications to deposit either the metal component of the silicide (e.g., PVD Ti from a Ti target for a Ti-salicide process) or the silicide itself (e.g., PVD $MoSi_x$ from a compound Mo–Si target for a polycide bilayer). While many metals form silicides with a combination of desirable properties (e.g., metallic conduction with resistivity much lower than heavily doped polysilicon, stable contact formation to Si, high-temperature stability, and self-passivation in oxidizing ambients), the principal silicides exploited in VLSI devices have been MSi_x where M = Mo, W, Ta, and Ti. Interconnect roadmaps suggest that Co will join this list for ULSI devices due to the

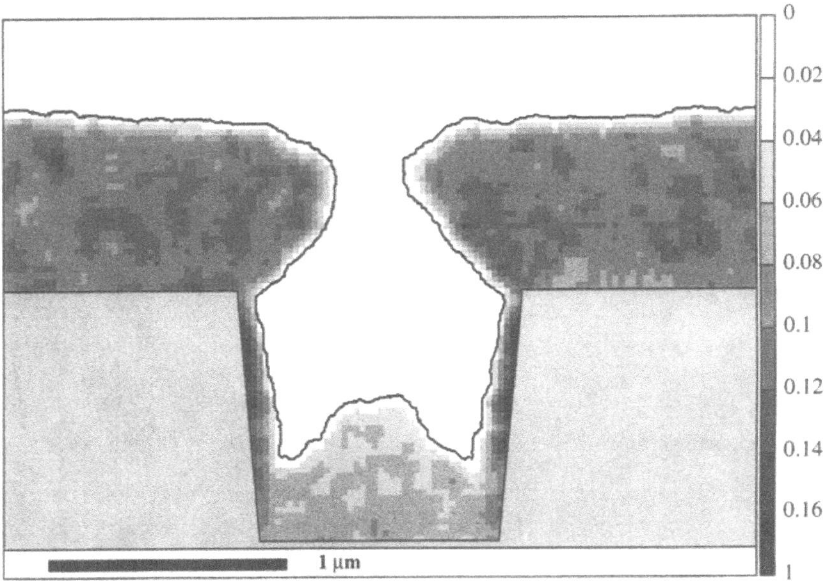

FIG. 9.31 Simulated Ti_xW_{1-x} film composition in a 1:1 aspect ratio trench showing variation in stoichiometry with location [9.36]. The legend bar indicates relative Ti concentration in weight percent.

attractive scaling properties of $CoSi_2$. Reviews of the physics and chemistry of silicides used in IC processing are provided in refs. 9.39–9.42, and in this section we simply point out several PVD-related aspects of silicide use.

9.7.1 MSi_x WHERE M = TA, MO, OR W

Although these refractory silicides can be cosputtered using separate metal and Si targets, they are more often sputtered from a single, composite target with ultrahigh density (≥ 95% of bulk density). Such targets are produced by mixing and pressing together fine particles of refractory metal and Si and then vacuum-sintering at high temperature and pressure to give a bulk density close to the theoretical maximum for the metallurgical alloy. Since target microvoids in a low-density target can trap gases and contamination that are subsequently released during sputtering, high-density targets have allowed silicide films with improved purity and reduced particulates. Also, since a greater in-plane atom density is exposed to Ar^+

bombardment, a higher deposition rate is achieved at a given magnetron power.

Another aspect of composite silicide target sputtering is that the target composition must be controlled to give the desired MSi_x film on the substrate. For IC applications, a film with [M]:[Si] ratio close to the disilicide composition ($TaSi_2$, $MoSi_2$, and WSi_2) is favored. However, a film with an over-stoichiometric amount of Si is often used to optimize stress and resistivity and to provide an in-film source of Si for a surface-passivating oxide. Due to the different sputter yields of the alloy constituents in the target and the difference in gas-phase scattering of the higher-mass metal atoms (^{181}Ta, ^{96}Mo, ^{184}W) and the ^{28}Si atoms, the as-deposited films often have a slightly different stoichiometry than the target — typically metal enriched. If an RF bias is applied to the substrate during PVD, the deposited film can be further depleted in Si that is preferentially resputtered relative to the high-mass metal. As a result, extra Si is often blended into the target during its manufacture to allow the user some flexibility in film stoichiometry. For example, a WSi_x film with $x = 2.0$ to 2.5 might be produced by PVD from a given $WSi_{2.7}$ target, depending on process conditions and postdeposition sintering. Since the as-deposited silicides tend to be amorphous, a high-temperature post-PVD sintering step (e.g., 900–1000°C) is often used to increase and homogenize grain size so that low-resistivity polycrystalline films can be obtained that, depending on the specific silicide and its stoichiometry, are in the range of about 30–100 $\mu\Omega$-cm.

9.7.2 $TiSi_2$ AND $CoSi_2$

These material systems are selected to illustrate the use of metal sputtering followed by annealing for silicidation. For example, PVD Ti is widely used at the Si contact level as a chemical cleaning agent to reduce native silicon oxide. After annealing, this same Ti film reacts with and consumes Si to form a low-resistance $TiSi_2$ layer of much greater thickness ($\approx$ 2.7 times thicker) than the starting Ti film. The amount of Si consumed during Ti-silicidation must not be so great that it affects the junction (which can be quite shallow (< 100 nm) in advanced ULSI devices), and as a result $TiSi_2$ layers are typically < 40-nm thick for this application with even thinner Ti starting thickness (< 15 nm). Also, since a 1-Å change in Ti thickness results in a 2.7-Å change in $TiSi_2$ thickness, excellent control of starting PVD Ti film thickness is required for process repeatability.

Both Ti and Co are well suited for the salicide process in which the elemental metal is sputter deposited as a blanket film and then processed to selectively leave a low resistivity film of $TiSi_2$ or $CoSi_2$ on oxide-defined windows [9.43]. Although salicide processing with Ti is established in present VLSI devices, a future concern is that the $TiSi_2$ formed after low-temperature annealing ($< 600°C$) has the C49 phase with base-centered orthorhombic crystal structure and relatively high resistivity ($\approx$ 60–70 $\mu\Omega$-cm). Therefore, a second anneal at higher temperature ($> 700°C$) is required to transform the C49 phase into the desired C54 phase with face-centered orthorhombic crystal structure and low resistivity ($\approx$ 15–20 $\mu\Omega$-cm). While this has not ruled out Ti-salicide processing for VLSI devices, the C49-to-C54 transition temperature increases with decreasing line width, and the higher temperatures may not be compatible with the reduced thermal budget of ULSI devices. Also, scaling down of the $TiSi_2$ layer thickness further increases the transformation temperature and/or annealing time. As a result, increasing attention is being given to Co-salicide processing.

Although Co offers a number of potential benefits [9.43], its application in multilevel metal interconnection is relatively new and poses some interesting challenges. First of all, care must be taken when using PVD Co for IC processing, since it is a midgap trap in Si and will affect MOS properties if it is allowed to cross-contaminate processing equipment. The other major issue relates to Co being a ferromagnetic material with high magnetic permeability. To efficiently sputter such a material with a magnetron, the ferromagnetic target must not shunt the magnetron cathode's plasma-enhancing DC magnetic field or act as a magnetic pole piece of the source. Even if sufficient electron confinement can be obtained in the presence of the target to sustain a magnetron discharge, the magnetic field perturbation may alter the desired target erosion profile with adverse impact on target utilization and film uniformity. Therefore, a magnetron with very strong permanent magnets and/or a very thin target must be used to ensure that the target material is magnetically saturated. Although very thin targets are generally at odds with high target life, the amount of Co needed for this application is sufficiently small (tens of angstroms per deposition) that rather thin Co targets can be used. In this regard, target suppliers have begun to fabricate Co targets with smaller grains and more preferred orientation. This microstructural engineering serves to reduce the relative permeability of the target and increase the pass-through flux of magnetic field for a given target thickness. Finally, the basic PVD source technology has already been developed to deposit cobalt alloys for hard disk drives (e.g., CoCrTa, CNiCrTa, and CoCrPt), although advanced magnetic hard disks are much smaller in diameter ($\leq$ 3 inch) than production Si wafers. Other than the case of Co, however, sputtering of magnetic materials is rarely attempted in semiconductor processing.

While the ferromagnetic nature of Co makes it more difficult to sputter than Ti, Co offers a number of process advantages over Ti, including the ability to deposit epitaxial $CoSi_2$ on Si(100). Ironically, PVD Ti has played an enabling role in this process, which uses an ultrathin intermediate layer of Ti (1–5 nm) sandwiched between a thicker Co film (15–20 nm) and the Si substrate [9.44]. The Ti cleans the native oxide from the Si surface and, during subsequent annealing in N_2, moderates the diffusion rate of Co to the Si leading to the growth of a single epitaxial $CoSi_2$ phase as opposed to a mixture of polycrystalline CoSi and $CoSi_2$ phases. Following annealing, a TiN/$CoSi_2$/Si(100) structure is formed. The process requires good uniformity of ultrathin Ti films and suggests that PVD Ti will continue to play an important role in silicide formation, even though the silicide being formed is $CoSi_2$.

9.8 Copper

9.8.1 Metallurgical Issues for PVD

PVD of Cu is well established in multilevel metallization. For example, sputtered Cu is sometimes used as part of the back-end-of-line metallization process for chip assembly and packaging. Also, alloy targets of Al with a few percent of Cu (e.g., Al-Cu, Al-Si-Cu) are commonly used to deposit interconnect wiring, and even though the Cu is alloyed with Al in the target matrix, the sputter process results in a flux of elemental Cu atoms at the substrate. On the other hand, serious interest in using pure copper interconnects as a material replacement for Al and its alloys is a relatively new development.

In particular, Cu has lower intrinsic resistivity than Al (1.7 $\mu\Omega$-cm at 20°C vs $\approx$ 3 $\mu\Omega$-cm for Al alloys) that should enable higher-speed, deep submicron devices ($\leq$ 0.18 μm) through reduced RC time-constant delays as well as greatly reduce the number of metal levels required to route the wiring of an advanced microprocessor (Fig. 9.33). Cu also has a higher melting point than Al and a better thermal expansion match to that of SiO_2 (linear coefficients of expansion for Al, Cu, and SiO_2 are $\approx$ 23.6, 16.5, and 0.5 ppm per °C, respectively), leading to superior resistance to stress migration, hillock formation, and electromigration — although both Cu and Al exhibit their best electromigration resistance when they are strongly textured in the (111) direction.

The primary clock-speed advantage of using Cu over Al is derived from its lower resistivity and is achieved by introducing Cu at the upper interconnect levels, where conductor lengths can be of the same order as the chip size. On the other hand, the increased reliability advantage of Cu over Al is realized by introducing Cu at the lower interconnect levels where the current density in the fine lines can be large enough to induce electromigration failure in traditional Al-based interconnects. In order to take advantage of Cu for upper or lower-level interconnect applications, integration issues such as oxidation (Cu does not form a self-passivating oxide like Al), corrosion, and poor adhesion to oxide need to be resolved. Also, suitable diffusion barriers need to be developed to prevent the rapid movement of Cu into both SiO_2 and Si. For example, Cu migration into SiO_2 can create electrical leakage paths between adjacent metal lines and/or layers. Also, Cu forms deep-level traps in Si and can consume Si via formation of Cu_3Si at temperatures as low as 200°C. Fortunately for PVD, candidate barriers for Cu such as Ta and TaN can be deposited with rather good conformality by sputtering.

In addition, a commercially viable anisotropic plasma etch process for Cu has been notoriously difficult to develop because the vapor pressure of Cu halides are very low at room temperature. This means that one cannot use the conventional "subtractive" process for MLM in which a blanket PVD metal film is deposited over a planarized oxide and then patterned and etched into separate metal lines by reactive ion etching (RIE). As discussed in Chapter 6, the industry is expected to switch to a damascene type of patterning (Fig. 9.32) in which Cu is first deposited into trenches that were first etched into the dielectric and is subsequently planarized, e.g., by chemical-mechanical polishing (CMP). As a further refinement, dual-damascene wiring can be used in which both vias and trenches are first etched into the dielectric and then are filled with Cu in the same deposition step [9.45]. This will be a particularly challenging application for PVD (analogous to simultaneously filling a rain gutter and a down spout) and will have to be carried out at low process temperatures consistent with the organic dielectrics being considered for ≤ 0.18-μm devices. In addition, filling must be done without leaving buried voids or forming seams where the sidewall deposits meet. The risk is that such features could be exposed after the Cu is chemomechanically polished back, producing topographical surface defects.

Whether or not PVD is up to the challenge of dual-damascene Cu wiring, it is worth noting that other candidate Cu deposition methods such as CVD Cu and electroplating may require either a PVD barrier/adhesion layer (e.g., PVD Ta or TaN) and/or a PVD Cu seed/nucleation layer. Also, in many advanced multilayer designs, the upper wiring layers are so-called fat levels in which the lines and vias are both relatively thick and of low

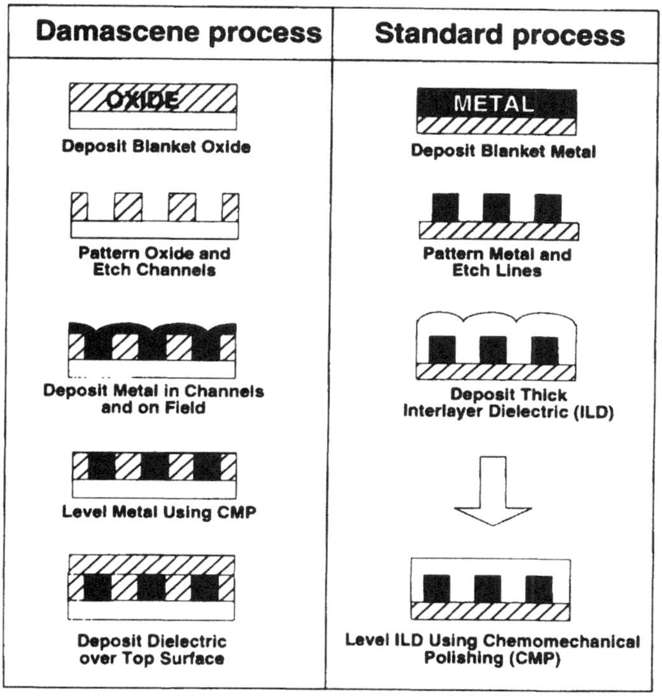

FIG. 9.32 Comparison of conventional and damascene metal wiring processes.

aspect ratio ($\approx$ 1:1). For these layers, PVD barriers and liners are expected to be applicable even beyond 0.18 μm, although the process used for Cu filling could be CVD or electroplating.

9.8.2 Sputtering and Self-Sputtering of Cu

Oxygen-free highly conductive (OFHC) elemental Cu targets are readily available with purity of 5 nines and above and, as with Al-Cu alloy targets, can be used for high-power magnetron deposition on 200-mm Si wafers with comparably high specific deposition rates ($\approx$ 20 Å/kW-sec). Cu has a higher sputter yield than Al under Ar^+ bombardment (e.g., 2.3 Cu atoms/ion versus 1.2 Al atoms/ion at 600 V) and a greater mass than the Ar working gas ($^{64}Cu > {}^{40}Ar > {}^{27}Al$). Hence, all things being equal, we would expect about 2 times more sputtered flux from a Cu target than from an Al target, with reduced gas-atom scattering and increased directionality in the deposited film.

Probably the most interesting aspect of PVD Cu is the possibility of using self-sputtering to completely do away with the inert Ar working gas and associated gas-atom scattering, residual gas impurities, and Ar incorporation that are of concern at conventional mTorr-type PVD pressures [9.46–9.49]. In this self-sputtering mode, the magnetron discharge is initiated with Ar gas but is sustained under ultralow pressure (e.g., 5×10^{-5} Torr) with sputtered Cu atoms that are ionized in the DC plasma region and accelerated by the electric field to the target. In conventional magnetron sputtering, the high density of secondary electrons leaving the target in crossed electric and magnetic fields ($\mathbf{E} \times \mathbf{B}$) gives rise to a high ionization rate of inert gas ions, which subsequently sputter-erodes the target While sputtered metal can also become ionized near the target, this ionization fraction is typically very small. On the other hand, if the production of metal ions and their self-sputter yield are sufficiently greater than unity, then a discharge can be sustained in the absence of any inert gas — provided that the magnetron fields are designed to redirect a large fraction of the metal ions back at the target. These conditions are discussed in refs. 9.46 and 9.49, and the basic idea is illustrated in Fig. 9.33.

The self-sputter yield Y_{ss} as a function of atomic number Z and ion energy E has been given by Zalm [9.50] as

$$Y_{ss} = \left(\frac{1.9Z^{1/2}}{U_0}\right)(E^{1/2} - 0.009U_0^{1/2}) \tag{9.4}$$

where U_0 is the sublimation energy of the element. Applying Eq. (9.4) to Cu and Al at E = 600 V, we see that Cu is an outstanding candidate for self-sputtered PVD since it has a yield of $Y_{ss} \approx 2$, which is several times larger than that of Al. On the other hand, the first ionization potential of Cu (E_I = 7.68 eV) is higher than Al (E_I = 5.96 eV) so it is more difficult to produce Cu^+ than Al^+ by electron-impact ionization. The net result, however, is that Cu can be self-sputtered at high rates (> 1 μm/min) at low pressure ($\approx 2 \times 10^{-5}$ Torr) in the absence of any inert working gas [9.46]. However, the power density (80 W/cm^2) in this case was several times higher than what could be sustained with a conventional large-area planar magnetron (e.g., 20 kW into a 12-inch-diameter target is ≈ 30 W/cm^2). It remains to be seen whether self-sputtering of Cu will enter mainstream IC manufacturing; however, the exploitation of metals ions to improve PVD performance is an established trend. For example, as discussed in Chapter 8,

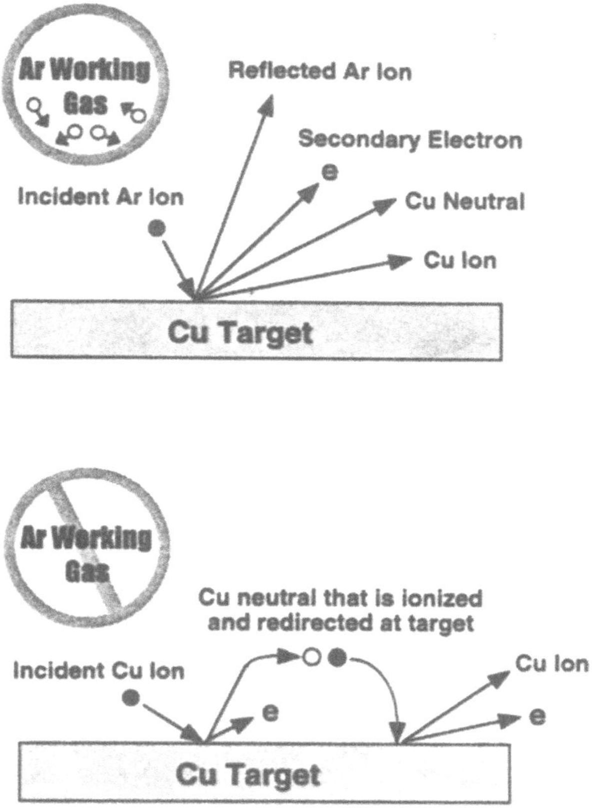

FIG. 9.33 Comparison of conventional PVD Cu with self-sputtered PVD Cu in which Cu^+ ions replace Ar^+ ions, allowing the Ar working gas to be eliminated from the process.

directional PVD of Cu and other metals based on plasma ionization of sputtered-metal neutrals (albeit at a relatively high pressure of inert working gas) is under active development.

9.8.3 PVD Ta and TaN Barriers

As noted earlier, integration of Cu into IC processing will require deposition of suitable barriers to prevent its diffusion into Si and electric-field assisted drift into SiO_2. An important difference in this regard between Cu and Al is that Al can be directly deposited onto oxide with good adhesion and formation of a self-limiting interfacial Al_2O_3 while Cu requires a barrier. Therefore, replacement of a PVD Al interconnect with PVD Cu also requires additional barrier layers on the sides of the metal line to prevent its contact with the intermetal dielectric. Since the resistivity of the barrier is much greater than that of Cu, the effective interconnect cross-sectional area is reduced, which offsets to some extent the RC gain expected from going to Cu over Al. This effect is illustrated in Fig. 9.34 for a via hole and a 20-nm-thick barrier layer. In the case of Cu, the bottom and sidewall barrier can use up 50% or more of the volume of a 0.18-μm hole. Assuming the barrier resistivity is much greater than that of Cu, this leads to an effective conductivity of the filled feature of 2.8 $\mu\Omega$-cm — much higher than the 1.7 $\mu\Omega$-cm one would obtain if the feature were filled with pure Cu. In contrast, the bottom-only barrier for Al takes up less than 5% of the via hole volume. From the resistivity standpoint, there is clearly a reason to use as thin or as conductive a barrier as possible. On the other hand, it is questionable just how thin such a material can be made before its barrier properties are no longer sufficient to prevent Cu diffusion and drift into the adjacent regions of the hole.

A wide variety of materials have been investigated as diffusion barriers for Cu, including refractory metals (e.g., W, TiW, Ta), nitrides (e.g., W_2N, TiN, TaN, Ta_2N), silicides (e.g., $TiSi_2$), and ternary amorphous alloys such as Ta-Si-N [9.51]. The Cu barrier/adhesion layer of choice is an open issue at this time. It may be possible to simply extend older materials that have performed well as barriers for Al metallurgy, such as Ti/TiN and TiW. However, two "new" materials — Ta and TaN_x — have shown considerable promise as thermally stable barriers with Cu metallurgy [9.52–9.54]. Ta is thermodynamically stable with respect to Cu, while Ta nitrides are chemically inert with Cu because of the absence of any compounds between Cu and Ta or Cu and N. It has been found that a small amount of contamination in a Ta film — e.g., a few atomic percent of oxygen or nitrogen — improves its ability to withstand Cu penetration. Hence, ultrapure Ta films are not preferred for this application.

In analogy with the Ti/TiN system, Ta can be deposited by the sputtering of an elemental Ta target in Ar, and TaN can be deposited by the reactive sputtering of Ta in an Ar/N_2 admixture. However, an important difference is that unlike the Ti/TiN case, there are two phases of PVD Ta films (a high-resistivity β-Ta phase with $\rho \approx 180$ $\mu\Omega$-cm and a lower-resistivity, body-centered cubic phase of bcc-Ta with $\rho \approx 40$ $\mu\Omega$-cm) as well as two nitridation states (TaN and Ta_2N). The impact of this complexity is illustrated in Fig. 9.35, where data on reactive sputtering of Ta onto room

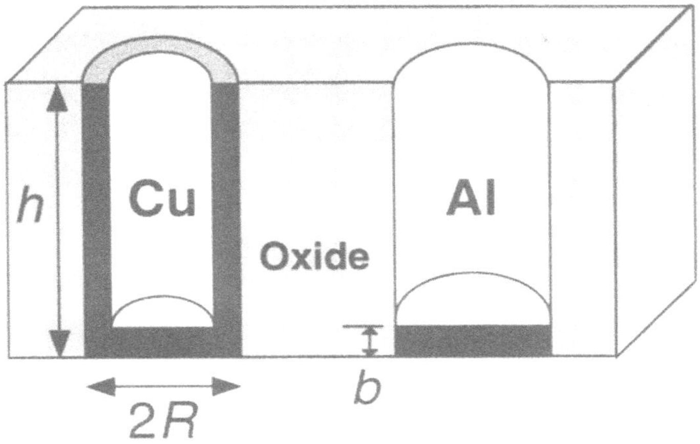

(a)

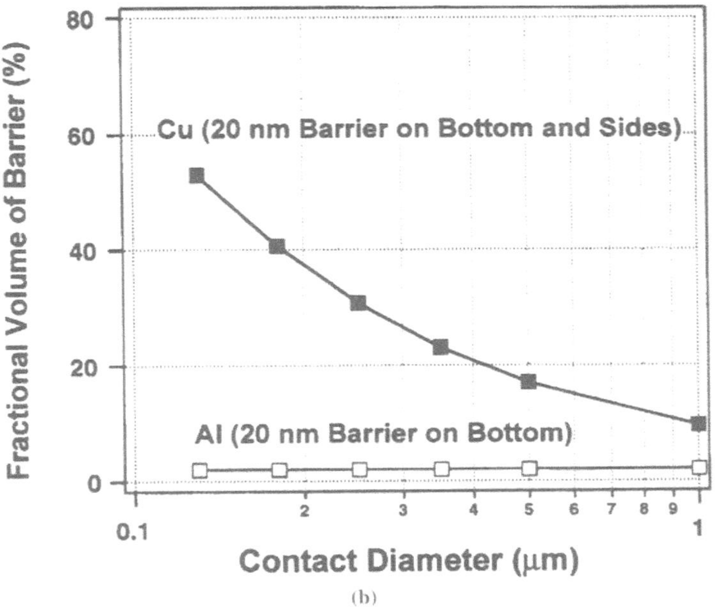

(b)

FIG. 9.34 (a) Different barrier requirements of PVD Cu and PVD Al lead to (b) different fractional volume of barrier in a via hole.

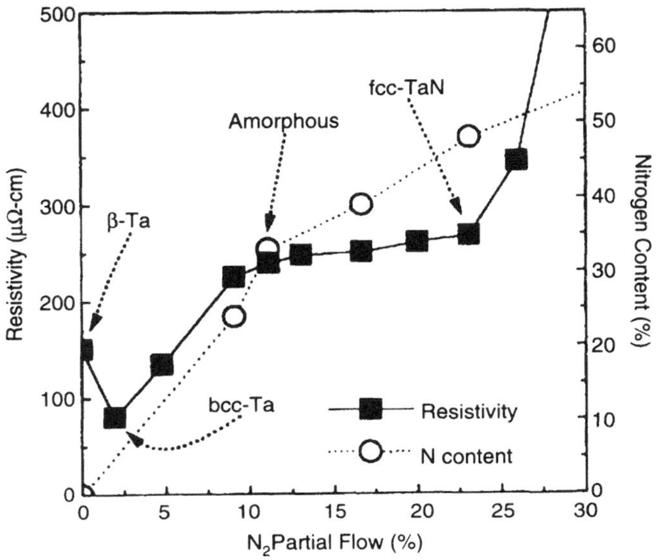

FIG. 9.35 Resistivity and nitrogen content of reactively sputtered TaN as a function of N_2-to-Ar flow ratio [9.52]. (Reprinted with permission from K.-H. Min et al, *J. Vac. Sci & Tech.* **B14(5)**: 3263–3269 (1996). Copyright 1996 American Institute of Physics.)

temperature Si(100) substrates is shown [9.52]. Based on a detailed analysis of these films, it was found that Ta sputtered in pure Ar gave predominantly β-Ta films (with a small admixture of bcc-Ta) whereas addition of a very small amount of N_2 to the working gas ($\approx$ 3% N_2/Ar flow ratio) led to bcc-Ta. Higher N_2/Ar ratios led progressively to a mixture of amorphous and crystalline Ta_2N, and finally fcc-TaN. Clearly the evolution of phase and microstructure make reactive PVD of TaN a more complicated system than that of PVD TiN and suggest that very good control of process conditions (e.g., flow ratio and substrate temperature) may be required to implement the process in a production environment. This advice holds true even in the case of elemental Ta, which can deposit in either the α- or β-phase depending on sputter deposition temperature, pressure, and substrate type. For example, workers have reported that sputtering Ta in 3 Torr of Ar onto Si wafers produced films of β-Ta for heater set-point temperature below 550°C (i.e., wafer temperature below 400°C) at which point α-Ta began to form. By either raising the Ar pressure to 8 mTorr or depositing on SiO_2, the temperature of the α-to-β transition could be reduced to a heater set point of 400°C and 350°C, respectively [9.55].

On the positive side, PVD Ta films have been found to be more directional than PVD Ti films and highly conformal in high aspect ratio structures [9.56]. Due to the much greater mass of ^{181}Ta compared to ^{48}Ti, one would expect Ta atoms to be deflected less by scattering with the Ar working gas, and hence to retain their as-sputtered directionality. This is reflected in Fig. 9.36, which shows the improved bottom coverage of 1:1 collimated PVD Ta over PVD Ti deposited with even greater (1.5:1) collimation. In addition, the deposition rate of Ta has been found to be several times that of Ti under comparable collimation conditions [9.55]. The improved conformality has been attributed to the reflection coefficient of Ta atoms increasing very steeply with decreasing angle of incidence — a dependence that may reflect a relatively high population of energetic atoms in the incident Ta flux [9.56]. As a consequence, near-normal-incidence Ta atoms that hit at the top of the via sidewalls have an effective sticking coefficient less than unity and are reflected and redistributed much deeper into the structure. The net result is more conformal sidewall coverage and possibly increased bottom coverage as well. In addition, a highly directional incident flux of Ta (e.g., using collimation or longer throw distance) can then provide quite good sidewall coverage, which is contrary to what one observes for atoms such as Ti.

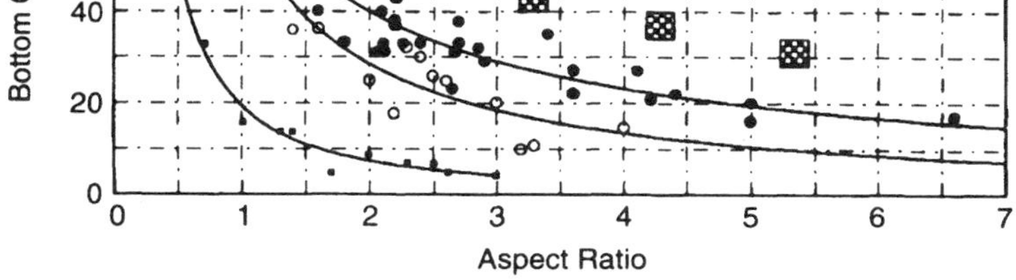

FIG. 9.36 Comparison of step coverage of PVD Ti and PVD Ta (checkered boxes) based on data in ref. 9.55. The PVD Ta was deposited using a 1:1 aspect ratio collimator.

9.9 PVD and CVD

PVD has historically been used to deposit conductors for contacts, barriers, liners, and wiring in multilevel metallization. On the other hand, with the notable exception of CVD W plugs (which dominate VLSI device wiring), CVD has been used to deposit insulators for dielectric isolation between the lines (so-called gap fill) and between the levels. Although PVD has been used to deposit thin high-dielectric constant insulators for DRAM storage capacitors (e.g., $Ba_xSr_{1-x}TiO_3$ = BST), it is unlikely that PVD will be used for thick dielectric isolation in advanced devices due to its limited step coverage and low deposition rate. Hence, there is growing interest in CVD metallization for ULSI devices due primarily to its improved conformality and fill capability.

Figure 9.37 uses TiN deposition to show in a simple schematic way the difference between PVD and CVD, and the two methods are further compared in Fig. 9.38. Of note, PVD is carried out in a process chamber having high or ultrahigh vacuum base pressure ($< 10^{-8}$ Torr) and mTorr-type

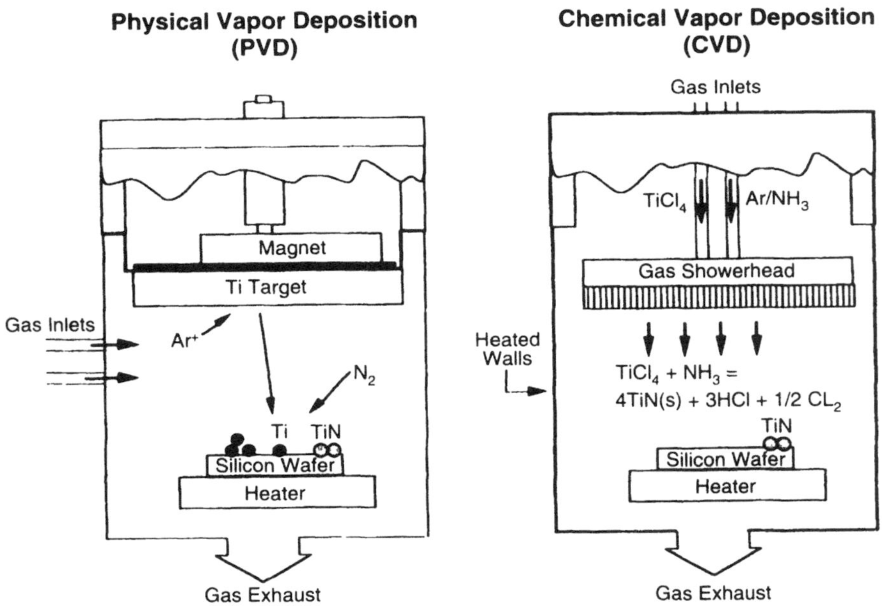

FIG. 9.37 Schematic representation of the deposition of TiN using PVD and CVD methods.

PVD	CVD
Physics-based (sputtering).	Chemistry-based (often organic).
Blanket, nonselective deposition. Modeling is relatively straightforward.	Selective deposition in some cases. Modeling is complex in general.
Base pressure ≈ 10^{-9} Torr (UHV). Operating pressure ≈ 1 mTorr.	Base pressure usually >10^{-6} Torr. Operating pressure ≈ 1 Torr.
Suitable for single-wafer cluster tool. Proven at 200 mm; scalable to 300 mm.	Suitable for single-wafer cluster tool. Proven at 200 mm; scalable to 300 mm
NOT used for insulators (e.g., ILD). Suitable for all conductors used in microelectronic interconnect.	Method of choice for ILD. Suitable for some conductors; pure metals and alloys are difficult.
High film purity. Step coverage limited for high AR.	Film purity can be an issue. Conformal step coverage possible.
Simple working gas (e.g., Ar, Ar/N_2). Nontoxic targets and noncorrosive gases.	Chemically complex precursors. Reactive gases and reaction products (sometimes hazardous and/or corrosive).

FIG. 9.38 Comparison of PVD and CVD metal deposition for microelectronic applications.

operating pressure (0.5–5 mTorr). By contrast, CVD requires a much cruder vacuum, with operating pressure in the Torr range (e.g., pressures for CVD W are ≈ 40 Torr). This allows CVD to make use of less costly, less complex vacuum pumping (e.g., cryopumps and cyropump regeneration cycles are avoided) and to avoid the need for separately injected backside gas to control wafer temperature transfer (see Section 5.3.4). Higher operating pressure also allows a simple vacuum chuck to be used in some cases in which the pressure behind the wafer is controlled by active pumping and kept enough below the operating pressure to provide a suitable pressure gradient, i.e., a suitable holding force per unit area. This avoids the need for a more costly, complex electrostatic chuck. On the other hand, the purity and electrical conductivity of CVD films are often compromised by the relatively poor vacuum ambient and by incorporation of impurities such as C and O that are common in organometallic precursors.

With regard to wafer temperature, the CVD process is often exponentially activated (deposition rate goes as $e^{-Ea/kT}$) so that uniform film thickness requires excellent uniformity of wafer temperature. On the other hand, the deposition rate of a PVD film is relatively insensitive to temperature (as is the sputter yield of the target); however, wafer temperature can have a strong influence on step coverage, film purity, and microstructure. Hence, global control of wafer temperature can be equally important in PVD.

Finally, we mention the issue of chamber coatings that are common to both technologies. Even in a well-engineered PVD chamber, a majority of the sputtered flux from the target intercepts and adheres to the sputter shields and other chamber fixtures (e.g., collimator). The goal therefore is not so much to prevent PVD films on these structures as to guarantee that the films do not cause particles or flake off. Therefore, promoting film adhesion and reducing film stress are important concerns, and conditions are chosen so that shields and other coated surfaces can last 50% or more of target life before replacement or cleaning is required. In CVD, on the other hand, line-of-sight deposition is not relevant. In this case, the goal is to prevent chemical gases from condensing or reacting on surfaces or from reacting prematurely in the showerhead with another precursor gas. Both chamber walls and showerheads are generally kept at an appropriate temperature with this goal in mind. Inevitably, some deposition occurs, necessitating the periodic use of in-situ reactive plasma or reactive gas cleaning. By contrast, in-situ cleaning of PVD chambers is rarely done.

It is not the purpose here to review CVD metallization, but simply to point out where CVD is likely to replace and/or augment PVD in microelectronic applications. The basic technical issue relates to the difference between the "directional" nature of PVD and the surface-activated film growth of low-pressure CVD (see Fig. 9.39). In conventional PVD, the combined effect of the broad angular distribution of sputtered flux, finite target size, and gas-phase scattering gives rise to a large fraction of low-angle material, leading in turn to low bottom and sidewall coverage and concerns about keyhole voids. Hence, this situation is not optimum for lining or filling high aspect ratio features. Although coverage would be improved if the sticking coefficient of the metal adatoms were very low, this is typically not the case for the metals and process conditions used in PVD. The use of collimation and/or lower pressure can improve directionality, but even at very low pressures (< 0.1 mTorr) and strong collimation, the angular spread of incident flux can be rather broad (e.g., FWHM $\approx \pm 20°$ for PVD Ti with a 1.5:1 aspect ratio collimator). If a highly anisotropic method such as RF-ionized PVD is used, the bottom coverage can be quite good (e.g., $> 80\%$ in AR = 5:1 holes), and ideally one could fill the hole from the bottom up or increase incident metal ion energy to resputter material from the bottom onto the sidewalls for a liner application (see Chapter 8). However, bottom-up filling requires the removal of the material deposited on the field by use of a subsequent etchback step. Also, get-

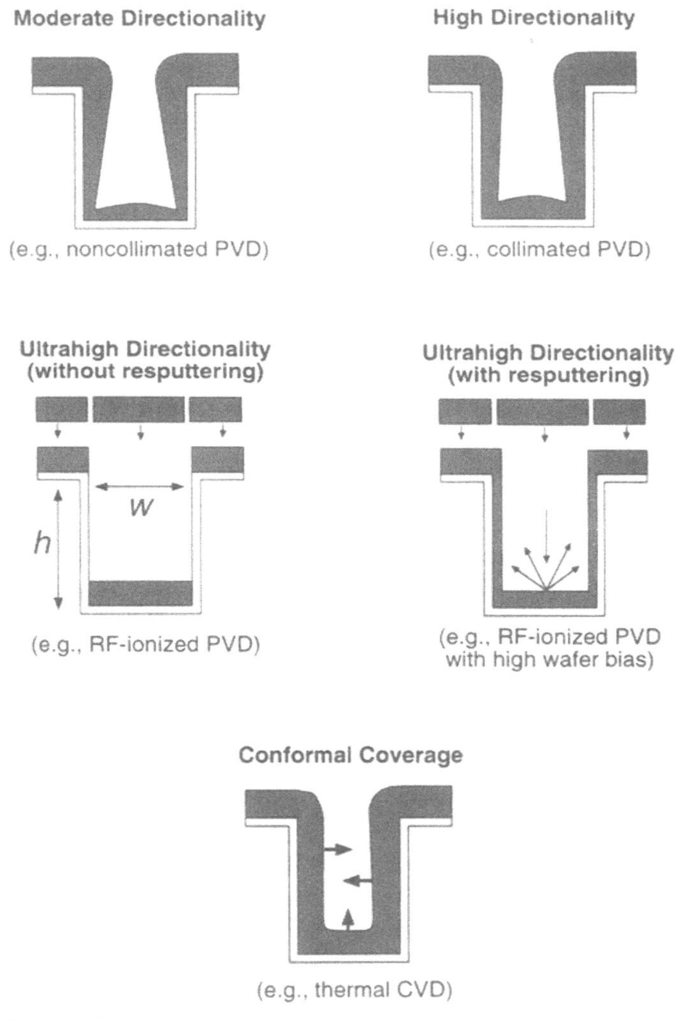

FIG. 9.39 Representation of film coverage by CVD and PVD with varying degrees of directionality.

ting a liner with high step coverage in a high aspect ratio hole (height h and width w with $h/w \gg 1$) is problematic because the cylindrical plug of material entering the via hole (cross-sectional area $A = \pi w^2/4$) must be redistributed over the much higher, interior surface area of the hole ($A = \pi w^2/4 \times [1 + 4h/w]$). A simple calculation based on conservation of mass shows that a uniformly thick, conformal PVD layer in a hole of aspect ratio $h/w \gg 1$ will at best have step coverage $\approx 1/(2\ h/w)$, which is 10% for a 5:1 aspect ratio hole. One solution is to utilize a hot PVD process such as reflow Al in which material from the adjacent field is moved into the hole by surface diffusion, the field itself being simultaneously planarized. Another approach is to abandon PVD for this application in favor of a surface-activated CVD process.

As a consequence of the relatively high CVD pressure ($\approx$ 0.1–10 Torr is typical for CVD metal deposition) and chemistry used, both the mean free path of the precursor molecules and their sticking coefficients are often reduced relative to that of the sputtered atoms in PVD. As a result, the precursor molecules undergo a large number of collisions both in the reactor and with the surfaces of microstructures on the wafer, which, coupled with surface diffusion and the proper process conditions, can give highly conformal films with uniform thickness and $\approx$ 100% step coverage. This mode of deposition is also well suited to rapid filling of narrow, high aspect ratio features since it requires only that deposited film thickness be greater than one-half of the via width — more or less independent of the via height. For example, a 1-μm-deep $\times$ 0.18-μm-wide via could be filled by depositing a fully conformal 1/2 x (0.18 μm) = 0.09-μm-thick film. Finally, CVD is sometimes capable of coating even reentrant features — a feat that is not possible with a more line-of-sight method of deposition such as PVD.

> The term *conformal* is used in different ways in the literature and relates to how uniformly a nonplanar structure is coated with a thin film. With respect to a contact, via, or trench, a conformal film is one whose sidewall coverage is of uniform thickness and equal to its bottom coverage. In general, this thickness will be less than the thickness on the field. A conformal film whose thickness is equal to the thickness deposited on the field is said to be "fully conformal" or to have "100% step coverage."

In spite of its compatibility with steep topography, CVD metallization has its own set of challenges. For example, it has been difficult to develop appropriate CVD chemistries to deposit high-purity Ti, Cu-doped Al alloys, or TiN films having both high conformality and purity comparable to PVD TiN. Also, when ultrahigh aspect ratio topography is not an issue (e.g., deposition of 1 μm of slab Al on a planarized oxide), PVD offers a mature and highly cost-effective solution. There are, however, situations in which the complementary benefits of each method can be used to advantage — e.g., the well-established use of PVD Ti/TiN as a contact-barrier layer-adhesion liner for CVD W contacts and plugs.

As shown schematically in Fig. 9.40, there are many other possibilities for combining CVD and PVD in the future. For example, Cu-doped CVD Al via plugs (AR = 5:1, 0.25 μm) have been produced by partially filling the via with undoped CVD Al and then, without a vacuum break, depositing a PVD Al-Cu(1%) overlayer [9.57, 9.58]. A brief anneal (e.g., 30 sec at 300°C) was then used to diffuse the Cu from the PVD film uniformly throughout the filled via. When a vacuum break was used between the CVD Al and PVD Al-Cu deposition, an amorphous interfacial layer formed

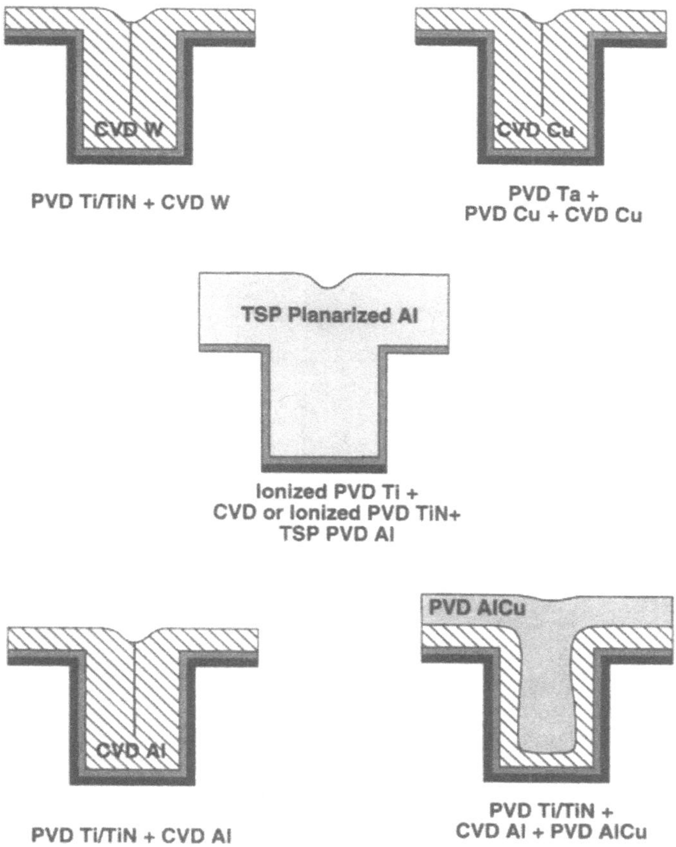

FIG. 9.40 Illustration of selected processes combining PVD and CVD.

that acted as a diffusion barrier to subsequent Cu diffusion. This in turn necessitated a much higher thermal budget anneal to diffuse the Cu (30 min at 450°C). Hence, a vacuum-integrated cluster tool is required to optimize the process.

Other PVD/CVD combinations that are under active development include the use of an RF-ionized PVD Ti layer combined with a CVD TiN barrier, and the use of a PVD Cu/Ta bilayer to facilitate subsequent filling with Cu by CVD or electroplating. Also, the cold-hot PVD Al process might be improved by use of a more conformal Al seed layer deposited by CVD Al or directional, ionized PVD Al with sufficient resputtering to satisfy sidewall coverage requirements. Finally, we mention a novel CVD-PVD approach to Al plug fill — known as polysilicon-aluminum substitution (PAS) — that relies on the same thermodynamic driving force that causes Al junction spiking [9.59]. In its simplest form, the PAS process involves filling of an oxide via with CVD poly-Si, removal of the poly-Si on the field regions by chemomechanical polishing, and then blanket deposition of PVD Al. A lengthy anneal of the vias (e.g., 500°C for 3 hours) allows the poly-Si to diffuse up into the bulk of the Al and the Al to diffuse down into the via where it substitutes for the poly-Si. Unlike the departed silicon, however, the Al in the via after annealing is single crystal. Done properly, PAS has been shown capable of filling 0.18-μm, 10:1 aspect ratio vias with single-crystal Al.

Assuming that such hybrid process flows are cost-effective, the major concern is how to avoid cross-contamination when PVD and CVD are done on the same vacuum-integrated process tool, given the high CVD operating pressures ($\approx$ 1–10 Torr for CVD vs 1 mTorr for PVD) and the exotic organometallic precursors that are likely to be used — such as DMAH (dimethyl-aluminum hydride) for Al, Cu(II)-$(hfac)_2$ for Cu (hfac is the ligand 1,1,1,5,5,5-hexafluoro-acetyl-acetonato), and TDMAT (tetrakis-dimethyl-amido-Ti) or TDEAT (tetrakis-diethyl-amido-Ti) for TiN. The use of such chemistry on a PVD–CVD cluster tool is a particular concern for hot PVD Al steps that are highly sensitive to even trace amounts of oxidants and hydrocarbon contamination, and will require good vacuum buffering and possibly multiple pump/purge cycles of the CVD module (e.g., pump to 10^{-6} Torr + backfill with Ar to 1 Torr + pump-down) to reduce the precursor partial pressure below a critical level before opening the isolation valve to the central handler.

Finally, we note that cost-of-ownership applies equally to both PVD and CVD technology. However, in the case of CVD the additional cost of pumps and maintenance to deal with reactive and/or corrosive gases and effluents needs to be considered. Also, cost and availability of the precursor can be an issue. For example, although CVD W plugs are widely used today ($3H_2 + WF_6 = W + 6HF$), the low availability and high cost of suitably pure tungsten hexafluoride (WF_6) in the 1980s delayed widespread deployment of both CVD W and WSi_2 by many years. In this regard, it is instructive to estimate chemical cost associated with the Al sources in PVD Al and CVD Al. The cost of the relatively common organometallic DMAH = $(CH_3)_2$-Al-H used in CVD Al is about \$25/gm (in 1997 dollars). Since the weight of a 1-μm film of Al ($\rho = 2.7$ gm/cm^3) deposited over a 200-mm wafer is about 0.08 gm, the cost per wafer associated with this Al precursor is $\approx$ \$2.10 — assuming that the CVD process is 100% efficient at converting the gaseous precursor into solid-phase Al films. In practice, the utilization efficiency is less than 100%, and a large fraction of the precursor ends up being pumped out of the chamber. Hence the actual cost per wafer could be several times greater — e.g., \$5–10. By comparison, the cost of a high-purity Al target is about \$5000 and produces about 8000 1-μm Al films over target life — a cost of $\approx$ \$ 0.65 per wafer.

9.10 Upper-Level Metallization

The topmost layer(s) of metal interconnection in a multilevel stack is typically reserved for power distribution and the transmission of global signals that run the length of the chip. To minimize voltage drops over these long runs, the wiring usually has large cross-sectional area A — e.g., very thick lines with very wide pitch — which reduces the resistance per unit length: $R/L = \rho/A$. The top layer of metal also provides the large-area bonding pads that are used to connect the chip to the outside world. PVD metal films between the bonding pad and the bonding wires can then serve as diffusion barriers, promote adhesion, and/or provide a better thermal expansion match to the wire bonding metallurgy. For example, in the so-called bump fabrication method used with tape automated bonding (TAB), an Al alloy bonding pad might be coated with a bilayer of PVD Au on PVD TiW. The TiW serves as a diffusion barrier between the Au and Al, and the Au provides an adherent surface for a subsequent electroplated bump of metallic Au. PVD is also used for backside conductive coating of the chip. For example, a PVD gold or Pt film might be deposited onto the backside of the Si. For this application, care must be taken in the design of the wafer holder and shielding to prevent any metal contamination from reaching the wafer frontside since metals such as Au diffuse rapidly in Si and can reduce minority carrier lifetime. Gold is extensively used in GaAs-based ICs and actually has a lower resistivity than Al (≈ 2.2 $\mu\Omega$-cm). However, high diffusion coefficients in both Si and SiO_2 and poor adherence to oxide have discouraged its use as an interconnect in silicon microelectronics.

> Roughly 1% of the metal in a typical personal computer is gold, which is found in the individual chips, circuit boards, and connectors. This corresponds to about 0.01 oz of Au in a 20-lb computer, which pound-for-pound is three times the amount of gold in the average gold mine (1 ton of ore yields about 0.3 oz of Au).

The requirements of advanced high-speed chips such as microprocessors and telecommunications devices are driving IC packaging to smaller sizes and larger numbers of input/output (I/O) connections. The established method of packaging VLSI devices is to wire bond connections from the die to a leadframe; however, this method is less attractive for advanced ULSI chips due to the cost and performance of a densely-packed, wire-bonded assembly. For example, cross-talk between adjacent, closely-spaced wire bonds could degrade the intrinsic signal propagation speed expected from a high-speed chip. This is analogous to the RC time constant delay introduced by a multilevel metal interconnect stack due to capacitive coupling between adjacent metal lines in a given level.

Based on such concerns, IC industry roadmaps show a trend away from wirebonded leads and toward flip chip packaging technology in which the full area of the chip is directly attached to the package [9.60]. In most flip-chip applications, the wire bonds are replaced by small metallic bumps on the circuit side of the die. The connections from die to substrate are then made by flipping the chip circuit-side down onto the bonding pads and reflowing the underlying bump metallization. Flip chip technology has been used in microelectronic production for many years; however, its increased use in advanced packaging affords a market opportunity for PVD. For example, workers have reported [9.61] that PVD is an attractive way of forming a popular, under bump metallurgy (UBM) that is typically carried out in a batch mode using sequential e-beam evaporation of a chromium underlayer (Cr), a chromium-copper interlayer, and a Cu overcoat layer. Using an S-Gun™ sputter source with two independently powered, ring cathodes — one made of Cr and the other made of Cu — it was possible to deposit the same Cr/Cr-Cu/Cu stack within a single PVD module on a cluster tool. As with the e-beam evaporated stack, the PVD stack was a seamless structure without any interface evident between the layers.

References

9.1 B. Roberts, A. Harrus and R. L. Jackson, "Interconnect metallization for future device geometries," *Solid State Tech.*, 69–78 (Feb. 1995).

9.2. L. C. Feldman and J. W. Mayer, *Fundamentals of Surface and Thin Film Analysis*, North-Holland, New York, 1986.

9.3. C. G. Masi, "Semiconductor surface analysis: Fifty flavors, and counting," *R&D Magazine*, 24–26 (June 1996).

9.4. P. K. Chu and R. S. Hockett, "New ways to characterize thin films," *Semicond. Int.*, 142–146 (June 1994).

9.5. M. Dax, "Thin film metrology: Visible and UV based techniques," *Semicond. Int.*, 81–88 (Mar. 1996).

9.6. L. Savage, "National semiconductor metrology program travels roadmap to future needs," *Solid State Tech.*, 46–51 (Dec. 1996).

9.7. C. J. Morath, G. J. Collins, R. G. Wolf, and R. J. Stoner, "Ultrasonic multilayer metal film metrology," *Semicond. Int.*, 85–92 (June 1997).

9.8. W. Tsai, M. Delfino, J. A. Fair, and D. Hodul, "Temperature dependence of the electrical resistivity of reactively sputtered TiN films," *J. Appl. Phys.* **73**(9): 4462–4467 (1993).

9.9. K. Ramkumar, S. K. Ghosh, and A. N. Saxena, "Aluminum Based Multilevel Metallizations in VLSI/ULSICs," in *Handbook of Multilevel Metallization for Integrated Circuits*, pp. 97–201, S. R. Wilson, C. J. Tracy, and J. L. Freeman Jr., Eds., Noyes Publications, Park Ridge, NJ, 1993.

9.10. B. A. Movchan and A. V. Demchishin, "Study of the structure and properties of thick vacuum condensates of nickel, titanium, tungsten, aluminum oxide, and zirconium dioxide," *Fiz. Metal. Metalloved.* **28**(4): 653–660 (1969).

9.11. J. A. Thornton, "Influence of apparatus geometry and deposition conditions on the structure and topography of thick sputtered coatings," *J. Vac. Sci. & Tech.* **A11**(4): 666 (1974).

9.12. J. A. Thornton, "High rate thick film growth," *Ann. Rev. Mat. Sci.* **7**: 239–260 (1977).

9.13. Y. Pauleau, "Interconnect materials for VLSI circuits," *Solid State Tech.*, 101–105 (June 1987).

9.14. R. W. Bower, "Characteristics of aluminum-titanium electrical contacts on silicon," *Appl. Phys. Lett.* **23**(2): 99–101 (1973).

9.15. I. Krafcsik, C. J. Palmstrom, J. Gyulai, E. Colgan, E. C. Zingu, and J. Mayer, "Thin film interactions of Al and Al(Cu) on W and Ti," in *Proc. of the Electrochem. Soc.*, vol. 83-1, extended abstract no. 436, p. 681 (1983).

9.16. V. Hoffman, T. Wang, and D. Reedy, "Monitoring wafer temperature and wafer temperature uniformity in cluster tools using Ti/Al monitors," *Varian Associates Internal Technical Report* (July 1995).

9.17. R. Wilson, T. Hulseweh, and W. Krolikowski, "Development of a wafer level technique for monitoring and control of deposition temperature in high-vacuum physical vapor deposition technology," *J. Vac. Sci. & Tech.* **B15**(1): 122–126 (1997).

9.18. S. M. Rossnagel, "Sputtered atom transport processes," *IEEE Trans. on Plasma Science* **18**(6): 878–882 (1990).

9.19. S. K. Dew, "Processes and simulation for advanced integrated circuit metallization," Ph.D. thesis (University of Alberta, Edmonton, Canada, Fall 1992).

9.20. D. Liu, S. K. Dew, M. J. Brett, T. Janacek, T. Smy, and W. Tsai, "Experimental study and computer simulation of collimated sputtering of titanium thin films over topographical features," *J. Appl. Phys.* **74**(2): 1339–1344 (1993).

9.21. S.-Q. Wang, J. Schlueter, C. Gondran, and T. Boden, "Step coverage comparison of Ti/TiN deposited by collimated and uncollimated physical vapor deposition techniques," *J. Vac. Sci. & Tech.* **B14**(3): 1846–1852 (May/June 1996).

9.22. J.-E. Sundgren, B. O. Johansson, A. Rockett, S. A. Barnett, and J. E. Greene, "TiN_x ($0.6 < x < 1.2$): Atomic Arrangements, Electronic Structure, and Recent Results on Crystal Growth and Physical Properties of Epitaxial Layers," in *Physics and Chemistry of Protective Coatings*, pp. 95–115. W. D. Sproul, J. E. Greene, and J. A. Thornton, Eds., AIP, New York, 1986.

9.23. P. J. Martin, R. P. Netterfield, and W. G. Sainty, "Optical Properties of TiN_x produced by reactive evaporation and reactive ion beam sputtering," *Vacuum* **32**(6): 359–362 (1982).

9.24. I. Petrov, A. Myers, J. E. Greene, and J. R. Abelson, "Mass and energy resolved detection of ions and neutral sputtered species incident at the substrate during reactive magnetron sputtering of Ti in mixed Ar + N_2 mixtures," *J. Vac. Sci. & Tech.* **A12**(5): 2846–2854 (1994).

9.25. S. Berg, H.-O. Blom, T. Larsson, and C. Nender, "Modeling of reactive sputtering of compound materials," *J. Vac. Sci. & Tech.* **A5**(2): 202–207 (1987).

9.26. V. A. Koss, I. V. Ioffe, and A. Belkind, "Computational model of reactive sputtering," *J. Vac. Sci. & Tech.* **A11**(3): 701–703 (1993).

9.27. W. Tsai, J. Fair, and D. Hodul, "Ti/TiN reactive sputtering: Plasma emission, X-ray diffraction and modeling," *J. Electrochem. Soc.* **139**(7): 2004–2007 (1992).

9.28. P. Carlsson, C. Nender, H. Barankova, and S. Berg, "Reactive sputtering using two reactive gases, experiments and computer modeling," *J. Vac. Sci. & Tech.* **A11**(4): 1534–1539 (1993).

9.29. C. E. Wickersham and J. E. Poole, "The effect of target temperature on reactive sputtering target parameters," Tosoh SMD, Technical Note 9.006A.

9.30. M. Biberger, S. Jackson, G. Tkach, and L. Ouellet, "Adhesion and barrier layers for CVD tungsten and PVD aluminum filled contacts and vias of various aspect ratios," *Semicond. FABTECH* **1**: 197–203, 1994.

9.31. W. Tsai, D. Hodul, T. Sheng, S. Dew, K. Robbie, M. J. Brett, and T. Smy, "Variation of composition of sputtered TiN films as a function of target nitridation, thermal anneal, and substrate topography," *Appl. Phys. Lett.* **67**(2): 220–222 (1995).

9.32. T. Perera, "Antireflective coatings — An overview," *Solid State Tech.*, 131–136 (July 1995).

9.33. C. Bencher, C. Ngai, B. Roman, S. Lian, and T. Vuong, "Dielectric antireflective coatings for DUV lithography," *Solid State Tech.*, 109–114 (Mar. 1997).

9.34. M. Rocke and M. Schneegans, "Titanium nitride for antireflection control and hillock suppression on aluminum silicon metallization," *J. Vac. Sci. & Tech.* **B6**(4): 1113–1115 (1988).

9.35. S. Chen, C. L. Chen, and S. Tsou, "Sputtered TiN performance as an anti-reflective coating in backend sub-μm i-line lithography process," in *Proc. of the VLSI Multilevel Interconnection Conf.*, pp. 393–395 (1991).

9.36. D. Liu, S. K. Dew, M. J. Brett, T. Smy, and W. Tsai, "Compositional variations in Ti-W films sputtered over topographical features," *J. Appl. Phys.* **75**(12): 8114–8120 (1994).

9.37. R. N. Tait, W. Tsai, D. Hodul, D. Su, S. K. Dew, M. J. Brett, and T. Smy, "Compositional Variation of Sputtered Ti-W Thin Films on Topography: TEM/EDX Measurements and SIMBAD Simulations," in *Advanced Metallization and Interconnect Systems for ULSI Applications in 1995*, pp. 311–316, R. C. Ellwanger and S.-Q. Wang, Eds., Materials Research Society, Pittsburgh, PA, 1996.

9.38. D. B. Bergstrom, F. Tian, I. Petrov, J. Moser, and J. E. Greene, "Origin of compositional variations in sputter-deposited Ti_xW_{1-x} diffusion barrier layers," *Appl. Phys. Lett.* **67**(21): 3102–3104 (1995).

9.39. S. P. Murarka, *Silicides for VLSI Applications*, Academic Press, New York, 1983.

9.40. F. Mohammadi, "Silicides for interconnection technology," *Solid State Tech.*, 65–72 (Jan. 1981).

9.41. M.-A. Nicolet and S. S. Lau, "Formation and Characterization of Transition Metal Silicides," pp. 329–464, in *VLSI Electronics: Microstructure Science*, Vol. 6, N. G. Einspruch and G. B. Larrabee, Eds., Academic Press, New York, 1983.

9.42. J. Winneri, "Silicides for high density memory and logic circuits," *Semicond. Int.*, 81–86 (Aug. 1994).

9.43. K. Maex, "$CoSi_2$: An attractive alternative to $TiSi_2$," *Semicond. Int.*, 75–80 (Mar. 1995).

9.44. S. Ogawa, M. Lawrence, A. Dass, J. A. Fair, T. Kouzaki, and D. B. Fraser, "Epitaxial $CoSi_2$ film formation on (100) Si by annealing of Co/Ti/Si structure in N_2," *Materials Research Society Symp. Proc.* **312**: 193–198 (1993).

9.45. C. W. Kaanta, S. G. Bombardier, W. J. Cote, W. R. Hill, G. Kerszykowski, H. S. Landis, D. J. Pindexter, C. W. Pollard, G. H. Poss, J. G. Ryan, S. Wolff, and J. E. Cronin, "Dual damascene: A ULSI wiring technology," in *Technical Proc. of the VLSI Multilevel Interconnection Conf.*, pp. 144–152 (1991); see also P. Singer, "Making the move to dual damascene processing," *Semicond. Int.*, 79–82 (Aug. 1997).

9.46. W. M. Posadowski and Z. J. Radzimski, "Sustained self-sputtering using a direct current magnetron source," *J. Vac. Sci. & Tech.* **A11**(6): 2980–2984 (1993).

9.47. T. Asamaki, R. Mori, and A. Takagi, "Copper self-sputtering by planar magnetron," *Jpn J. Appl. Phys.* **33:** 2500–2503 (1994).

9.48. N. Hosokowa, T. Tsukada, and H. Kitahara, "Effect of discharge current and sustained self-sputtering," *Le Vide*, supplement 201 (*Proc. of the 8th Int. Vacuum Congress*, Cannes, France), pp. 11–14 (Sept. 1980).

9.49. Z. J. Radzimski, O. E. Hankins, J. J. Cuomo, W. P. Posadowski, and S. Shingubara, "Optical emission spectroscopy of high density metal plasma formed during magnetron sputtering," *J. Vac. Sci. & Tech.* **B15**(2): 202–208 (1997).

9.50. P. C. Zalm, "Some useful yield estimates for ion beam sputtering and ion plating at low bombarding energies," *J. Vac. Sci. & Tech.* **B2**(2): 151–152 (1984).

9.51. S.-Q. Wang, "Barriers against copper diffusion into silicon and drift through silicon dioxide," *Materials Res. Soc. Bull.* **19**(8): 30–40 (1994).

9.52. K.-H. Min, K.-C. Chun, and K.-B. Kim, "Comparative study of tantalum and tantalum nitrides (Ta_2N and TaN) as a diffusion barrier for Cu metallization," *J. Vac. Sci. & Tech.* **B14**(5): 3263 3269 (1996).

9.53. M. Takeyama, A. Noya, T. Sase, A. Ohta, and K. Sasaki, "Properties of TaN_x films as diffusion barriers in the thermally stable Cu/Si contact systems," *J. Vac. Sci. & Tech.* **B4**(2): 674–678 (1996).

9.54. B. Mehrotra and J. Stimmell, "Properties of direct current magnetron reactively sputtered TaN," *J. Vac. Sci. & Tech.* **B5**(6): 1736–1740 (1987).

9.55. M Biberger, S. Jackson, and E. Klawuhn, "Low pressure sputtering of copper and related barriers for seed layers and complete planarization," in *Technical Proc. of the ISSP Conf.*, June 1997; see also "Processing and integration of copper interconnects," R. L. Jackson, E. Broadbent, T. Cacouris, A. Harrus, M. Biberger, E. Patton and T. Walsh, *Solid State Tech.* 59–59 (March 1998).

9.56. S. M. Rossnagel, C. Nichols, S. Hamaguchi, D. Ruzic, and R. Turkot, "Thin high atomic weight refractory film deposition for diffusion barrier, adhesion layer and seed layer applications," *J. Vac. Sci. & Tech.* **B14**(3): 1819–1827 (1996).

9.57. I. Beinglass and M. Naik, "Advanced interconnect and via technology utilizing integration of CVD Al and PVD AlCu," *Technical Proc. of SEMICON-Korea* session 2, part II, pp. 73–78 (February 1997).

9.58. G. A. Dixit, A. Paranjpe, Q.-Z. Hong, L. M. Ting, J. D. Luttmer, R. H. Havemann, D. Paul, A. Morrison, K. Littau, M. Eienberg, and A. K. Sinha, "A novel 0.25 μm via plug process using low temperature CVD Al/TiN," *Technical Digest of the International Electron Devices Meeting*, pp. 1001–1003 (1995).

9.59. H. Horie, M. Imai, A. Itoh, and Y. Arimoto, "Novel high aspect ratio aluminum plug for

logic/DRAM LSIs using polysilicon-aluminum substitute (PAS)," *Technical Digest of the International Electron Devices Meeting*, pp. 946–948 (1996).
9.60. A. J. Babriarz, "Key process controls for underfilling flip chips," *Solid State Tech.*, 77–83 (Apr. 1997).
9.61. D. R. Marx, A. Lateef, and A. Clarke, "Sputtering deposits evaporation-quality UBM for flip-chip," *Semicond. Int.*, 97–102 (Mar. 1998).

Chapter 10 Process Modeling for Magnetron Deposition

Over the past 10 years, considerable effort has been put into the computer simulation of the sputter deposition process. The underlying goal has always been to perform a virtual experiment that predicts some aspect of the deposition process without actually conducting the physical experiment. This goal has been met in some cases, and yet each of the modeling approaches in use has various assumptions or shortcuts that make the results less than ideal. In this chapter we will present a brief overview of some of the major approaches to the modeling of sputter deposition. More detailed information is available in a recent companion volume [10.1] and other references at the end of this chapter.

There are three areas of interest for modeling in sputter deposition, and models of each region have been developed. At the cathode surface, it is interesting to examine the sputtering process as well as the emission or reflection of various other species. In the region between the cathode and the sample, modeling is appropriate to examine the transport and/or scattering of sputtered atoms as well as geometrical filtering effects due to distance or collimation. Finally, at the sample, modeling can be appropriate to examine the topographical nature of the deposition, the microstructure of the film, and other related physical characteristics.

Models for each of these three areas generally come in two forms: analytic and Monte Carlo. Many of the analytic models are two-dimensional, but some of the Monte Carlo models have been developed in three dimensions. In the first two regions (cathode and transport), the Monte Carlo models are atomic in nature: The model simulates single particles (mostly atoms) and deals with simple physical interactions. At the substrate location, all models of both types treat the film growth process in terms of large groups of atoms rather than single particles. The physics of these ensembles of particles is generally tied to atomic or traditional solid state physics, but it is clearly not a first-principles atomic interaction approach. This is mostly a computer-capacity problem, not a physics problem. As an example, a via with an aspect ratio of 3:1 and a diameter of 0.35 microns has a net volume of 9.5×10^{-14} cm^3. To fill this volume requires 5×10^9 atoms, plus a similar amount or more on the surrounding field region near the via — clearly a number that will strain most practical computer models. It is anticipated that the widespread advent of parallel-processing computers along with smaller feature dimensions should lead to a more rigorous, atomic approach to film modeling. As an alternative example, a feature 1000-Å wide (0.1 microns) is equivalent to roughly 400 atoms in width. If one were to consider the deposition of very thin films, such as diffusion barriers that were only a few tens of atoms thick, and also make use of the intrinsic symmetries of trenches, the computational problem becomes orders of magnitude more manageable, and it is possible to consider a first-principles, single-atom-based model.

10.1 Cathode Surface Models

The modeling of physical sputtering is generally done for the energy range of interest to magnetron sputtering by using a Monte Carlo approach. The most widely used program is known as TRIM (transport of ions in matter), and this program was originally developed to examine the world of energetic ion implantation (i.e., at much higher energies [10.2–10.4]). However, the basic physics at low energies are essentially the same, and TRIM has been widely used to model and predict the sputter yield, the energy dependence of the sputtered atoms, and the angular distribution of the sputtered atoms, as well as the probability, angular, and energy distributions of inert gas ions that are neutralized and reflected from the surface at moderate energies.

As an example, Eckstein has examined the case of 1 keV Ar^+ onto Ni. The energy distributions of the sputtered Ni atoms are shown in Fig. 10.1 at two different angles of incidence: 0° (normal incidence) and 75°, which is close to grazing [10.4]. The simulation shows that the majority of sputtered atoms have energies of under 10 eV and that a high-energy tail exists up to about 200 eV for normal incidence almost all the way to the incident energy (1 keV) for grazing bombardment.

This general approach to modeling has been quite successful at predicting the various physical properties of the emitted species from the cathode, and the program has been adapted to run on a conventional PC. The results from TRIM and its many progeny are often now accepted as legitimate data. This data can then be used without experimental confirmation is other aspects of the film deposition process.

The TRIM approach is also useful in exploring the physical processes that occur at the sample surface. For example, a fractal variant of TRIM has been used to model reflection of depositing metal atoms as a function of the inclination angle of the surface [10.5–10.6]. TRIM is also appropriate for use in modeling the resputtering of surface atoms during either inert gas ion bombardment of a growing film (e.g., bias sputtering) or energetic deposition, such as may occur during I-PVD (see Chapter 8). The results of the sputtering and ion-impact models are in the form of particle fluxes emitted from the surface with known energy and angular distributions. For

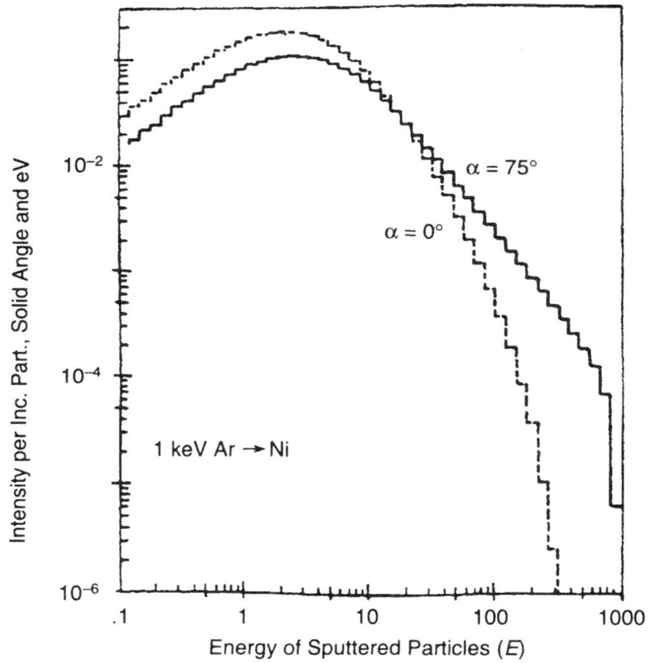

FIG. 10.1 Energy distributions of sputtered Ni atoms for Ar^+ bombardment at 0° (normal incidence) and 75° (near grazing) [10.4].

the case of sputtering and sputter emission, the TRIM-related models can predict the average yield, the average kinetic energy of the sputtered particles, and whether the angular distribution is a cosine distribution or some variant, such as an over-cosine distribution that has a larger relative fraction of the emitted particles emitted normal to the surface.

The TRIM models are also useful in exploring other relevant ion-impact processes. The most important to sputter deposition is the neutralization and energetic reflection of sputtering inert gas species. This occurs primarily in cases where the substrate atomic mass is greater than the atomic mass of the incoming ion. The incident ions, which are neutralized as they approach close to the cathode surface, are nearly elastically reflected from the surface and leave the surface region with a significant fraction (20–50%) of their initial energy. As these energetic reflected neutrals can impact the growing film, the energy and angular distributions of this flux can be critical to film properties.

10.2 Transport Modeling

Sputtered atoms emitted from the cathode must pass through the background gas of the chamber, as well as potentially such geometrical obstacles as a collimator, on their way to the sample. Collisions with background gas atoms will result in a general reduction in the kinetic energy of the sputtered atoms as well as a scattering or loss of directionality. Sputtered atoms that impact on physical surfaces such as a collimator or the chamber walls are obviously no longer part of the depositing flux at the sample, and their loss will alter the net deposition process. Although there have been some analytic approaches to modeling gas-phase transport [10.7–10.9], most groups have used a Monte Carlo approach [10.10–10.13]. In this scenario, single atoms of some known energy and direction are ejected from the cathode, and their collisions with various other atoms and surfaces are followed until the atoms are lost by condensing on a surface. After many tens of thousands of atoms are followed, it is possible to make some statistical assessments of the net transport trends.

An example of this approach given in Chapter 6 showed the average kinetic energy of sputtered Cu atoms as a function of chamber gas pressure (Fig. 10.2). The species of background gas is critically important due to its relative mass compared to the sputtered atom. Effectively, the higher the mass of the gas compared to the sputtered atom, the more rapidly the sputtered atom is scattered. Somekh showed, in a classic early study, how increasing the mass of the working gas results in a more rapid quenching of the initial kinetic energy as a function of distance from the source [10.12]. The results for the energetic fluxes are shown in Fig. 10.3. Similar studies have focused on the reduction in the net number of energetic particles along with the increase in diffusing, thermalized atoms. Early work by Motohiro and Taga shows both of these species as a function of the distance away from the source at constant pressure (Fig. 10.4) [10.13].

In addition to determining the average kinetic energy of the sputtered atoms, two other features can be explored by modeling: geometrical filtering and species-dependent transport. An example of geometrical filtering is shown in Fig. 10.5, where the flux is modeled at it passes through the cell of a collimator. As atoms impact the surface of the collimator, they stick and are removed from the transmitted distribution. The angular divergence of the transmitted flux is significantly narrowed for the transmitted flux, compared to the case without a collimator [10.14]. The deposition rate is also strongly reduced (Fig. 10.5). It is then appropriate in this case to explore the situation of a multicelled collimator. Since the walls of each

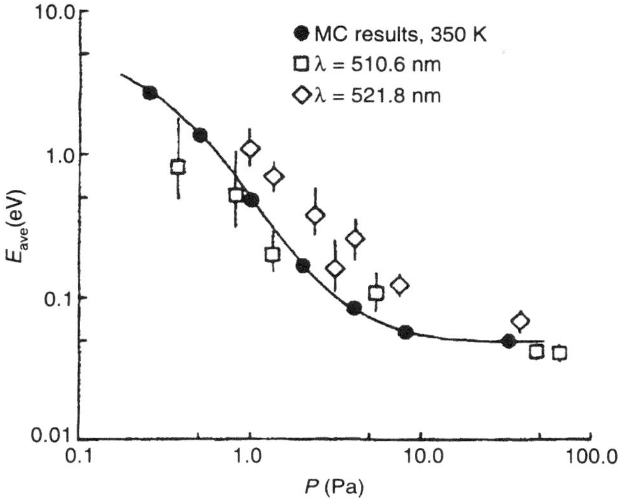

FIG. 10.2 Average kinetic energy of sputtered Cu atoms as a function of chamber pressure. The results of the Monte Carlo calculation are the filled circles. Experimental measurements are from [10.10] and are shown as open squares and diamonds [10.11].

cell or hole in a collimator collect material, a sample position close to the collimator should show the shadowing effect of these walls. This is shown in Fig. 10.6 for the case of a square collimator and a short sample distance below the collimator [10.15].

Since most transport codes are Monte Carlo in nature and track the dynamics of individual atoms, it is possible to compare the transport probability for atoms of different masses sputtered from the same cathode. This is most important in cases where the two components of an alloy target have widely differing masses, as, for example, would TiW and AlCu. It is also relevant in these cases that the heavier species has a mass that significantly exceeds the working gas (Ar, 40 AMU), and the lighter species is significantly less than the working gas. This means that the lighter species will be more readily scattered and stopped by the background gas, whereas the heavier species will tend to push its way through more efficiently. It would then be expected that the transport for the heavier species would be higher and the angular distribution less isotropic. The result will be a smaller angular divergence for the heavier species compared to the lighter one (Figure 10.7) [10.16].

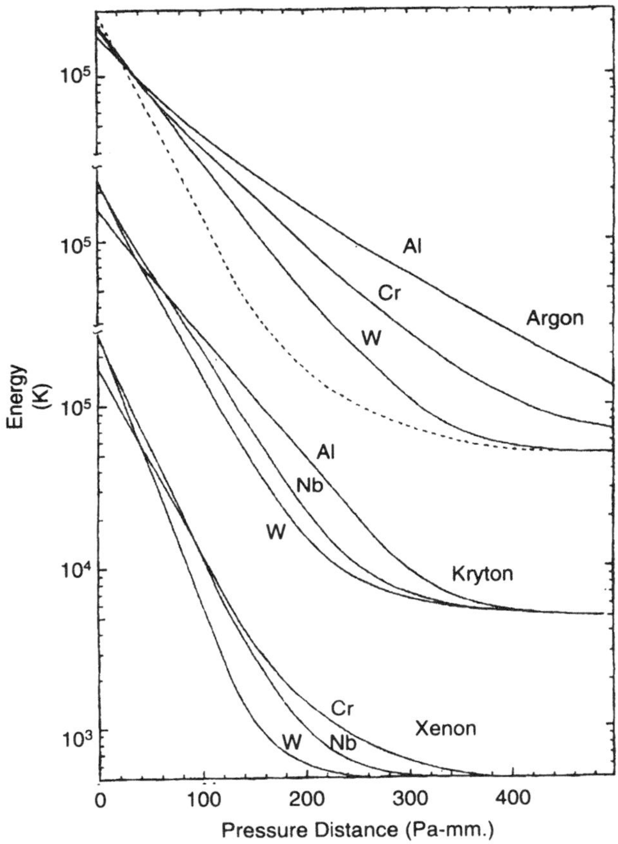

FIG. 10.3 Kinetic energy (in degrees K) for sputtered atoms as a function of the pressure–distance product [10.12].

The net result of this unequal transport and differing directionality for each species is potentially a pressure-dependent change in the film composition and, perhaps more importantly, a topography-dependent film composition. In this latter case, the film stoichiometry in deep features may be different from that on planar structures because of the difference in directionality of the two species. However, this is for transport alone. When other factors at the depositing film surface are taken into account, such as reflection or resputtering from the film surface, the compositional profile may change (see Section 9.7).

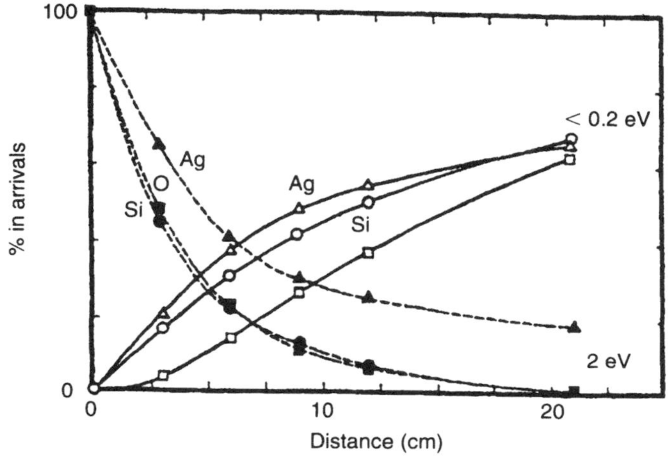

FIG. 10.4 Percentage of ballistic and thermalized (< 0.2 eV) particles as a function of distance [10.13].

10.3 The Wafer Surface

The physical processes that occur at the film and wafer surfaces are mostly well known and characterized. The fundamental problem with contemporary computer models of film deposition and growth is that, due to the limited, practical size of available computers, it has not really been possible to track each atom in the depositing film; there are simply too many. This has led to two general classes of models: one based on an approximation of a surface as a series of short line segments and another that uses aggregates of large numbers of atoms in the form of disks that then are used to construct the film. The line-segment approach is intrinsically two-dimensional and is thus limited to modeling surface features such as trenches, which are essentially infinite in one direction. The disk approach, also known as a molecular dynamics approach, can be extended to three-dimensions, although most published work seems to be in two dimensions.

The incoming flux to a surface can be approximated by an examination of the experimental conditions. For example, if the deposition was occurring at low pressure where the mean free path of the sputtered atoms was much longer than the cathode-to-sample distance, the arriving atoms

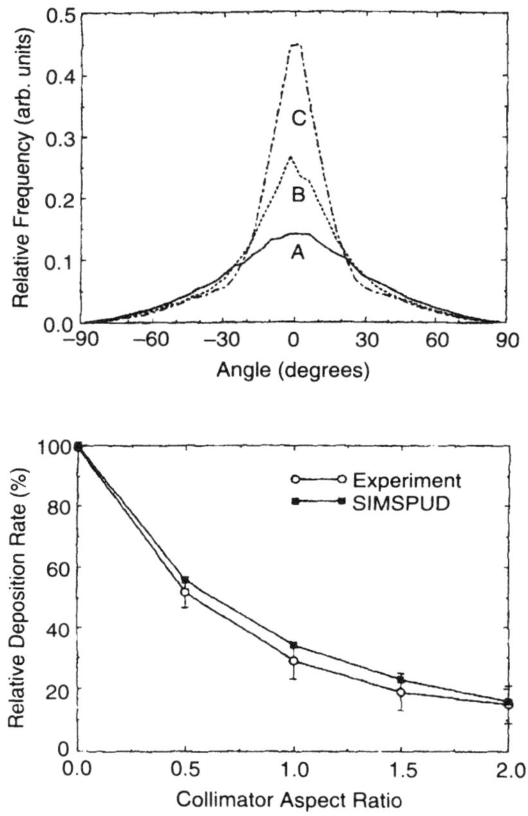

FIG. 10.5 (Top) Modeled angular divergence of sputter-deposited atoms for: A; no collimator, B; a collimator of aspect ratio 1:1, and C; a collimator of aspect ratio 2:1. (Bottom) The calculated net deposition rate compared to experimental values as a function of collimator aspect ratio [10.14].

would have the same angular and energy distributions that they left the cathode with. Increased pressure would result in a wider angular distribution as well as a reduced average energy.

Some models tie the transport part of the model to the film deposition part [10.16]. The transport model predicts a certain angular and energy distribution of atoms that arrive at the sample location, and this information is fed directly into the film deposition part of the model.

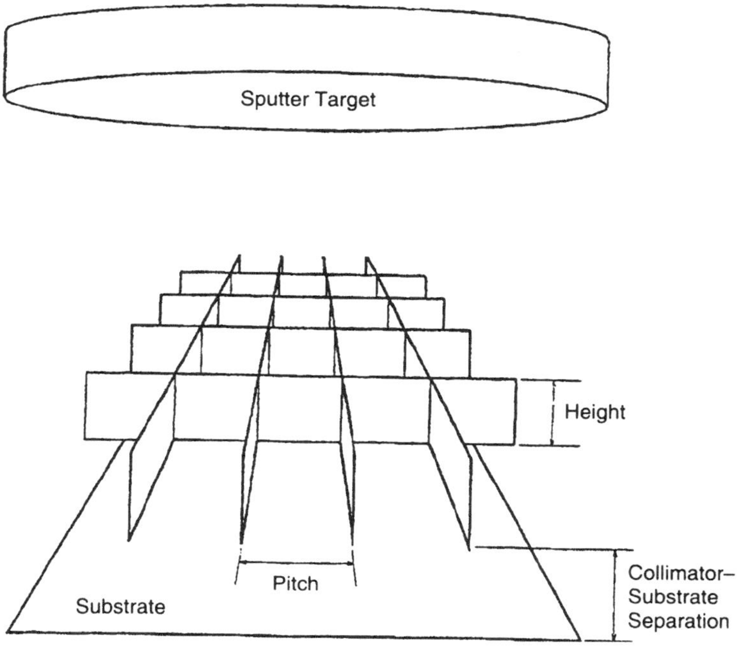

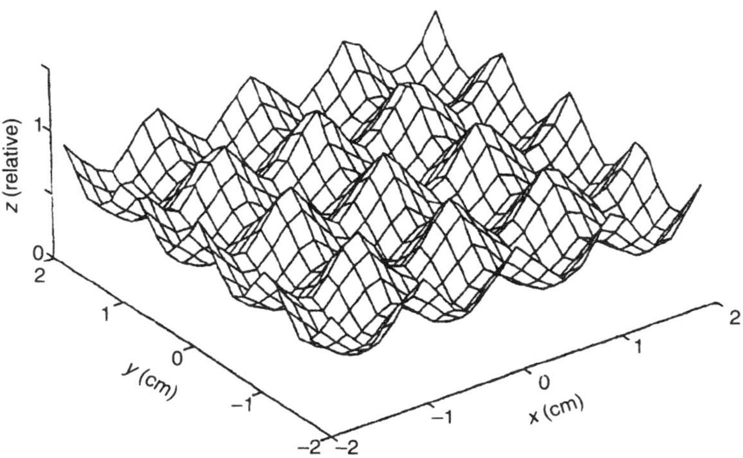

FIG. 10.6 Three-dimensional view of flux transmitted through a collimator at a short distance below the collimator [10.15].

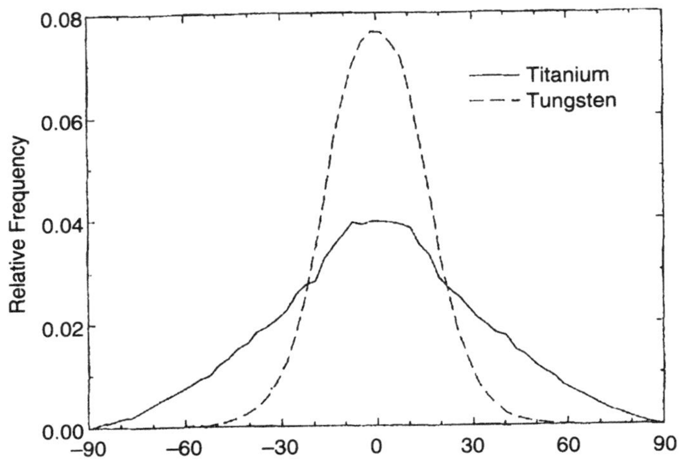

FIG. 10.7 The relative divergence of sputtered Ti and W species at 7 mTorr [10.16].

10.3.1 Line-Segment Models

This approach is intrinsically two-dimensional, which is appropriate for long trenches or perhaps circular holes. The substrate surface is broken up into many hundreds of short line segments. The length of the line segments is arbitrary, but typically models use a few hundred line segments to model a feature several microns wide, resulting in approximately 60 atoms per line segment. The deposition process occurs by randomly choosing a line segment and allowing some amount of film material to land on the segment: The amount is determined by consideration of the incident fluxes (perhaps a result of the transport models described above) as well as the position and angle of the line segment. In addition to deposition, it is also possible to include diffuse or specular reemission from the entire segment, which might be due to either nonsticking of the incident flux or evaporation of substrate material. Specular reemission would be most probable for grazing-angle deposition that might occur on the steep sidewalls of a feature. In many cases, it is also possible to consider resputtering of the film on the line segment. This might be due to the effects of inert gas bombardment, energetic neutral bombardment, or energetic, depositing metal species such as might be present during ionized PVD. The material resputtered from the film can be tracked and redeposited on nearby line segments. The line-segment approach uses bulk values for such things as the sputter yield, reflection probability, sticking coefficient, etc., which is an obvious limitation from a physics point of view.

The overall film topography is then determined by tracking these processes at the randomly chosen deposition/etching sites over a period of time. In general, it is also necessary to develop certain criteria for dealing with discontinuities in the film profile, and the details of these effects are described in many articles [10.17–10.20].

As an example of the two-dimensional, line-segment approach to deposition modeling, Fig. 10.8 shows the sequence of conventional sputter deposition into three trenches varying in aspect ratio from 0.5 to 2.0 [10.21]. The mostly isotropic nature of the deposition can be seen, particularly with the higher aspect ratio feature, in the build-out of the upper sidewalls of the deposit and the eventual closure and void formation. This modeling approach has also been used to combine the effects of a neutral, isotropic deposition with an ionized, directional deposition. This is shown in Fig. 10.9, in which the relative ionization is varied from 33% to 67% of the total depositing flux [10.19].

A final example of this approach explores the effect of resputtering of the depositing film and the subsequent deposition of the resputtered material (Fig. 10.10). This is particularly important in I-PVD cases where the depositing metal ion energy has a measurable sputter yield (> tens of eV). As can be seen from the figure, increased depositing ion energy results in significant resputtering of the deposited film. This forms bevels on the upper sidewalls of the trenches, and the resulting redeposition of this sputtered material on the opposite sidewall tends to close off the feature, forming a void. (Similar studies for resputtered liners were shown in Section 8.3.)

The line-segment approach to modeling of the depositing film has several limitations, although it can be very useful for the prediction of topography in trenches. The model is intrinsically two-dimensional, which means it is inappropriate for vias or complicated geometries (vias under lines, etc.) that are intrinsically three-dimensional. Since it is not really an atom-based model, it is not possible to introduce physical effects such as surface diffusion, grain formation, or film structure. For those features, it is necessary to go to the molecular dynamics approach.

However, even with its limitations, line-segment modeling has been found to be very useful in predicting the topography development during complicated deposition conditions and can be used as a diagnostic tool either to understand and calibrate the degree of directionality and/or resputtering, or to imply other physical properties, such as effective sticking coefficients [10.22]. In the former case, the use of this type of model in

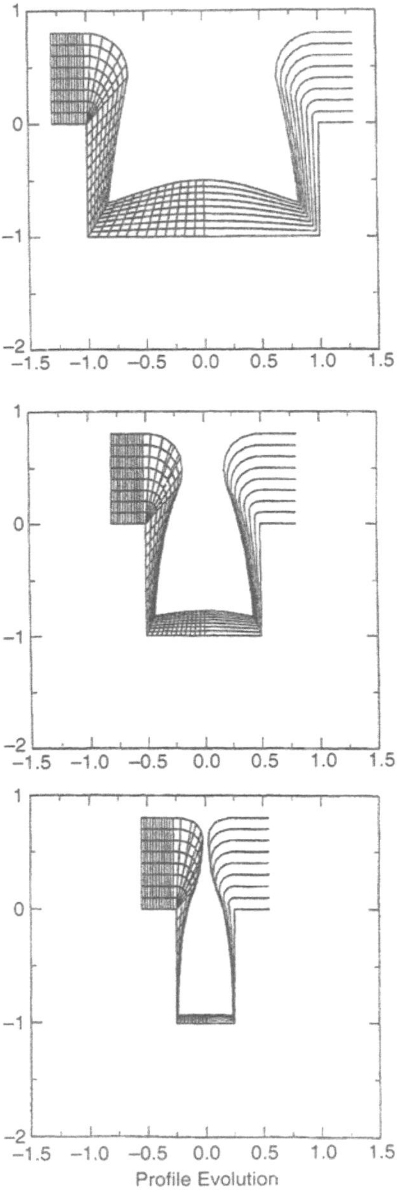

FIG. 10.8 Line-segment model showing conventional magnetron deposition (cosine flux) onto trenches of aspect ratio 0.5 to 2.0 [10.21].

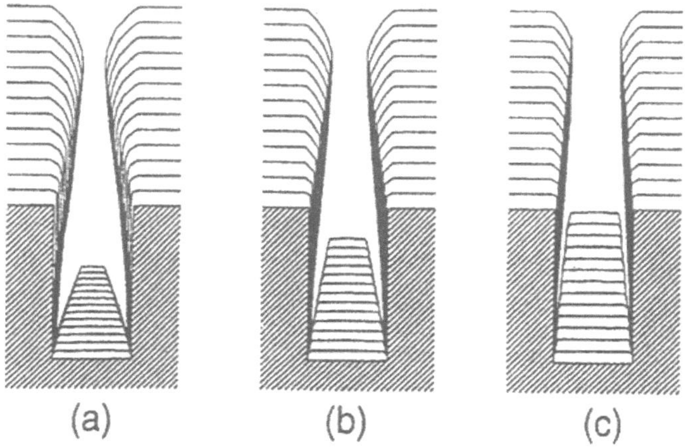

FIG. 10.9 Model of ionized PVD deposition for the case of (left-to-right) 33% ions and 67% neutrals, 50% ions and 50% neutrals, and 67% ions and 33% neutrals [10.19].

conjunction with experimental samples has been used to "measure" such plasma properties as the degree of ionization of the depositing atoms or the functional density and temperature of the neutral species.

10.3.2 Molecular Dynamics Film Growth Models

This class of models approximates both the arriving flux and the already deposited film as an array of small disks. Pioneering work by Mueller at CSIRO mapped out a wide variety of physical deposition phenomena using

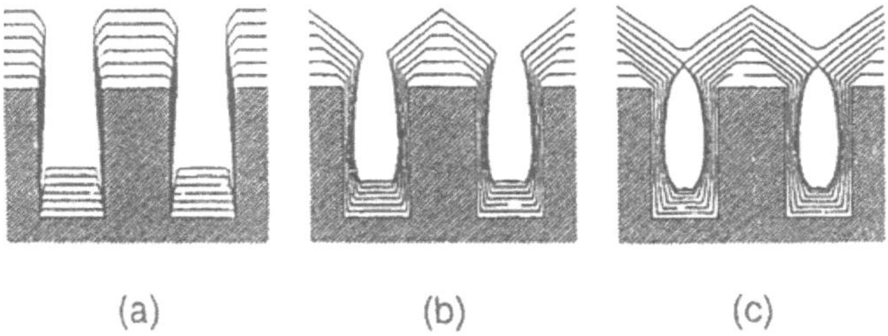

FIG. 10.10 Modeling of increased ion energy, resulting in a higher etch-to-deposition ratio for I-PVD deposition [10.19].

this approach [10.23]. An example of the microscopic surface is shown in Fig. 10.11 [10.16]. In this case, a new disk is incident on an existing film structure. Potential sites for adsorption are shown in the figure adjacent to gray disks, which represent nearest potential neighbors. The incident disk hits the surface and relaxes into some nearby site based on a calculation of the lowest surface energy.

One of the intrinsic advantages of this approach is that it allows the development of a physical structure to the film. If, for example, the substrate/film surface is cold and no rearrangement due to surface diffusion occurs, the film may show a very columnar structure. As the substrate temperature is increased, adatom diffusion occurs and the resulting grains are much wider and less columnar. (This was described in detail in Section 7.2.1 and Fig. 7.3.)

This approach can also explore deposition at angles other than 90°. Fig. 10.12 [10.16] shows the tilting of the intrinsic columnar structure of a cold deposition as the incident flux is inclined by 30 and 60°. This could also readily show the effect of an asymmetric deposition into trench features, similar to what might occur at a deep trench near the edge of a wafer using a long-throw, low-pressure deposition [10.16, 10.22]. (This was described in Section 6.2.) A related example shows the film coverage and structure for deposition over a step when the depositing flux is 5° from normal. This would be similar to deposition in the middle to outer regions of a 200-mm wafer during a long-throw deposition. The simulations show

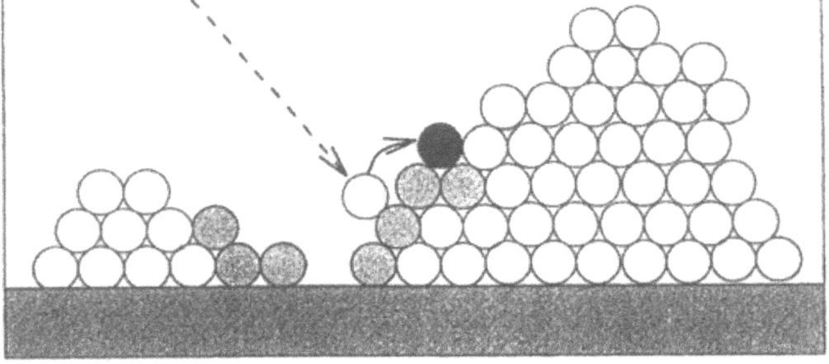

FIG. 10.11 Schematic of deposition using a surface composed of disks approximating groups of atoms [10.16].

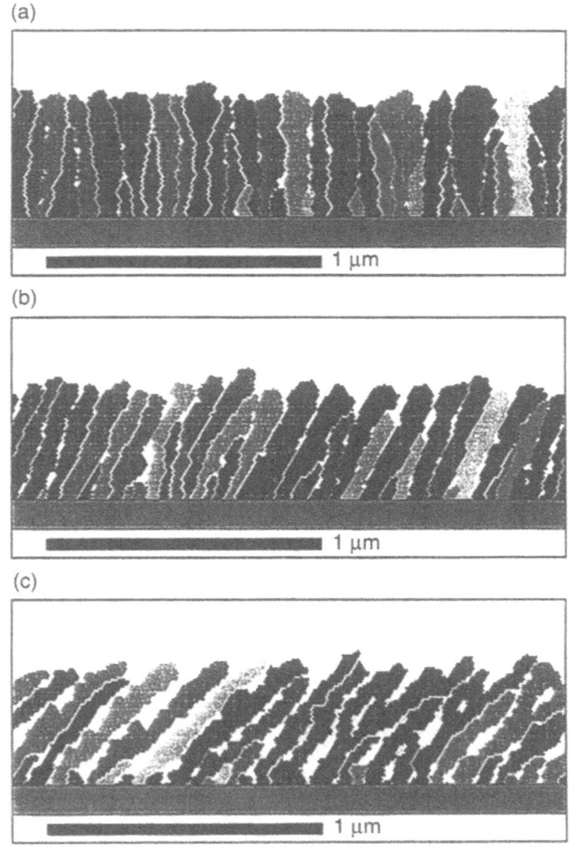

FIG. 10.12 Deposition at 0, 30, and 60° angles of incidence (from the surface normal) with a fixed diffusion length of 0.02 microns (similar to Fig. 7.3d) [10.16].

a shadowing effect on the side of the feature away from the depositing flux (Fig. 10.13) [10.16].

This same approach can be used to examine the topography of deposits into deep features such as trenches. One of the first things that is seen with these simulations is the low density of columns on the steep sidewalls within a feature (Fig. 10.14a). This is consistent with experimental observations (see, for example, Fig. 6.20) of up-tilted columns of significantly lower density than on the planar areas. From a functional point of view,

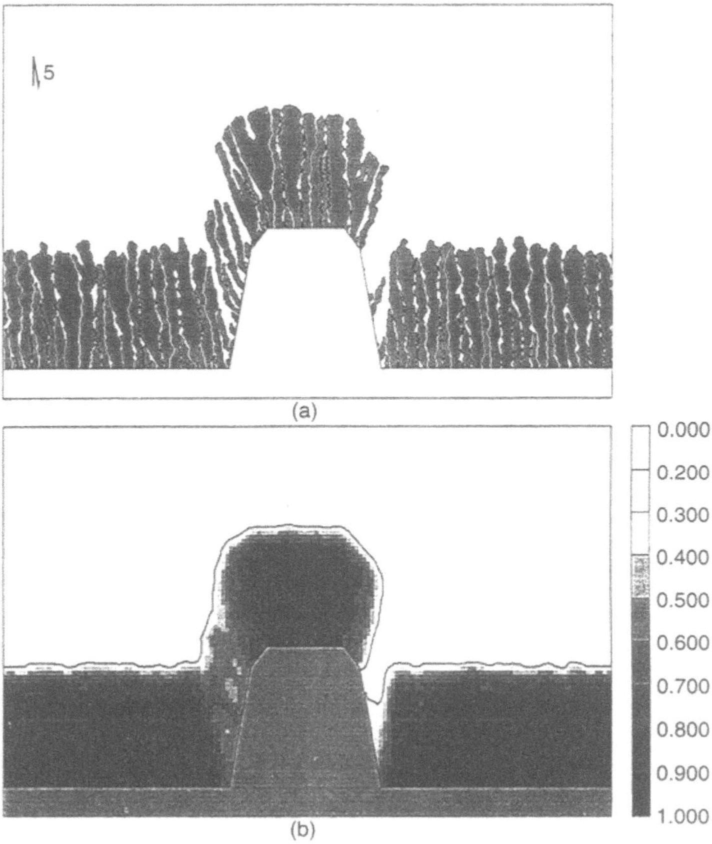

FIG. 10.13 Microstructure (a) and film density (b) for deposition of a thin film over a step with a depositing flux which is 5° from normal incidence [10.16].

this sort of low-density coverage is not desirable for diffusion barrier or seed layer applications. The model can also be extended to the asymmetry intrinsic to long-throw deposition at the wafer edge. Figures 10.14a and 10.14b show the observed film structures at the wafer center (Fig. 10.14a) or perpendicular to the wafer edge at the outer edge of the wafer (Fig. 10.14b). Experimentally, similar results were seen in Fig. 6.8.

A related characteristic of the molecular dynamics approach is the development of a grain structure to the deposited film. This would typically occur in an elevated-temperature deposition process where there

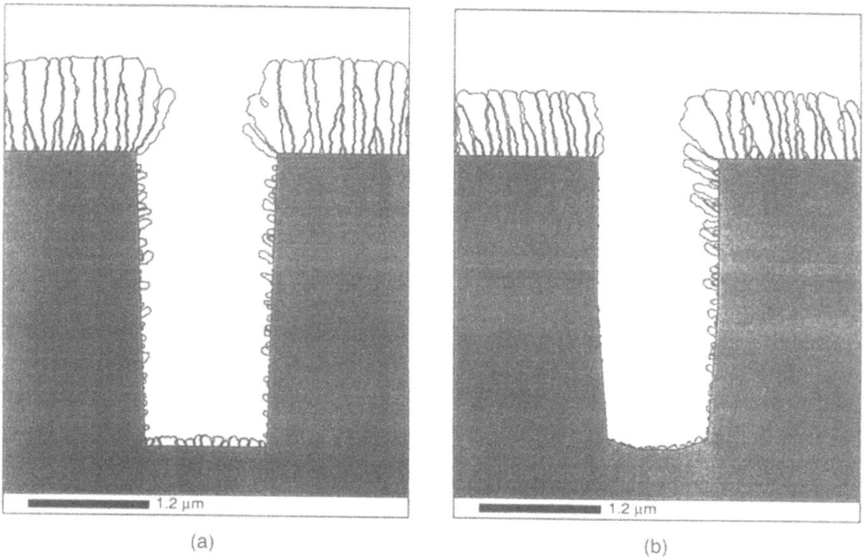

FIG. 10.14 Simulation using SIMBAD of deposition (a) near the wafer center and (b) near the wafer edge for long-throw deposition [10.16].

was significant surface diffusion and mobility of the deposited species (see Section 7.3). The modeling allows predictions to be made of the number of grains formed as well as related features such as the dependence of the reflow time on the number of grains present. In effect, with more grains present, the surface diffusion is slowed down somewhat because of the inhibiting presence of extra grain boundaries, which are sinks for depositing atoms [10.24].

10.3.3 Monte Carlo Models

Monte Carlo modeling techniques can also be used to help describe the physical properties of the deposited film, including the density and the stress. As a result, for example, it is possible to correlate the film stress with the depositing kinetic energy of the film atoms [10.24] (Fig. 10.15). This shows trends similar to those observed experimentally: At low sputtering pressures and presumably high kinetic energies for the nonthermalized depositing atoms, the films tend to be in compressive stress. This is

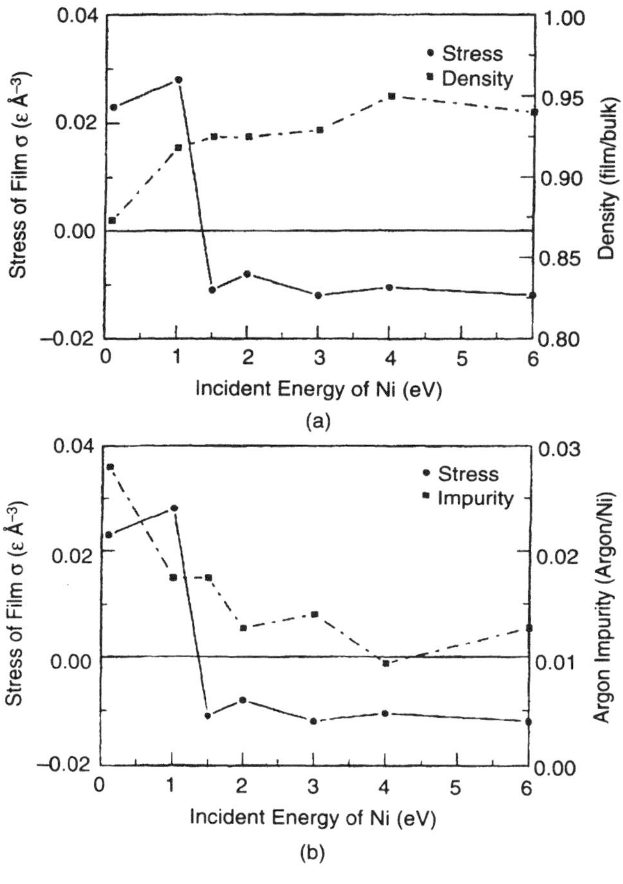

FIG. 10.15 (a) The combined effect of ion bombardment with 50-eV Ar and gas impurities on the stress and density of a condensing Ni film as a function of the incident energy of the Ni. (b) The stress and impurity levels for the same case. The ion-to-atom ratio was 0.76 and the gas-to-atom flux ratio was 0.134 [10.24].

partly due to the defects introduced by the more energetic deposition (higher kinetic energy, more energetic bombardment by reflected neutrals) and may also be related to incorporated inert gas impurities in the film.

Another option available in this type of modeling is the examination of the chemical nature of the film. One example of this is the case of alloy sputtering described in Chapter 6, in which the transport of one of the species was different from the other. In general, the heavier species was scattered less during transport and was more directional than the lighter, more easily scattered species. As a result, the composition in a deep feature can be shown to be dependent on the depth into the feature, becoming enriched in the heavier, more directional species at the bottom [10.26].

It is also possible to track the reaction level during reactive sputtering with a Monte Carlo model. Figure 10.16 shows the relative nitrogen level in a deep trench for the reactive deposition of TiN from a nonnitrided Ti target. In this figure, the darker regions are close to 1:1 stoichiometry (TiN), but the lighter areas at the bottom of the trench are clearly under-nitrided, which may have implications on the quality of the diffusion barrier [10.16].

10.4 Conclusion

Process modeling can be a versatile way to perform a virtual experiment, modeling a sputtering, transport, or deposition process without ever turning on a sputtering system. Commercial versions of several of these

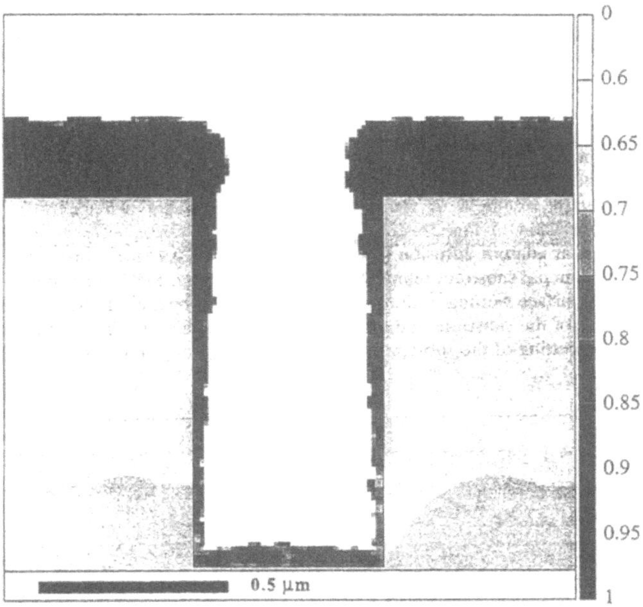

FIG. 10.16 The SIMBAD-predicted chemical composition for the case of reactive deposition of TiN into a deep trench. The target in this case was nonnitrided. The legend at the right side of the figure indicated the relative nitrogen concentration.

models also exist, either for sale to the end user or as a service, so that the user can gain access to the prediction capabilities without generating a new model. The brief description and examples in this chapter are a tiny fraction of the available, published studies in this technology. For example, a very similar modeling approach can be used to describe chemical transport and reactions at a wafer surface, modeling a CVD or PECVD deposition process. Also, close examination of some of the subtle features in the molecular dynamics approach can lead to insight into physical characteristics of the deposit such as stress or defect density [10.24–10.26].

At the time of this chapter (1998), progress is also underway with an even more physical approach to film modeling, one that tracks the arrival and behavior of individual atoms rather than ensembles of atoms. The low-energy physics of these atoms is relatively well understood, but it has simply not been practical using available computers to develop single-atom-based film models. However, with the much wider availability of parallel computing coupled with the increasingly smaller dimensions of relevant film structures, a full-atomic three-dimensional model should be developed soon. As described earlier, as trench widths move down toward 1000 angstroms (0.1 micron) — characteristic of the 4 or 16 gigabit DRAM generations — the trench is now only 400 atoms or so wide. Therefore, it is much more conceivable that a computer will be able to model the development of films in three-dimensional features of this size range with available computing resources.

References

10.1. S. M. Rossnagel, *Modeling of Film Deposition for Microelectronic Applications*, Academic Press, San Diego, 1996.

10.2. J. P. Biersack and L. G. Haggmark, "A Monte Carlo computer program for the transport of energetic ions in amorphous targets," *Nucl. Instrum. & Meth.* 174: 157 (1980).

10.3. J. P. Biersack and W. Eckstein, "Sputtering studies with the Monte Carlo program TRIM.SP," *Appl. Phys. A* **34**:73 (1984).

10.4. W. Eckstein, "Energy distributions of sputtered particles," *Nucl. Instrum. & Meth. in Phys. Res.* **B18**: 344 (1987).

10.5. D. Ruzic, "Fundamentals of Sputtering and Reflection," in *Handbook of Plasma Processing Technology*, pp. 70–87 S. M. Rossnagel, J. J. Cuomo, and W. D. Westwood, Eds., Noyes Publications, Park Ridge, NJ, 1990.

10.6. D. Ruzic, "The effects of surface roughness characterized by a fractal geometry on sputtering," *Nucl. Instrum. & Meth. in Phys. Res.* **B47**: 118 (1990).

10.7. W. D. Westwood, "Calculation of deposition rates in diode sputtering systems," *J. Vac. Sci. & Tech.* **15**: 1–9 (1978).

10.8. I. Abril, A. Gras-Marti, and J. A. Valles-Abarca, "Energy transfer processes in glow discharges," *J. Vac. Sci. & Tech.* **A4**: 1773–1778 (1986).

10.9. J. A. Valles-Abarca and A. Gras-Marti, "Evolution towards thermalization and diffusion of sputtered particle fluxes: Spatial profiles," *J. Appl. Phys.* **55**: 1370–1378 (1984), and J. A. Valles-Abarca and A. Gras-Marti, "Slowing down and thermalization of sputtered particle fluxes: Energy distributions," *J. Appl. Phys.* **54**: 1071–1075 (1983).

10.10. G. M. Turner, I. S. Falconer, B. W. James, and D. R. McKenzie, "Monte Carlo calculation of the thermalization of atoms sputtered from the cathode of a sputtering discharge," *J. Appl. Phys.* **65**: 720 (1986).

10.11. L. T. Ball, I. S. Falconer, D. R. McKenzie, and J. M. Smelt, "An interferometric investigation of the thermalization of copper atoms in a magnetron sputtering discharge," *J. Appl. Phys.* **59**: 720 (1986).

10.12. R. E. Somekh, "The thermalization of energetic atoms," *J. Vac. Sci. & Tech.* **A2**: 1285–1291 (1984).

10.13. T. Motohiro and Y. Taga, "Monte Carlo simulation of the particle transport process in sputter deposition," *Thin Solid Films* **112**: 161–173 (1984).

10.14. D. Liu, S. K. Dew, M. J. Brett, T. Janacek, T. Smy, and W. Tsai, "Experimental study and computer simulation of collimated sputtering of Ti thin films over topographical features," *J. Appl. Phys.* **74**: 1339–1344 (1993).

10.15. C. Sorlie, M. J. Brett, S. K. Dew, and T. Smy, "Advanced process simulation of metal film deposition," *Solid State Tech.*, 101 (June 1995).

10.16. M. J. Brett, S. K. Dew, and T. J. Smy, "Thin Film Microstructure and Process Simulation Using SIMBAD," in *Modeling of Film Deposition for Microelectronic Applications*, S. M. Rossnagel, Ed., Academic Press, San Diego, 1996.

10.17. S. M. Rossnagel and R. S. Robinson, "Monte Carlo model of topography development during sputtering," *J. Vac. Sci. & Tech.* **21**: 790 (1982).

10.18. S. Hamaguchi, M. Dalvie, R. T. Farouki, and S. Sethuraman, "A shock-tracking algorithm for surface evolution under reactive ion etching," *J. Appl. Phys.* **74**: 5172 (1993).

10.19. S. Hamaguchi and S. M. Rossnagel, "Simulations of trench-filling profiles under ionized magnetron sputter metal deposition," *J. Vac. Sci. & Tech.* **B13**: 183 (1995).

10.20. S. Hamaguchi and S. M. Rossnagel, "Liner conformality in ionized metal sputter deposition processes," *J. Vac. Sci. & Tech.* **B14** (1996).

10.21. S. Hamaguchi, unpublished work, IBM Research, 1995.

10.22. S. M. Rossnagel, C. A. Nichols, S. Hamaguchi, D. Ruzic, and R. Turkot, "Thin, high atomic weight refractory film deposition for diffusion barrier, adhesion layer and seed layer applications," *J. Vac. Sci. & Tech.* **B14**: 1819 (1996).

10.23. K.-H. Mueller, "Stress and microstructure of sputter deposited thin films: Molecular dynamics investigations," *J. Appl. Phys.* **62**: 1796–1799 (1987).

10.24. C.-C. Fang, V. Prasad, R. V. Joshi, F. Jones, and J. J. Hsieh, "A Process Model for Sputter Deposition of Thin Films Using Molecular Dynamics," in *Modeling of Film Deposition for Microelectronic Applications*, S. M. Rossnagel, Ed., Academic Press, San Diego, 1996.

10.25. D. Liu, S. K. Dew, M. J. Brett, T. Smy, and W. Tsai, "Compositional variations in Ti-W films sputtered over topographical features." *J. Appl. Phys.* **75**: 8114 (1994).

10.26. T. S. Cale and V. Mahadev, "Feature Scale Transport and Reaction During Low Pressure Deposition Processes," in *Modeling of Film Deposition for Microelectronic Applications*, S. M. Rossnagel Ed., Academic Press, San Diego, 1996.

Chapter 11 Sputtering Targets

Simply stated, PVD is the controlled erosion and transfer of material from a target to a substrate by means of the sputtering process. The sputtering source initiates the process and provides the needed control to turn bulk targets into thin films suitable for microelectronic applications. Such films must be deposited economically (i.e., low cost-of-ownership) and with tight tolerances on film uniformity, chemical purity, microstructure, and in-film particles. Each of these film attributes is in turn strongly influenced by the sputtering target itself — often in an interrelated way. For example, to deposit uniform films on the wafer, a planar magnetron source is often engineered to give relatively high erosion at the edge of the target, which compensates for its finite geometric size. This in turn impacts target utilization and PVD cost-of-ownership since the entire target must be replaced when the preferentially eroded region at the edge reaches the target backing plate. As another example, Al_2O_3 inclusions can affect the purity of an Al target, and these same insulating inclusions can give rise to electrical arcing at the target surface with resulting particle generation. As a result, targets are no longer regarded as passive elements in a PVD system, and increasing attention is being placed by both target manufacturers and PVD users on target purity, target metallurgy (e.g., grain size and crystallographic orientation), and the design of targets tailored for both a given PVD cathode design and process application.

The focus of this chapter will be predominantly on metallic sputtering targets. These can be used either for deposition of metals when sputtered with an inert gas species or for the reactive deposition of nitrides or oxides (see Chapter 3 for a discussion of reactive PVD). Compound targets, such as nitrides or oxides, are rarely if ever used for the deposition of most semiconductor materials such as TiN, TaN, WN, SiO_2, TiO_2, etc., for several reasons. For example, these targets are generally insulating, which requires the use of RF power. This adds significant complexity to the sputter tool in the form of RF matchboxes, tuning circuits, additional electromagnetic shielding, filtering of other circuits, etc. Another primary constraint is that oxide targets, and to some extent nitride targets, are difficult to bond and handle. They are brittle and have poor thermal conductivity, which makes them susceptible to cracking and structural failure — particularly when used under high-power, high-deposition rate conditions.

The primary exceptions to the metals-only approach to PVD are piezoelectric materials and superconducting thin films — both of which are rarely, if ever, extended to production manufacturing systems — as well as a class of complex oxides with very high dielectric constant that are beginning to be applied to IC applications. Members of this latter class include barium strontium titanate (BST = $Ba_xSr_{1-x}TiO_3$), strontium bismuth tantalate (SBT = $Sr_xBi_{1-x}TaO_3$) and lead zirconium titanate (PZT = $Pb_xZr_{1-x}TiO_3$), which can have dielectric constant value k of several hundred compared to $k \approx 4$ for SiO_2 and $k = 1$ for vacuum. The primary interest in these high-k materials is their use as the dielectric of a small, planar capacitor that could then replace more complex designs such as the deep trench capacitors or multilevel capacitors currently used to store electrical charge (i.e., to store digital information) on advanced memory chips such as a 1 Gb DRAM. Since this application involves a very thin, planar film (< 1000 Å) with only moderate step coverage, PVD is a suitable deposition method. On the other hand, CVD methods continue to be developed for high-k film deposition and may offer a more cost-effective solution as well as extendability to very advanced devices (> 4 Gb DRAM) with more stringent step coverage needs.

11.1 Target Fabrication

A sputtering target for microelectronic application consists of an appropriately shaped and dimensioned slab of target material, which is then attached (typically bonded) to a simple backing plate or to a more complex mechanical assembly for use with a given supplier's sputtering cathode. Figure 11.1 shows a variety of targets and assemblies used in advanced IC production. The target might be formed or machined into a large-diameter circular disk to be used with a company-specific planar magnetron cathode such as the AMAT Durasource™, MRC RMX™, or Varian Quantum™. Alternatively, the target could consist of a single piece with complex topography like the ring source of a conical magnetron (e.g., Varian ConMag™) or separate pieces intended for a multiple-cathode source such as Sputtered Films' dual-cathode S-Gun™.

The lateral dimension of the target is primarily set by the need to uniformly coat a stationary large-diameter wafer, which favors large targets, while at the same time minimizing deposition of expensive target material on anything other than the wafer, which favors small targets. The thickness of the target is primarily set by the desire to maximize the time for any part of the target to erode to the backing plate without needlessly wasting material since, regardless of starting thickness, the eroded front surface of the target will eventually become sufficiently nonuniform that film uniformity is out of spec. These considerations applied to planar magnetron coating of 6- and 8-inch wafers have led to the use of circular targets with diameter

FIG. 11.1 A variety of PVD target shapes — planar, conical, and ring-shaped — are used in advanced microelectronic manufacturing and depend on cathode design (courtesy of Tosoh, SMD, Grove City, OH).

≈ 12 inches and thickness up to ≈ 0.5 inch, excluding a possible backing plate that might add another 0.25 inch of thickness to the assembly.

Regarding target diameter in the future, one supplier (C. Wickersham of Tosoh SMD) has plotted the historical increase in target diameter ϕ_{target} with time as wafer diameter has increased in production from ϕ_{wafer} = 4 inch to 200 mm. He finds that these two diameters are strongly correlated and track each other through an empirical relationship of the form $\phi_{target} = 1.46\ \phi_{wafer} + 35$ mm. Using this formula, we would expect the diameter of targets needed to coat next-generation 300-mm wafers to be ≈ 475 mm (18–19 inch).

PVD targets are machined from a solid block of material; how one fabricates this starting block depends on the specific metallurgy in question. For example, Al alloys (Al-Si, Al-Cu, Al-Si-Cu) are made by first casting the metal and then working it by rolling or forging with various annealing treatments. Ti targets are manufactured in much the same fashion as the Al alloys; however, Ti-W targets utilize powder metallurgy. In this case, powders of W and Ti or TiH_2 are mixed and then compacted under high pressure and high temperature (ranging from ≈ 750–1500°C). The result is a lump of metal at near 100% of bulk density that is then machined to the final target shape and surface finish. For example, hot isostatic pressing (HIP) at 800°C and 400 MPa can be used to produce W-10 wt % Ti targets with greater than 99% of bulk density [11.1]. Depending on process conditions, the Ti-W target may contain a single phase or two phases, with the single phase being very brittle and difficult to work.

11.2 Target Cooling

DC magnetrons consume a great deal of energy relative to what they deliver to the wafer and have been described somewhat cynically as an expensive wafer heating device. For example, planar magnetron deposition of Al at 1 μm/min from a 700-cm^2 sputter target might have a specific deposition rate (i.e., film deposition rate normalized to magnetron power) of $\approx$ 15 Å/sec-kW, requiring a cathode power of 12 kW (e.g., cathode current of I_{dc} = 25 A and discharge voltage of V_{dc} = 500 V). As calculated in Section 5.5.4, the flux of Al atoms to the wafer represents $\approx$ 0.2 W/cm^2 or $\approx$ 60 W over a 200-mm wafer. Even assuming that several times this power is delivered to nonwafer surfaces such as the PVD shields, the total represents only a small fraction of the 12 kW applied to the PVD source. This is not to say that the power delivered into the process chamber is unimportant. For example, it can lead to large temperature rises of the wafer and other exposed chamber parts. However, from the standpoint of energy use, almost all of the magnetron discharge power (the product of I_{dc} and V_{dc}) is consumed by the cathode, which — for 12-kW and conventional diameter targets — represents > 10 W/cm^2. The relatively inefficient use of electrical energy in DC magnetron sputtering is due to two effects: (1) the nature of physical sputtering and (2) the nature of the diode plasma system.

As described in Chapter 2, physical sputtering is a momentum transfer process from the incident, energetic ion to the atomic lattice of the target. Under somewhat random collisional processes, one or more of these target atoms is ejected due to the bombarding particles. A relevant example is the case of Ar^+ sputtering of AlCu. In this case the operating or discharge voltage of the cathode is perhaps 500 V, which imparts 500 eV of kinetic energy to the incident Ar ion. The sputter yield of 500 eV Ar^+ on AlCu is about 1.0, and the average kinetic energy of the ejected Al atom might be 10 eV. Therefore, from a simply particle point of view, the emission process is only $(1.0 \times 10)/500 = 2\%$ efficient in terms of returned energy to the discharge. However, there are also additional processes to consider. The secondary electron yield for Ar^+ on Al might be $\approx$ 5%, and each secondary electron picks up the full discharge potential as it returns back into the plasma. There are also other minor sources of energy from the cathode:

thermal blackbody radiation from the slightly heated surface, reflected neutrals (which are a very small effect for Ar^+ on Al), and some optical emission. The net result is that only 10% or so of the incident energy returns from the cathode in the form of energetic particles or photons; the remaining 90% is absorbed as heat by the cathode and must be taken away by water cooling of the cathode (air cooling is not sufficient to deal with the heat load on production DC magnetrons but can be used in smaller, research-oriented sources).

Target cooling, then, is a key engineering problem for magnetron sputtering. A standard 12- to 13-inch-diameter target is generally rated at 20–25 kW, which means that it must have sufficient water cooling to absorb nearly 20 kW of thermal energy. The water flow requirements are on the order of 5 gallons or more per minute, requiring water lines of about 1-inch diameter (comparable to the flow when filling a car's 15-gallon gas tank in 3 minutes from a service station pump). As a practical matter, the power capability of a magnetron sputtering cathode is limited primarily by cooling and not by plasma issues.

The water flow can be calculated as follows. Taking the heat capacity of water as 1.0 cal/gm-°C, using chilled water near 0°C, and using the maximum output water temperature of 100°C, the incoming power P in watts (joules/sec) can be written as

$$P = \left(\frac{1\ \text{cal}}{\text{gm-°C}}\right) \times \left(\frac{4.184\ \text{J}}{\text{cal}}\right) \times (100°\text{C}) \times \frac{dM}{dT}$$

where dM/dT is the mass flow rate of the water (1 gal/min = 65 gm/sec). At an applied power of $P = 1$ kW, we calculate a minimum water flow of ≈ 3 gm/sec = 1/20 gal/min. This is an absolute minimum, however, since it is preferable for safety reasons not to have an output water temperature exceeding 35°C. This consideration leads to a practical value about 5 times higher, or an effective flow rate of about 1 gal/min per 4 kW of applied power.

The power density on the cathode of a swept-field magnetron varies with time as the moving etch track revolves across the cathode space. Even though the instantaneous power density in the etch track might be 10 times greater than the average power density (time-averaged over many rotations of the magnet array), thermal calculations typically use the average power density. For example, a 12-inch diameter AlCu cathode operating at 20 kW has an average power density of 20 W/cm^2. Assuming a backing plate of

0.25 inch and a target thickness of 0.5 inch, this leads to the surface temperature at the cathode being a few tens of degrees higher than the backside water temperature. However, if the magnet rotation is stopped and the etch track remains stationary, the power density would be > 200 W/cm^2, resulting in potential local melting of the AlCu target. Needless to say, production sputtering tools have interlocks that detect both adequate water flow as well as magnet rotation. This situation is not unique to PVD. For example, in a high-current batch ion implanter, each wafer in the batch may be rotated through the high-power density ion beam with only a few msec spent under the beam per rotation. Should the beam be allowed to dwell too long on a given wafer, excessive photoresist heating or even catastrophic melting of the Si wafer could occur.

At the PVD shield and wafer locations there are also concerns about thermal status during deposition. For example, a conventional-diameter AlCu cathode (12- to 13-inch diameter) might be operated at 20 kW to obtain a deposition rate of 1 μm/min. Approximately 10% of this 20 kW, or 2 kW, is delivered to the discharge chamber, where it eventually reaches the wafer, chamber walls, and fixturing. At the wafer, the high deposition rate can cause significant heating (see Section 5.3.4). Assuming an approximate atom size of 2.5 Å, a deposition rate of 1 μm/min is equivalent to about 67 atomic layers/sec. Each arriving atom brings along its kinetic energy plus its heat of condensation, which — along with other minor contributions from the plasma — might amount to 12–15 eV per adatom. Integrating this over a 200-mm wafer leads to a deposited energy flux of 50–70 W. The power deposited on the shields is significantly higher than this because they are located closer to the cathode and also function electrically as the de facto anodes in the plasma circuit. Shields can easily reach temperatures of 200–300°C during continuous operation. Collimators too, as described in detail in Chapter 6, have been measured to exceed 450°C during extended, high-power operation. The thermal cycling of all of these chamber parts is important in that the resulting stress in the films inevitably deposited on them can result in subsequent delamination and flaking.

Target heating can affect the PVD process in a variety of ways. For example, excessive target heating can cause undesired outgassing of impurities or induce thermal stress resulting in particle emission and possibly cracking. Thermomechanical damage to magnetron parts, harmful effects to permanent magnets, or even loss of a critical dimension by thermal expansion are possible. It is even possible for solder-bonded targets to physically fall off of their backing plates due to thermal-stress-induced delamination. More subtle effects of heating might include changes in target microstructure — such as increased grain size — that can affect sputtered film properties. For example, Al films are observed to have the best sheet resistance uniformity when the Al target has both a preferred (100) orientation and fine grains ($< 100\ \mu m$), so recrystallization of a fine grain target is to be avoided. As a rule of thumb, recrystallization is a concern when the operating temperature of the target exceeds one-half of its melting point. For a pure Al target ($T_{mp} = 660°C = 933$ K), this means that temperature should be kept below $\approx 194°C$. In cases of reactive sputtering of TiN, where the degree of target nitridation is of direct concern, increases in Ti target temperature can cause the process to shift from a non-nitrided, metallic mode to a nitrided mode. Radiant heat from a hot sputtering target can also increase substrate temperature, affecting film microstructure and composition. Therefore, active cooling of the target is critical.

A variety of methods for controlling cathode temperature have been employed, most of which rely on water cooling with particular attention to reducing the thermal resistance between the target and a cooled backing plate. For example, an *expansion contact method* has been successfully used with conical magnetrons in which the thermal expansion of the target brings it into physical contact with a water-cooled ring [11.2]. As the target expands, the contact pressure can become large enough for effective conduction cooling. When sputtering power is removed, the target cools and contracts so that it is no longer in contact with the ring, which makes eventual replacement very easy.

Planar magnetrons typically employ a bonded contact between the target and water-cooled backing plate. The trend is away from solder or epoxy and toward more reliable diffusion bonding in which solid state diffusion at elevated temperature and/or pressure is used to "cement" the target and backing plate. For example, concerns over debonding have prevented very high power operation of Ti targets solder bonded to Cu backing plates. However, Ti targets that have been diffusion bonded to high-strength, Al backing plates (tensile strength of bond ≈ 1.8 kpsi) can reliably operate at powers up to 20 kW [11.3, 11.4]. Whatever method of bonding is used, verification of bonding integrity is often provided by the target supplier using nondestructive ultrasonic imaging (≈ 10 MHz) of the completed target–backing plate assembly. Finally, we note that bonding can be avoided totally if the target and backing plate are machined from the same monolithic block of material; however, this can lead to significantly increased costs.

To cool the backing plate and other cathode components, a bathtub-like arrangement can be used as shown in Fig. 11.2. In this case, cathode components such as permanent magnets need to be encased (potted) in a

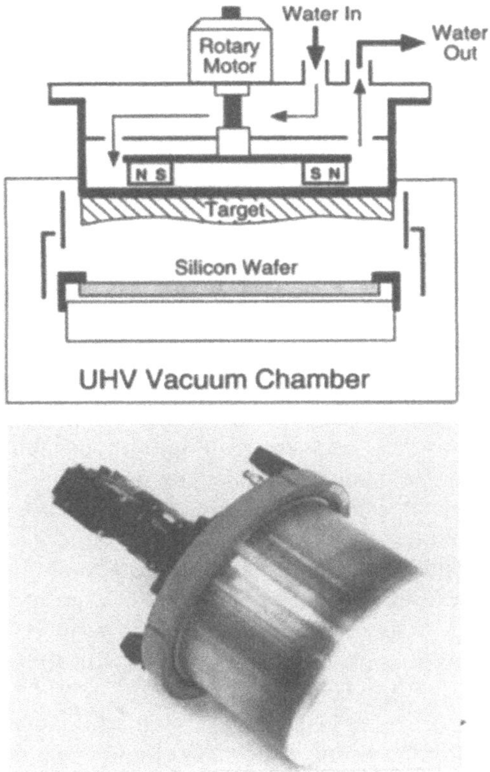

FIG. 11.2 Conduction cooling of a planar magnetron target by use of a bathtub-type arrangement located behind the backing plate.

water-resistant material to prevent corrosion, and deionized water should be used to prevent electrolytic corrosion between the electrically biased backing plate and the grounded water supply. The electrical conductivity of any cooling fluid in contact with the cathode should also be low enough to minimize current leakage to ground when maximum voltage is applied. Assuming a conservative design in which only a small increase in cooling-water temperature is allowed (output temperature $< 35°C$), we showed earlier that a relatively high water flow rate of > 1.0 gal/min per 4 kW of applied power is required. The combination of atmospheric pressure (the target front surface is at mTorr vacuum) plus the water pressure needed for proper flow rate can produce bowing of the target, similar to what happens to an edge-clamped wafer with backside gas (see Section 5.5.4). For example, 14.7 psi of atmospheric pressure plus 35 psi of water pressure translates into a load force of 5600 pounds over a 12-inch-diameter target used for coating 200-mm wafers. Larger diameter magnetrons of the sort being developed for 300-mm wafers and flat panel display applications will be even more susceptible to pressure-induced target bowing or deformation unless significantly thicker target/backing plate assemblies are used. To deal with this scale-up issue, large-diameter magnetron designs utilize water cooling channels between the target and the backing plate (Fig. 11.3). In addition to cooling large targets without bowing, the water-channel approach addresses reliability concerns with bathtub-type cooling, such as magnet corrosion and rotating water seals.

11.3 Target Burn-In

Dielectrics on the surface of a metal target can cause electrical arcing and particle generation in the PVD source, while surface contaminants can poison PVD films. Therefore, whenever a PVD module is vented to atmosphere (e.g., to change target and/or shields or to perform maintenance), the

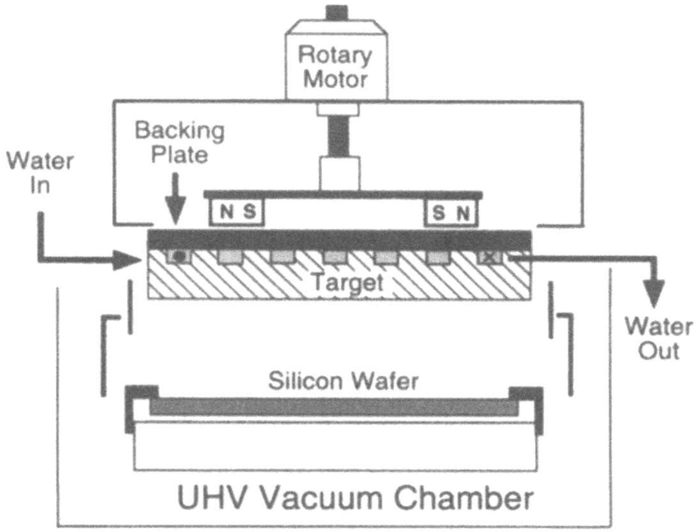

FIG. 11.3 Conduction cooling of a planar magnetron target by use of water-cooling channels between the target and the backing plate.

target must be reconditioned to remove adsorbed oxides, nitrides, and contamination before it can be used to deposit high-quality PVD films. For this purpose, a "target burn-in and conditioning" process is used whereby the source is gently ramped up in power to provide in-situ cleaning while avoiding such things as arc tracks or damage from thermal stress. Improper burn-in can be particularly severe for powder metallurgy targets since they are often not fully densified and can literally blow apart from gases trapped within microvoids. Burn-in is also important for a target that has been sitting idle in the process chamber. For example, at a base pressure of 1×10^{-8} Torr, residual gas arrival rates are $\approx$ 0.01 Å/sec. Assuming a sticking coefficient of only 0.5, the resulting layer of contamination formed on the target in one 8-hour shift can be over 150-Å thick.

In a typical burn-in procedure, a dummy wafer is placed on the substrate holder and the source is turned on at low power and high pressure, where the target voltage is low. Power is then progressively increased and pressure lowered. The highest burn-in power is typically higher than the process recipe power to ensure that the target sees its highest temperature prior to film processing — similar to the wafer degas strategy discussed in Section 5.3.2. For example, the Al target burn-in for a PVD Al process at 9.6 kW might consist of 4.0 kW-hr of deposition with the source ramped up as follows: 15 min each of 1 kW, 2 kW, and 4 kW power deposition at 5 mTorr, followed by 10 min of 8 kW deposition at 2 mTorr, and finally 5 min of 11 kW deposition at 2 mTorr.

11.4 Target Composition

We begin this section by noting that atomic percent, i.e., the ratio of numbers of atoms, is the common method of specifying chemical composition and materials purity. For example, $MoSi_2$ is a compound with 33% (1/3) Mo atoms and 67% (2/3) Si atoms. Many surface analytical techniques, such as auger electron spectroscopy (AES), X-ray photoelectron spectroscopy (XPS), and secondary ion mass spectrometry (SIMS) also use atomic percent, which can then be converted into chemical formulas. Unfortunately, sputtering target composition is almost always given in weight percent, which can be quite different from atomic percent. Weight percent is simply the normalized ratio of the weights of the constituent components of the target. For example, fabrication of a 100-gm target containing equal weight percents of W and Ti might begin with a mixture of 50 gm of W and 50 gm of Ti, which are then sintered into a composite matrix. However, in atomic mass units (AMU), W has an atomic weight of

183.9 AMU and Ti has an atomic weight of 47.9 AMU. This means that 50 gm of W is only 27% of a mole of W (1 mole = 6.02×10^{23} atoms = an Avogadro's number of atoms), whereas 50 gm of Ti is 104% of a mole. Hence, the atomic ratio of the target turns out to be 80% Ti and 20% W ($Ti_{0.8}W_{0.2}$) which is much different from its 1:1 weight ratio.

Weight percent is often written in parentheses after the elemental symbols. Hence, a W(30)Ti(70) target has 30% W by weight and 70% Ti. In the case of Al alloys containing very small amounts of Cu and/or Si, the weight percent is usually omitted from the Al. Hence, an Al alloy with 1.0 weight percent of Si and 0.5 weight percent Cu would be written as AlSi(1.0)Cu(0.5) and not as Al(98.5)Si(1.0)Cu(0.5). The relationship between atomic and weight percent is straightforward to calculate by simply counting atoms and knowing the atomic weight of each species. Consider a binary target A_xB_{1-x} consisting of material A with atomic mass m (in AMU) and material B with atomic mass M. From the target chemical formula, the atomic percent of element A is $100x$. It is then easy to show that

$$\text{Weight percent of A} = \text{wt }\%_{A} = \frac{100}{\left(1 + \left(\frac{M}{m}\right)\frac{(1-x)}{(x)}\right)}$$

$$\text{Weight percent of B} = \text{wt }\%_{B} = \frac{100}{\left(1 + \left(\frac{m}{M}\right)\frac{(x)}{(1-x)}\right)} \tag{11.1}$$

where

$$(\text{wt }\%_{A} + \text{wt }\%_{B}) = 100\%$$

As an example of the use of Eq. (11.1), consider a Ti_xW_{1-x} target with $x = 0.3$, i.e., $Ti_{0.3}W_{0.7}$. In this case, A is Ti with m = 47.9 AMU; B is W with M = 183.9. Since $x = 0.1$, the atomic percent of Ti is $100 \times 0.3 = 30\%$. Using formula (11.1), we calculate the weight percent of Ti as 100/(1 + (183.9/47.9)(0.7/0.3)) = 100/(1 + 8.96) = 10.04%. Hence, the atomic percent of Ti is 3 times greater than its weight percent, and the $Ti_{0.3}W_{0.7}$ target composition can be written by weight as Ti(10)W(90). Not surprisingly, the discrepancy between weight percent and atomic percent is greatest for species with significant differences in their atomic weights. For example, Al has an atomic weight of 27 AMU and Cu has an atomic weight of 63.5 AMU. Thus a commonly used Al alloy target with 0.5 weight percent of Cu — AlCu(0.5) — then turns out to have approximately 0.2 atomic percent of Cu.

Regardless of how one specifies target composition, it may gradually shift over time. This may be the result of the following effects.

Dissimilar Sputter Yields If two elements have different sputter yields, they are also likely to have different yields when present as the components of an alloyed target. The first approximation (often used for surface analysis measurements involving sputter erosion depth profiling) is simply to assume the bulk, elemental sputter yield for each alloy constituent. This results in a rapid depletion at the cathode surface of the higher-sputter-yield material and the possibility that the initial films deposited from the target may be enriched in the higher-sputter-yield material. This process is self-limiting in that the surface composition eventually adjusts to a slightly higher concentration of the lower-sputter-yield material — just enough to counteract the higher sputter yield of the other component. The net effect can be formation of an altered surface layer in the top few tens of angstroms of the target that has a different composition from the bulk. If target temperature is allowed to rise to the point where there can be sufficient volume diffusion ($\approx$ 500°C for AlCu), the entire target may eventually be depleted of the higher-yield material. Fortunately, for most material systems of interest to IC processing (such as AlCu), the altered layer is formed rapidly and will not be an issue following the burn-in procedure used for degassing.

Granular Targets Sputter targets that are fabricated with complex stoichiometry (e.g., ternary oxides such as BST or SBT) are usually made by mixing powders of their constituent materials, and then sintering and hot pressing. Since these grains are randomly oriented and the surface is not completely flat, there can be subtle changes in composition over time as the individual grains are slowly exposed and then sputtered. This, in a sense, is a microscopic analogy to the altered-layer problem described above in which each grain within the cathode functions as an individual, microscopic target.

Oxidation Issues Targets of compound materials, such as Ti-W, that are formed from hot pressing of powders are sensitive to oxidation of the individual powder grains. Depending on the fabrication process environment and control, this problem can be minimized.

Redeposition and Transport Most manufacturing-scale magnetron sputtering uses low operating pressure, typically 0.5–4 mTorr. At these pressures, gas-phase scattering is low and few of the sputtered atoms are scattered in-flight. However, these scattered atoms may be deposited back onto the cathode. In the case of alloy cathodes whose constituents have very dissimilar mass (e.g., TiW and AlCu), the lighter species may be preferentially redeposited onto the cathode — particularly at higher pressures. This can lead over time to changes in the composition of the target surface. This effect, though, is somewhat self-compensating in a way similar to the yield-related altered layer formation, and is rarely a concern. Conversely, the deposited film composition may be inversely related to this scattering issue, as the heavier atoms from the target pass through the background gas more easily and preferentially deposit on the sample [11.5].

Nonunity Sticking Coefficient Sputter deposition is generally considered to be characterized by a 1.0 sticking coefficient for the sputtered atoms. That is, the sputtered atoms hit a surface and stick immediately without bouncing off. There may be some cases, particularly with alloys having a large mass mismatch (e.g., TiW and AlCu) in which the sticking probability could be different for each species. This may depend on the relative masses involved — e.g., light atoms might bounce off of a high-mass film surface — or perhaps on the kinetics of the deposition. More massive materials (e.g., refractories such as Ta or W) tend to have higher kinetic energy and as such may be more likely to reflect from high-angle surfaces such as the sides of a trench.

11.5 Target Purity

All sputter targets — be they elemental (e.g., Ti), binary (e.g., WSi_2), or alloy (e.g., Al-Si-Cu, Ti-W) — contain impurities. It is neither practical nor cost-effective to require 100% purity in a production sputter target. On the other hand, even trace amounts of selected elements can adversely affect thin film properties and device performance. Therefore, suppliers have devoted much effort to reducing the level of critical impurities in their targets. This drive toward ultrahigh purity material is sometimes referred to as "the nines game," since target purity is generally stated in the language of "nines." For example, a "four-nines-five" or 4N5 Ti target would have 99.9995% purity, with the total level of all impurities being < 5 ppm $=$ 0.0005% by weight. Although 6N purity Al and Al alloy sputter targets are available, the purity of Al targets used in production are typically 5N to 5N5, while Ti target purity tends to be $\approx$ 4N5 to 5N. Since individual elements can be present at much lower levels than the total impurity content, an element-by-element analysis covering most of the elements of the periodic table is typically provided for a given target using analytical methods with detection limits $\approx$ 0.01–0.001 ppm. Established techniques such as spark source mass spectrometry (SSMS), glow discharge mass spectrometry (GDMS), and X-ray fluorescence (XRF) mapping are used to certify the purity of the starting target material or, for a nondestructive method such as XRF, of the finished target itself. Figure 11.4 presents a representative analysis of a high-purity 4N Ti target.

Since one pays a significant premium for an additional nine or even nine-five of materials purity, it is important to focus on the problem elements. Regardless of target chemical composition, alpha-particle emission from heavy elements such as uranium (^{238}U) and thorium (^{232}Th) turns out to be of general concern, since this can result in significant electron-hole pair generation in active device regions with subsequent "upsets" or even permanent damage to devices. As a result, the sum total concentration of these two elements is typically restricted to $<$ 1 ppb (0.0000001%) in PVD targets intended for metallization.

Element	ppm	Element	ppm	Element	ppm
H	< 10	Zn	< 0.1	Pr	< 0.03
C	60	Ga	< 0.05	Nd	< 0.1
N	10	Ge	< 0.05	Sm	< 0.1
O	500	As	< 0.05	Eu	< 0.07
Li	< 0.001	Se	< 0.05	Gd	< 0.1
B	< 0.002	Br	< 0.03	Tb	< 0.04
Be	< 0.002	Rb	< 0.04	Dy	< 0.1
F	0.1	Sr	< 0.02	Ho	< 0.04
Na	< 0.02	Y	< 0.02	Er	< 0.1
Mg	< 0.03	Zr	< 0.1	Tm	< 0.03
Al	30	Nb	< 0.2	Yb	< 0.2
Si	0.3	Mo	0.2	Lu	< 0.03
P	0.4	Ru	< 0.06	Hf	< 0.1
S	0.3	Rh	< 0.02	Ta	< 0.03
Cl	0.1	Pd	< 0.08	W	< 0.4
K	< 0.06	Ag	< 0.04	Re	< 0.04
Ca	0.06	Cd	< 0.08	Os	< 0.07
Sc	< 0.001	In	< 0.03	Ir	< 0.08
Ti	Matrix	Sn	< 0.4	Pt	< 0.3
V	4	Sb	< 0.03	Au	< 0.09
Cr	30	Te	< 0.05	Hg	< 0.2
Mn	< 0.02	I	< 0.02	Tl	< 0.1
Fe	30	Co	< 0.1	Pb	< 0.2
Co	< 0.02	Ba	< 0.02	Bi	< 0.08
Ni	1	La	< 0.02	Th	< 0.08
Cu	2	Ce	< 0.02	U	< 0.08

FIG. 11.4 Representative analysis of a high-purity 4N Ti target (Vacutec™ target from Atramet, Inc., Farmingdale, NY).

In the case of high-purity Al targets, the dominant impurities tend to be light elements such as H, C, and O, with oxygen of particular concern since even trace amounts of oxidants (e.g., H_2O partial pressure $> 10^{-7}$ Torr) can retard processes such as high-temperature Al reflow [11.6]. Assuming as a worst case that all of the impurity content in a 5N5 Al target is due to O, does this amount of sputtered O pose a problem? The deposition rate from an Al target is typically ≈ 1 μm/min $= 160$ Å/sec. Assuming that the relative sputtered flux of O and Al from the target are not too different from their relative bulk concentration leads to an oxygen "deposition rate" at the wafer of about 160 Å/sec $\times$ 5 ppm $= 8 \times 10^{-4}$ Å/sec. This degree of oxygen bombardment corresponds to an oxygen partial pressure in the chamber of $\approx 8 \times 10^{-10}$ Torr, which is 100 times lower than the partial pressure needed to affect Al reflow. Of course, the target will adsorb moisture, nitrogen, hydrocarbons, etc. between the time it is certified and the time when a PVD film is deposited. Proper target burn-in is therefore essential to recovering the low contamination level of the as-manufactured target. Also, reducing sources of contamination such as chamber outgassing, permeation of elastomer-seals, etc. may be a better return on investment than going to a higher purity target.

With regard to chamber outgassing, it is worth noting that achieving a low base pressure greatly reduces the impact of surface oxidation and/or contamination during the time of the PVD deposition. For purposes of illustration, assume that all of the chamber outgassing at a base pressure of 10^{-6} Torr is potential contamination and has a sticking coefficient at the surface of 1.0. At base pressure, about 0.5 monolayer of "crud" will form in about 1 sec — this formation time being inversely proportional to pressure. Assuming a deposition rate for Al of 1 μm/min ($\approx$ 67 monolayers/sec) and that the Ar working gas introduces no impurities of its own, then the residual gas contamination present during deposition will lead to an impurity level in the Al film of $\approx 0.5/67 = 0.75\%$. By reducing the base pressure in the PVD module to, say, 10^{-8} Torr, through better vacuum practices and improved vacuum hardware, the contamination so introduced into the film will be reduced by a factor of 10^{-6} Torr $/10^{-8}$ Torr $= 100$ — to 0.0075% or 75 ppm.

11.6 Target Utilization

A target is a consumable item, and how effectively it is utilized during PVD has a strong impact on cost-of-ownership. In this discussion, we define *target utilization* as the percent of starting target material that is used up before the target must be replaced at its "end-of-life." There are other ways of defining or discussing target utilization, such as the percent of material eroded from the target that is actually deposited on the wafer; however, as a practical matter the weight of target material remaining at the end of its useful life is easy to quantify. Target utilization of 100% is never achieved in practice; nevertheless, high-purity targets can be rather costly, and changing targets too frequently affects tool productivity. Therefore, one would like both high target utilization and long target life — which turn out to be interrelated issues, as shown in Fig. 11.5.

When one considers the angular spread of the sputtered material and the finite solid angle subtended by the target at the wafer, it becomes clear that uniform target erosion cannot produce uniform thin film deposition. For example, the center of the wafer "sees" a greater amount of target material than does the edge of the wafer, so that a uniformly eroded target would produce a deposition profile thicker at the center. Therefore, the magnet array of the DC magnetron source is designed to produce a nonuniform erosion profile across the target that compensates for the target-to-wafer geometry and takes into account such process-related variables as gas-phase scattering. For example, in some cathodes, a radially symmetric "W"-shaped erosion profile is utilized having relatively greater erosion near the edge (Fig. 11.6), whereas others utilize multiple, concentric etch tracks.

As a planar target is sputtered, the nonuniformity of the erosion profile increases because the parallel component of the magnetic field at the target surface (which determines the local plasma density and therefore the

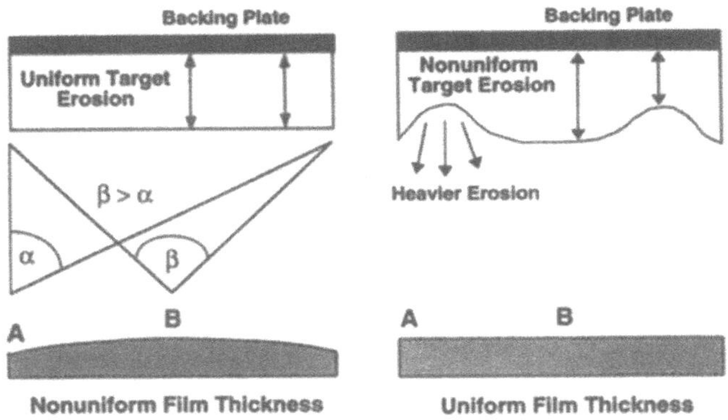

FIG. 11.5 Nonuniform target erosion can be used to produce more uniform films, but this has a negative effect on target utilization.

FIG. 11.6 Obtaining an optimum balance between film uniformity and target utilization has led to the use of tailored target erosion profiles in both circular and rectangular planar magnetrons (courtesy of Sierra Applied Sciences, Boulder, CO).

local Ar^+ flux) increases as the eroded target surface gets closer to the magnets behind the backing plate. Target life is then determined by the lesser of two times: (1) the time it takes the fastest eroding spot on the target to hit the backing plate, or (2) the time it takes the target profile to have evolved to the point that a key film property such as nonuniformity of thickness or step coverage can no longer be maintained. Advanced planar magnetron sources (c. 1995) designed for 200-mm wafer coating and film uniformity of $3\sigma \approx 5\%$ are capable of depositing > 6000 μm of Al with target utilization $> 50\%$. Since Al films are typically ≈ 0.8–1-μm thick, the number of processed wafers through Al target life would then be $\approx$ 6000–9000. In general, target life and film uniformity have an inverse correlation, with higher film uniformity requiring a less uniform erosion profile, leading to lower target utilization. For example, target life for $3\sigma = 3\%$ film deposition could be 2 times less than for $3\sigma = 5\%$. Therefore, whenever comparing target life of different PVD sources, the same film uniformity should be used. Similarly, since film thickness can change strongly in the vicinity of the wafer edge, the number of wafers during target life that meet a given film uniformity spec will depend on the specific edge exclusion used when measuring that uniformity.

Once the target has reached end-of-life and is removed, recycling of the spent target material and/or reuse of the target assembly with a new target is sometimes used to further reduce costs.

11.7 Microstructural Engineering

Target microstructure can have a direct influence on both the uniformity and quality of sputtered films. Thus target suppliers continue to engineer the preferred crystallographic orientation (texture), grain size, and grain distribution of the target to produce optimum sputtered film properties, improve repeatability, and reduce particle generation [11.7–11.10]. The general trend in advanced microelectronic applications of PVD is toward homogeneous, fine-grained sputter targets with random crystallographic orientation.

Crystallographic Orientation Sputtered atoms tend to leave the surface of a single-crystal target (or a single microcrystal in a polycrystalline target) preferentially along close-packed or nearest-neighbor directions that correspond to high-atomic density (see Fig. 11.7). For example, in Al the preferred emission is along the (110) crystallographic direction. The distribution of grain orientation within a polycrystalline metal target then affects the overall angular emission distribution of the sputtered atoms from the target. Generally, the thermomechanical process of target manufacturing imparts a preferred alignment to the grains, or "texture," which influences film uniformity. Figure 11.8 shows calculated and measured angular distributions of sputtered Al from (100) and (110) single crystal targets. Figure 11.9 presents the calculated film uniformity on a 200-mm wafer from these single-crystal Al targets and from polycrystalline Al with texture ranging from strongly (100) to strongly (110). The simulation suggests that a predominantly random spatial distribution of crystallites with a very slight (100) texture produces the best thickness uniformity. Highly randomized orientation offers another potential benefit. Namely, as a target with a preferred orientation erodes during sputtering, its surface topography evolves to expose a variety of new crystal planes at different areas over the surface (e.g., on the sides of the erosion grooves). This can result in an erosion-dependent film uniformity. Experimental work on both Al alloys and Ti confirms that a randomly oriented target texture is preferred for optimum blanket film uniformity and consistency through target life.

Grain Size Target grain size affects the deposited film in several ways. When the grain size is large (> 1 mm) and the target-to-wafer spacing is

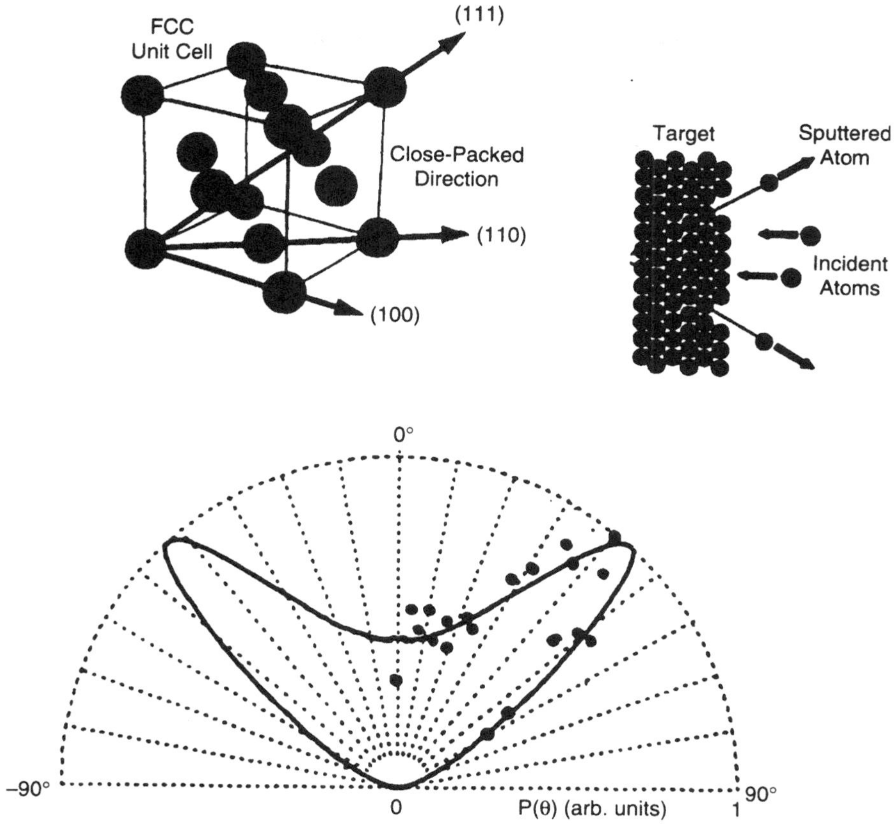

FIG. 11.7 Angular distribution of sputtered atoms from single-crystal (100) Al displaying lobes associated with close-packing directions [11.9].

small (≈ 5 cm), local differences in each grain's angular emission distribution can create local regions on the wafer that have different deposition rates, which affects global uniformity. This problem can be solved by reducing target grain size below 500 μm or using much larger source-to-substrate spacing. This is supported by data such as that in Fig. 11.10 (from ref. 11.7), which shows sheet resistance uniformity of Al-0.5%Cu films from Al alloy targets of different average grain size. Grain size can also influence particle generation (as illustrated in Fig. 11.11) for reactive sputtering of TiN from a 4N5 purity Ti target [11.7]. Figure 11.11 shows that

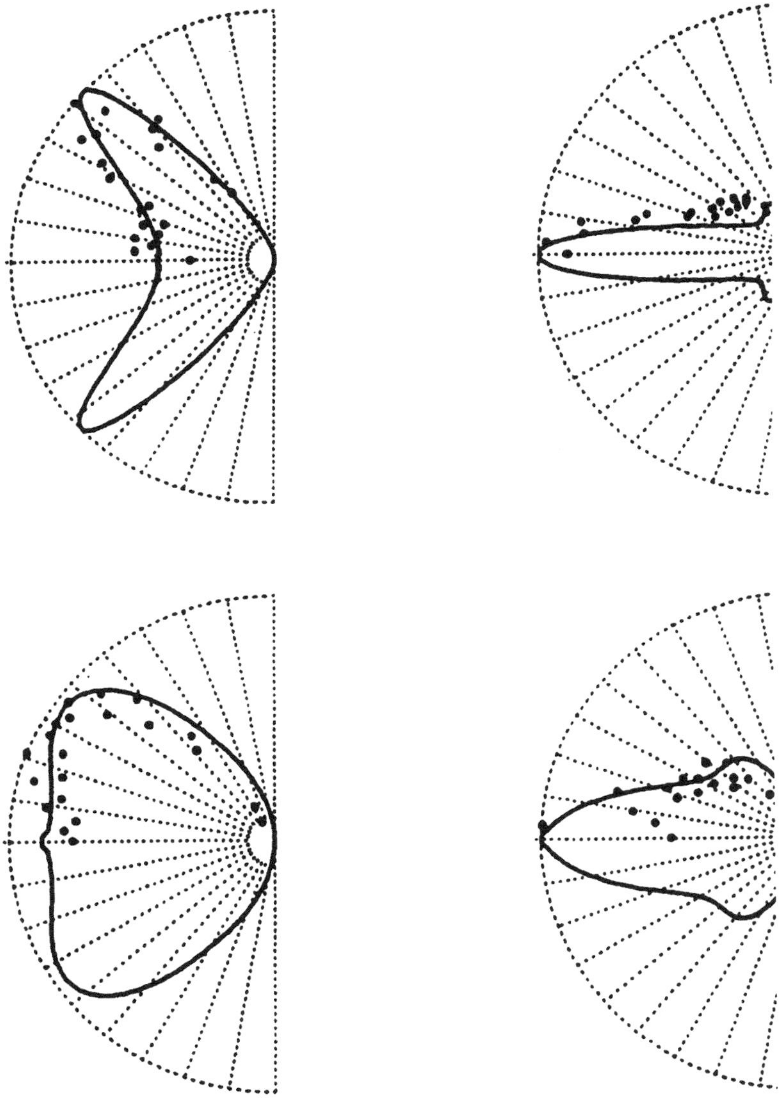

FIG. 11.8 Calculated and measured angular distribution of sputtered Al atoms from single-crystal (100) and (110) Al targets and similarly oriented polycrystalline Al targets [11.9].

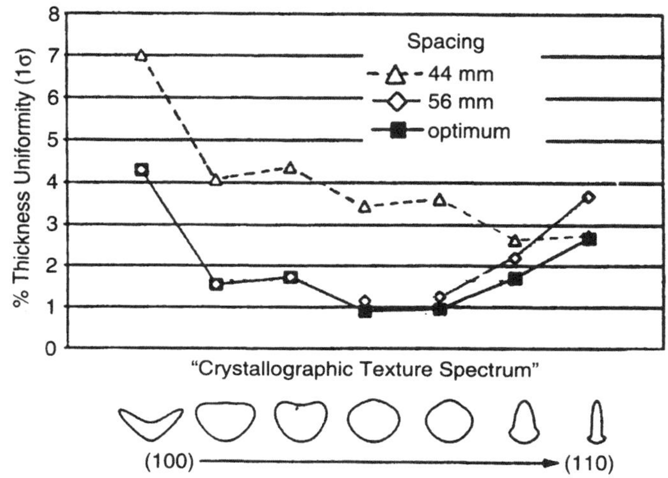

FIG. 11.9 Calculated Al film uniformity versus target-to-substrate spacing for single-crystal (100) and (110) Al targets and polycrystalline targets with texture ranging from strongly (100) to strongly (110) [11.9].

submicron particle levels (size range = 0.3–0.5 μm) could be greatly reduced in this case by going to fine and even ultrafine grain targets with average grain size ≈ 10 μm, and that a high-performance target can maintain particle levels of ≈ 0.01/cm^2 through life.

Second-Phase Microstructure Both the resistivity and reflectivity of Al alloy films can be influenced by the amount of second-phase precipitation (e.g., Al_2Cu) in the target [11.10, 11.11]. It is believed that the emission of second-phase molecular species (e.g., Si_2, Al_2Cu) from the target creates second-phase nucleation sites in the PVD film, which can influence how stress is relieved. For example, in Al-Si-Cu films at elevated temperatures (> 400°C), second-phase precipitates can suppress hillock formation in favor of the growth of single-crystal "whiskers."

Copper Segregation Segregation of copper in macroscopic regions of the target is also to be avoided since this can cause microarcing, nodules, or film segregation effects such as variation in electromigration resistance or dry etch rate across the wafer. Both of these effects depend on the local concentration of Cu in the AlCu alloy film.

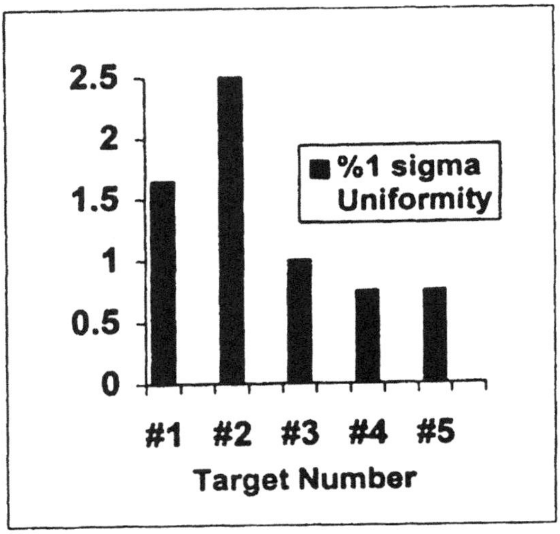

Target Grain Size (μm)				
Target #	Target Face	Mid-Radius	Center	Average
1	>300	120	263	>228
2	>300	>300	>300	>300
3	62	69	63	65
4	66	72	69	69
5	61	53	66	60

FIG. 11.10 Sheet resistance uniformity for Al-Cu alloy targets depends on average grain size (data from ref. 11.7).

11.8 Particle Generation

A major source of particles in a well-designed PVD module can often be the target itself. A general issue is that the nonuniform erosion can lead to material being sputtered from heavily eroding areas of the target and back onto more lightly eroding areas — typically near the center. Build-up of sputtered material can then lead to flaking. A related issue is the use of a Ti target exclusively for reactive TiN deposition — for example, dedicating a PVD module for deposition of a TiN antireflection coating (ARC) layer. TiN has a high compressive stress, and over the course of time a thick layer of TiN will build up on the shields that can spall and produce

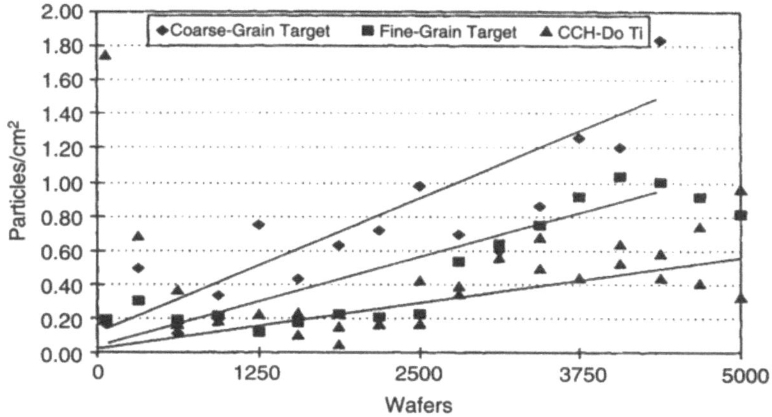

(a)

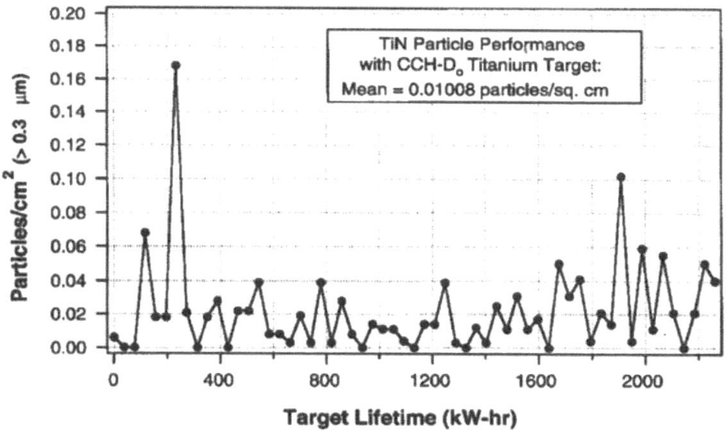

(b)

FIG. 11.11 (a) Density of 0.3–0.5-μm particles in TiN films deposited onto a 150-mm wafer from a Ti target with coarse grains, fine grains (< 100 μm), and "CCH™-D_0" grain size < 30 μm [11.7]; (b) Particle density in TiN films on a 200-mm wafer are maintained at levels of ≈ 0.01 particles/cm^2 through target life by using a high-performance Ti target and a high-vacuum PVD cluster tool (courtesy of P. Gilman, MRC).

flakes and particles. One way of dealing with this situation is to periodically perform so-called Ti pasting in which the target is sputtered without reactive gas to deposit a Ti layer on the TiN-coated shields. Ti has good adhesion to TiN and the Ti/TiN sandwich has lower stress than the TiN layer alone. By using the Ti as a stress-reducing paste, additional TiN depositions can be carried out without shield cleaning. In a production environment, pasting might occur after every 100 wafers or so. Conversely, if the sputter chamber is equipped with a mechanical shutter that can be positioned between the target and wafer, pasting for just a few seconds might be done after each wafer while the shutter is closed. In either mode, the cumulative effect of pasting on overall wafer throughput needs to be considered, since the practice can reduce useful kW-hr of Ti target life by 20–30%.

Other issues of particle generation relate to target quality and tend to be materials-specific. For example, areas of low density (microvoids) produced in Ti targets during target manufacturing can trap gases (see Fig. 11.12). As these gases are released during target erosion, they can cause local high-pressure regions that induce electrical arcing from the target to the plasma with related particle generation. Also, since the gases in the voids are at relatively high pressure compared to the mTorr process ambient, a microburst of released gas can ballistically launch a piece of target material that eventually impacts the wafer surface. Refractory metal targets such as TiW that have not been properly burned-in also have a tendency to grow dendrites or cones on the target surface (Fig. 11.13). These cones continue to grow until they reach a critical height above the target surface at which arcing will occur.

Contamination, on the target surface or included in the bulk, will also tend to arc and flake. For example, a relatively high density of Al_2O_3 inclusions in Al targets has been correlated with arcing events at the target

FIG. 11.12 SEM of a microvoid in a Ti target (from ref. 11.12).

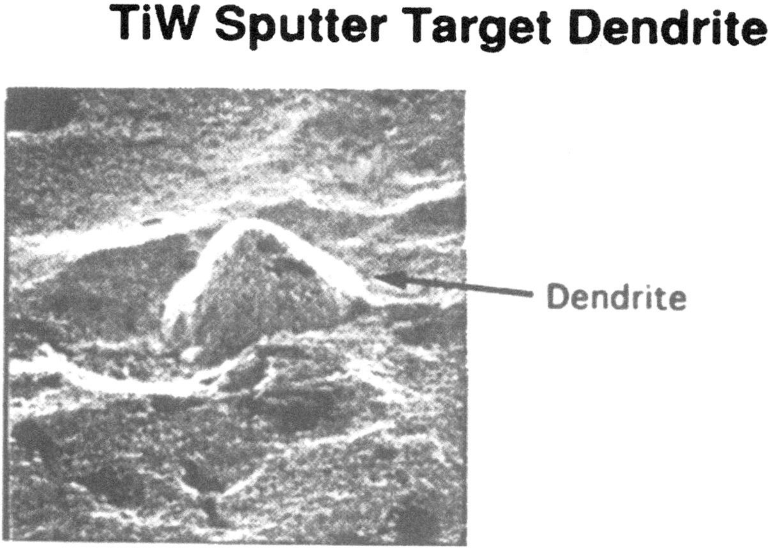

FIG. 11.13 SEM of dendritic growth on TiW film (from ref. 11.13).

surface. Because the inclusions are insulating, they charge up due to ion bombardment of the target, producing local electric fields that can exceed the dielectric strength of the insulator. An arc is then initiated by the "flashover" that occurs when dielectric breakdown is reached. The arc in turn can produce localized melting and explosive emission of Al droplets from the target and onto the wafer. This particular mechanism of particle generation is greatly reduced by the ultralow levels of oxygen (< 10 ppm by weight) found in high-purity Al targets [11.13].

References

11.1. C. E. Wickersham Jr., J. E. Poole, and J. J. Mueller, "Particle contamination during sputter deposition of W-Ti films," *J. Vac. Sci. & Tech.* **A10**(4): 1713–1717 (1992).
11.2. C. E. Wickersham Jr. and J. E. Poole, "Target Operating Temperatures in Conical Magnetron Type Sputtering Sources," Tosoh SMD, Technical Note TKN 9.008A.
11.3. MRC Technical Brief on "CCH™-Titanium PVD Targets," (1996).
11.4. MRC Technical Brief on "IntegraBond™ Diffusion Bonding," (1995).
11.5. S. M. Rossnagel, I. Yang, and J. J. Cuomo, "Compositional changes during magnetron sputtering of alloys," *Thin Solid Films* **199**: 59–69 (1991).

11.6. K. Kikuta and T. Kikkawa, *Extended Abstract of the 53rd Autumn Meeting*, Vol. 2, p. 586, Japan Society of Applied Physics, 1992.

11.7. P. S. Gilman, "Microstructurally controlled sputtering targets," *Semicond. FABTECH* **3**: 209–211 (1995). See also A. E. Braun, "Sputtering targets adapt to new materials and shrinking architectures", *Semicond. Int.*, 127–134 (June 1998).

11.8. R. S. Bailey and N. C. Hill, "Process, equipment, and materials control in integrated circuit manufacturing," *SPIE Proceedings* **2637**: 56–64 (Oct. 1995).

11.9. J. S. Fan, R. S. Bailey, and C. E. Wickersham Jr., "New developments and applications for sputtering targets at Tosoh SMD," submitted for presentation at SEMICON-China (November 1997).

11.10. C. E. Wickersham Jr., "Nondestructive testing of sputtering targets," *Solid State Tech.*, 75–80 (Nov. 1994).

11.11. S. Whitney, R. W. Lionetti, C. Wickersham Jr., L. Succo, J. Esposito, and M. Cleeves, "Influence on the propensity for whisker growth in sputter-deposited aluminum films," Tosoh SMD, Technical Note TKN 8.004A (1988).

11.12. K. J. Hansen, "Microcontamination from physical vapor deposition process and equipment," *Technical Proc. of SEMICON-Korea*, 139–152 (Nov. 1993).

11.13. K. S. Bailey, A. Leybovich, J. E. Poole, T. Kuniya, N. C. Hill, and C. E. Wickersham Jr., "Particle emission from Al_2O_3 doped aluminum targets during sputter deposition," *Technical Proc. of the VLSI Multilevel Interconnection Conf.*, p. 317 (June 1994).

Author Index

G

H

I

J

S

Z

Subject Index

C

D

Q

R

S

T

X

Y

Z

ISBN 0-12-533026-X

FSC
www.fsc.org
MIX
Papier | Fördert gute Waldnutzung
FSC® C083411

Kolibri 360 GmbH
Ferdinand-Jühlke-Straße 7
99095 Erfurt, Deutschland
produktsicherheit@kolibri360.de